Medical Cell Biology
Second Edition

Medical Cell Biology
Second Edition

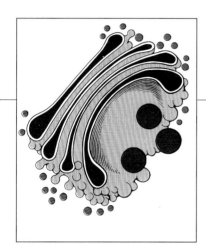

Edited by
Steven R. Goodman, Ph.D.

Professor and Chairman
Department of Structural and Cellular Biology
University of South Alabama
College of Medicine
Mobile, Alabama

Lippincott - Raven
P U B L I S H E R S
Philadelphia • New York

Acquisitions Editor: Richard Winters
Coordinating Assistant Editor: Erin O'Connor
Indexer: Susan Thomas
Art Director: Diana Andrews
Cover Designer: Jeane Norton
Production Manager: Maxine Langweil
Production Editor: Mary Ann McLaughlin
Compositor: Lippincott–Raven Electronic Production
Printer/Binder: Courier Kendallville

Library of Congress Cataloging-in-Publication Data
Medical cell biology / edited by Steven R. Goodman. — 2nd ed.
    p.   cm.
    Includes bibliographical references and index.
    ISBN 0-397-58427-X (softcover : alk. paper)
    1. Cytology. I. Goodman, Steven R.
    [DNLM: 1. Cytology. 2. Molecular Biology. QH 581.2 M4892 1998]
QH581.2.M43 1998
571.6'1—dc21
DNLM/DLC   97-26571
For Library of Congress   CIP

The Cover
Glial cell types in cultures of 5-day postnatal rat optic nerve revealed by immunofluorescence microscopy. Astrocytes (*shown in red*) are labeled with a fluorescent antibody to the cytoskeletal component, glial fibrillary acidic protein. A single oligodendrocyte (*shown in green*) is identified with a myelin protein–specific antibody. (Photograph courtesy of A. L. Gard and G. R. Dutton.)

The editor dedicates this book to:

The Lord: the source of resurrection power.

The love of my life: Karen Frances Goodman

and

Our children: Gena, Jessie, Scotty, and Laela

Contributing Authors

Ron Balczon, Ph.D.
Associate Professor, Department of Structural and Cellular Biology
University of South Alabama, College of Medicine

Anthony L. Gard, Ph.D.
Associate Professor, Department of Structural and Cellular Biology
University of South Alabama, College of Medicine

Steven R. Goodman, Ph.D.
Professor and Chairman, Department of Structural
and Cellular Biology
University of South Alabama, College of Medicine
Director, USA Comprehensive Sickle Cell Center

Stephen G. Kayes, Ph.D.
Professor, Department of Structural and Cellular Biology
University of South Alabama, College of Medicine

Warren E. Zimmer, Ph.D.
Associate Professor, Department of Structural and Cellular Biology
University of South Alabama, College of Medicine

Clinical cases contributed by Betty S. Pace, M.D.
Assistant Professor, Departments of Pediatrics
and Structural and Cellular Biology
University of South Alabama, College of Medicine

Preface

When we wrote the first edition of *Medical Cell Biology*, our goal was to create an affordable textbook that, in 300 pages, contained all of the information concerning cellular biology that a medical student would need to practice modern medicine and to perform well on standardized exams. We thank you all—educators and students—for making the first edition a tremendous success. I especially appreciate the kind letters and cards that we received from many of you, describing your gratitude for the quality of the textbook. Your comments were very helpful in making the second edition of *Medical Cell Biology* an even stronger educational tool. For example, several faculty members who are using our textbook in their courses requested that we supply the diagnoses for the clinical cases found at the beginning of each chapter. Because the majority of our readers preferred this approach, over having the students determine the diagnoses through self-study, we have taken their advice in this second edition. All new cases have been written by Dr. Betty Pace for the second edition and can be found at the beginning of the chapters. At the end of each chapter, the student is given the diagnosis, plus an explanation of the disease, and of how the disorder relates to the science of that particular chapter.

Another suggestion that I received came from students whom I helped to prepare for National Board Part I as part of the National Medical School Review. These students requested National Board–type questions at the end of each chapter. Therefore, in *Medical Cell Biology*, second edition, we have added National Board–type questions at the end of each chapter, with answers given at the end of the book. Then, a wonderful suggestion came from Richard Winters, Senior Editor at Lippincott–Raven Publishers. Richard suggested that we create a computer disk with approximately 250 National Board type questions that would be sold with the book. The result is the accompanying disk that we have been able to provide to our readership without raising the price of the book.

Finally, I had the privilege of chairing a committee from the Association of Anatomy, Cell Biology and Neurobiology Chairpersons that created the Cell Biology/Histology portion of the new U.S./Canadian Curriculum in the Anatomical Sciences. This document was made available to all anatomy, cell biology, and neuroscience departments in the United States and Canada in the spring of 1997. It will, no doubt, be a driving force for defining medical schools curriculum and questions on standardized exams for many years to come. As chairman of the working group that assembled the curriculum in Cell Biology/Histology, I have made sure that *Medical Cell Biology*, second edition, contains *all* of the information and topics defined as necessary by the U.S./Canadian Curriculum in the Anatomical Sciences. To meet this requirement, the authors have updated all chapters of the textbook, and Chapter 10, on Cell Motility, has been completely rewritten.

Medical Cell Biology, second edition, remains an affordable textbook that gives the medical student (or students in other health professions) all that they need to know on the subject in 300 pages. In addition, the second edition has been modified so that it faithfully represents the Cell Biology portion of the U.S./Canadian Curriculum in the Anatomical Sciences. Finally, by adding National Board type questions at the end of each chapter, and on disk, we have made the second edition an excellent Board preparation tool. As always, I welcome your comments on the fruits of our labor. I read all of your letters and, as you can tell, your suggestions are taken to heart in each new edition.

Preface to the First Edition

Cellular biology, a subject of tremendous import to the modern clinician and medical student, is taught in varied ways in medical schools around the world. It usually constitutes 10 to 20 lectures housed within either a histology, physiology, or biochemistry course. Given the limited amount of time to present cell biology within the increasingly compressed curriculum of medical school (as well as dental and nursing schools), many of us select the most essential information for presentation in our 10 to 20 lecture hours and then assign large amounts of supplemental reading in one of several outstanding cell biology textbooks. The problem that arises is that the most widely used cell biology textbooks were written for graduate students, not medical students, and contain roughly 1200 pages covering all of the cell biology needed for today's Ph.D. candidate. These textbooks are also expensive. The result is that many medical students do not purchase the assigned textbook and rely, instead, on their class notes or transcript service notes. Those who do purchase the textbooks become quickly frustrated in trying to keep up with long reading assignments that often stray into areas which, while interesting, are not directly relevant to their medical career.

The circumstances described above lead students to come to their course directors and lecturers and ask the common question: "What do I really need to read in the textbook?" The question, if you consider it carefully, is quite logical. What the students are telling us is that they do not have time in a compressed medical curriculum to read everything, and they want to know what is really important to their education as future clinicians (or dentists or nurses).

This textbook is our attempt to give the medical student an affordable textbook, of approximately 300 pages, which contains what we believe to be the content of cell biology that a clinician needs to know in the modern practice of medicine. We have limited the number of pages by focusing on the cell biology of humans and other animals, eliminating subjects that are of little interest to the medical student (e.g., plant cell biology). We have attempted to demonstrate the medical relevance of the subject by including many common diseases where the cellular and molecular basis of the disorder is understood. In addition, we begin every chapter with a clinical case that is directly linked to the material presented in the text. Schools that are using problem-based learning (PBL) techniques can utilize these cases in their small group sessions. Students in non-PBL curricula can utilize these cases for their own self-study or in student-formed study groups. The cases are meant to support and amplify the information stated in the text. While others have created shortened cell biology textbooks by eliminating cutting-edge science, we feel that this approach sacrifices the excitement of cell biology. We therefore have brought every subject up to the moment. We have utilized active subheadings that summarize the information of each section in the textbook. We find this a more useful tool for student review than a comprehensive summary at the beginning or end of each chapter.

In limiting our textbook to 300 pages, we decided not to include a free-standing chapter on the immune system. In most cases, immunology is currently taught by microbiology faculty as part of their medical courses, and not extensively within histology, physiology, or biochemistry courses. However, the cell biological aspects of the immune system are discussed within appropriate chapters of our textbook. Furthermore, we have included many light and electron micrographs of cell structure within this textbook. We feel that it is essential that the student come away with an understanding of how cell function and structure are linked. Many of the electron micrographs of organelles are taken from luteal cells within the corpus luteum, structures important to our current national interest in women's medicine.

We wrote this textbook because we needed it for our medical histology course, and believed that it would be of value to other schools teaching medical cell biology. We have attempted to write a genuine *medical* cell biology textbook. Your students will be the judges as to whether we have succeeded. If they buy the book, and you no longer hear the question "What do I really need to read in the textbook?", then we have made great progress. If they learn cell biology to your satisfaction, and become well-prepared physicians, then we will have succeeded. We welcome your feedback on our textbook.

Acknowledgments

The editor gratefully acknowledges Karen Frances Goodman for her typing of the textbook and Paula Spencer for her help in incorporating the equations in Chapters 1, 2, and 8. Further acknowledgment goes to Miss Gena Goodman for assistance in editing.

Contents

Chapter 4
Organelle Structure and Function   111

Chapter 5

Regulation of Gene Expression   165

Medical Cell Biology

Second Edition

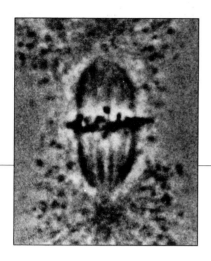

Chapter 1

Organization of the Cell

Clinical Case

A 25-year-old woman, Lorna Lane, complained of fatigue, weight loss, joint pains, and a petechial rash on her legs. She had thrombocytopenia (40,000/mm³), and a diagnosis of idiopathic thrombocytopenia purpura was made. After obtaining a bone marrow aspirate with an increased number of megakaryocytes, she was treated with prednisone (Deltasone, Prednicen, Sterapred). Her platelet count increased to 120,000/mm³ (normal range 150–250,000/mm³). At a follow-up visit she had developed an erythematous rash across the nose and cheeks that was worse in the sunlight, hepatomegaly, and hematuria. Her laboratory test showed pancytopenia, an elevated erythroid sedimentation rate, and hypergammaglobulinemia; her urine contained red cells, protein, and red cell casts. An autoantibody screen was positive for antinuclear and anti-DNA antibodies. A chest x-ray showed a small pleural effusion on the right side.

A question many medical students consider is: Why should we learn about cells? An obvious answer is that for the students to obtain passing grades in histology, pathology, biochemistry, physiology, and other medical school courses, they should have a basic knowledge of cell biology. A somewhat deeper and a bit more idealistic answer is found in an observation made almost 150 years ago by Rudolph Virchow. Virchow was a German pathologist who, in the 1850s, proposed that pathologic changes were due to malfunctions in cells. Virchow's ideas can be expanded to include infectious agents, which are bacterial cells, parasitic organisms, or viruses that take up residence in human cells. For a physician who is fighting disease, Virchow's proposal means that the arena in which those battles are fought is at the cell surface and within the cell.

Today, most modern clinical treatments involve either surgically removing aberrant or damaged cells, modulating the biochemical or metabolic activity of cells, specifically killing cells, or influencing the communications that occur among cells in the body. It is hoped that, in the future, increased knowledge of cells and disease will lead to clinical advances such as new vaccines, immunomodulatory drugs that either heighten or suppress the activity of the immune system, new drugs for fighting viral infections and specifically for destroying cancerous growths, more effective diagnostic reagents, and, perhaps, the ability to replace damaged or lacking genes within cells with healthy copies. For a physician to keep abreast of the rapid changes that occur in the clinical treatment of disease, a knowledge of cell biology will be essential.

Cell Structure

The Cell Is the Smallest Functional Unit of Tissues and Organs, and Eukaryotic Cells Contain Organelles

This textbook deals with the structure and function of eukaryotic cells, with special emphasis on human cells. First we must ask: What is a cell? Schleiden and Schwann, two nineteenth-century microscopists, first described cell theory as we currently know it. In 1858, Virchow stated, "Every animal appears as a sum of vital units, each of which bears in itself the complete characteristics of life." This early version of cell theory can be restated as follows: a cell is a living functional unit separated from its environment by a plasma membrane. Cells can be separated into two categories: prokaryotic and eukaryotic. We will be chiefly concerned with *eukaryotic cells*, which, by definition (the Greek word *karyon* means "nucleus"), contain a nucleus. Eukaryotic cells comprise protists, fungi, plants, and animals, and are generally 10–100 μm in linear dimension. Prokaryotic cells (bacteria and Cyanobacteria) are relatively small (1–10 μm diameter) and contain simple internal structure. The prokaryotic plasma membrane, often surrounded by a tough protective cell wall, encloses a single cytoplasmic compartment that contains DNA, RNA, proteins, and other small molecules.

The structure of a eukaryotic cell, as viewed through the light microscope, is presented in Fig. 1–1A. The smallest objects that can be visualized by the modern light microscope are approximately 0.5 μm wide. Therefore, in the light-microscopic view of the cell (see Fig. 1–1A), we can see a plasma membrane that defines the outer boundaries of the cell and surrounds the cell's protoplasm or contents. The protoplasm includes the nucleus where the eukaryotic cell's DNA is compartmentalized away from the remaining contents of the cell (the cytoplasm). In addition to the nucleus, the cytoplasm contains other organelles, which are better visualized with an electron microscope that can view components as small as 2 nm. An electron micrograph of a cell is presented in Fig. 1–1B. The eukaryotic cell's organelles include the nucleus, mitochondria, endoplasmic reticulum, Golgi apparatus, lysosomes, peroxisomes, cytoskeleton, and plasma membrane (see Fig. 1–1C). Mitochondria are similar to bacteria in size and shape. They contain their own mitochondrial DNA and make proteins. Like bacteria, they reproduce by dividing into two. Mitochondria contain a double membrane, with the inner membrane containing many infolds, referred to as the *cristae*, that contain the machinery for cellular respiration. The *endoplasmic reticulum* (ER) consists of flattened sheets, sacs, and tubes of membranes throughout the cytoplasm and is responsible for protein synthesis (rough ER) and lipid

metabolism (smooth ER and rough ER). The *Golgi apparatus* is a stack of flattened membranous sacs involved in the modification and transport of molecules made in the endoplasmic reticulum. The membrane-enclosed *lysosome* contains enzymes required for intracellular digestion. *Peroxisomes* are membrane-bound organelles in which hydrogen peroxide is generated and degraded. The organelles occupy approximately half the total volume of the cytoplasm. The remaining compartment of cytoplasm (minus organelles) is referred to as the *cytosol* or *cytoplasmic ground substance*. The cytosol comprises water, ions, various organic building-block molecules such as amino acids and nucleotides, numerous proteins, and important cellular molecules such as glucose and adenosine triphosphate (ATP) (Panel 1–1). Eukaryotic cells also differ from prokaryotic cells in having a cytoskeleton that gives the cell its shape, its capacity to move, and its ability to transport organelles and vesicles from one part of the cell cytoplasm to another. The cytoskeleton is composed of microfilaments (7–8 nm diameter), intermediate filaments (10 nm diameter), microtubules (24 nm diameter), and a spectrin-based membrane skeleton. Eukaryotic cells are generally larger than are prokaryotic cells and, therefore, require a cytoskeleton and membrane skeleton to maintain their shape, which is related to their functions. A schematic summary of cell structure is given in Fig. 1–1C.

The development of multicellular organisms required that cells closely related by ancestry become differentiated from one another, some developing certain features, others developing different traits. Although various cells within a human tissue (e.g., nervous tissue) are radically different from each other, they have descended from the same fertilized egg. Usually (one exception is human erythrocytes), cells retain all of the genetic material contained in the precursor fertilized egg. This raises the central question of developmental biology: Why does one cell become a muscle fiber and another cell a neuron if they contain the same genetic information? Simply stated, differentiation or specialization depends on gene expression. Eukaryotic cells contain a large amount of DNA, only approximately 1% of which encodes protein. The remaining DNA is structural (involved in DNA packaging) or regulatory (helping switch on and off genes). Therefore, whether a cell becomes a neuron or a muscle fiber depends on the expression of muscle-specific genes or neuron-specific genes (although both are present).

The remainder of this chapter will deal with the experimental methods that are used to study cells. Cell biology is an extremely broad field that encompasses cell structure, biochemistry, molecular biology, and immunology. As such, it will not be possible to describe all of the techniques that are used by modern cell biologists; nor will it be possible to provide much detail con-

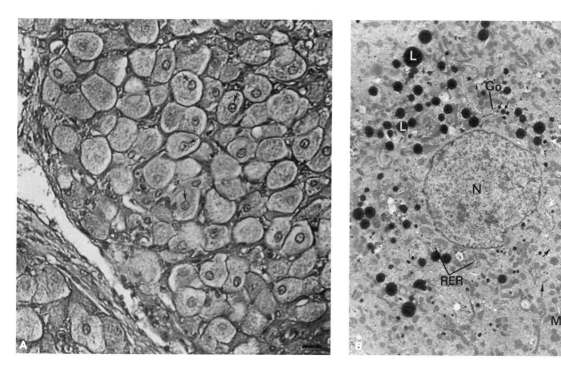

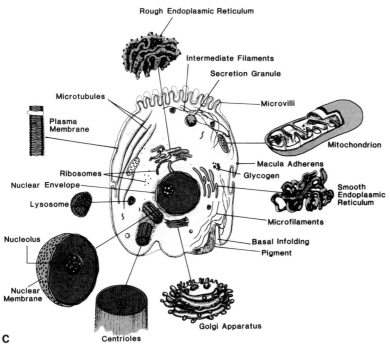

Figure 1–1. Cell ultrastructure. **A:** Light microscopy photomicrograph of a section of corpus luteum. This tissue contains large luteal cells (**arrows**, nuclei) that are responsible for both steroid and protein hormone synthesis (bar, 30.25 μm). **B:** Electron microscopy photograph of a large luteal cell. Note the nucleus (*N*), plasma membrane (*P*), mitochondria (*Mi*), cytoplasmic lipid (*L*), Golgi complex (*Go*) with forming secretory granules (*small arrows*), rough endoplasmic reticulum (*RER*), and stacks of smooth endoplasmic reticulum (*large arrows*) (bar, 1.25 μm). **C:** Artist's rendition of a typical human or animal cell with the organelles labeled. (**A** and **B** courtesy of Dr. Phillip Fields.)

cerning any of the individual techniques. However, entire books have been written on most of these procedures, and a list of pertinent references will be included at the end of the chapter.

Tools of the Cell Biologist

Various techniques are used today for studying cells. Because cells and organelles are too small to be examined by the naked eye, the backbone of cell biology is the microscope. However, the boundaries between the fields of cell biology and biochemistry, physiology, immunology, microbiology, and molecular biology are no longer clear-cut and distinct, and a modern cell biology laboratory often incorporates methods from each of these disciplines into the experimental analyses of eukaryotic cellular phenomena. Therefore, in this section, we will describe not only the various forms of microscopy, but also many of the other techniques that are contributing to the explosion of knowledge concerning cell structure and function.

The Resolving Power of a Microscope Is Limited by the Wavelength of the Illuminating Source

All microscopes, whether light or electron, work on the same principles and are governed by the same set of physical laws. All microscopes contain an illuminating source, which emits a form of radiation that is focused onto a sample to permit specimen viewing. In the microscope used in most laboratories, the illuminating source is a lamp that emits light in the visible range. The light is focused by a condensing lens onto a sample, then passes through the specimen and is captured by an objective lens. The light that is absorbed by the sample produces contrasts that help reveal structural detail. The light collected by the objective passes through an ocular lens in the viewing eyepiece that focuses it onto the retina of the eye (Fig. 1–2A). The principal reasons for using a light microscope for viewing cells and tissues are that microscopes magnify objects and, more importantly, allow the resolution of structural detail.

The *limit of resolution* is defined as a minimum distance between two closely positioned objects at which the two subjects can be observed to be distinct entities. The resolving power of the microscope is determined in part by the wavelength of the illuminating source and by the characteristics of the optical system, and can be calculated using the equation

$$d = \frac{0.61\, \lambda}{NA_{objective} + NA_{condenser}}$$

where *d* is the resolving power, λ is the wavelength of light used, and NA is the numerical aperture of the

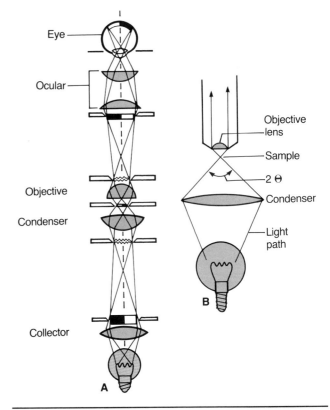

Figure 1–2. The characteristics of optical systems for light microscopy. **A:** A diagram of the basic bright-field light microscope. Light is focused on a sample by a condenser lens, then light that passes through the specimen is collected by an objective lens. The light then passes through an eyepiece or ocular lens and is focused onto the retina of the eye. **B:** The numerical aperture of a lens is dependent on the amount of light that can be picked up by the lens.

lenses. The numerical aperture for a lens is calculated by multiplying the sine of the half angle of light picked up by the lens by the refractive index of the medium between the lens and the specimen (see Fig. 1–2B). When viewing subjects through air, the maximum numerical aperture obtainable is 1.0, whereas the maximum numerical aperture possible with an oil immersion lens is 1.4. Therefore, under optimal conditions, the theoretical limit of resolution for a light microscope is approximately 0.1 μm. Objects that are closer than this distance to one another will be viewed as one. In reality, the resolving power of a microscope will not be at this optimal value because of flaws and aberrations in the optical system; the resolving power of the best light microscopes is closer to 0.2 μm. This value of resolution is approximately 1,000-fold better than the unaided human eye. Resolving power and magnification are not the same: An image can be magnified, or enlarged, to any size, but objects that are closer than 0.2 μm from one another will be viewed as a single entity in standard light microscopes.

Panel 1-1
The structures of important biomolecules and the formation of important chemical bonds

A. Amino acid structures with three-letter and one-letter codes.

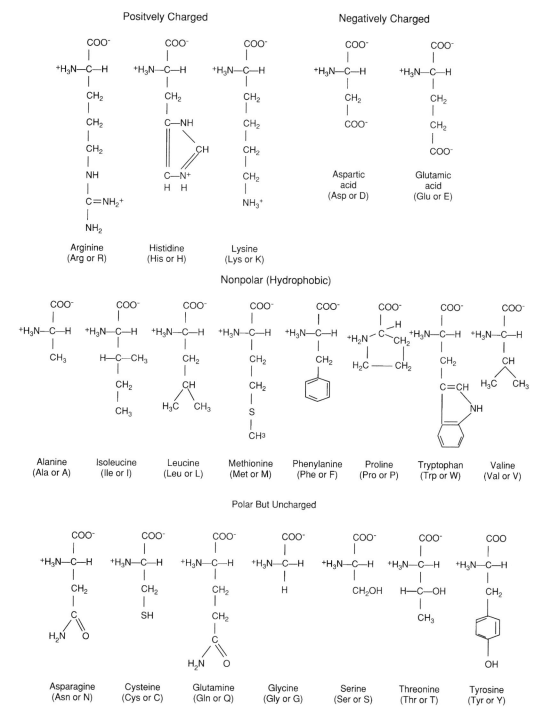

Positively Charged

Arginine (Arg or R), Histidine (His or H), Lysine (Lys or K)

Negatively Charged

Aspartic acid (Asp or D), Glutamic acid (Glu or E)

Nonpolar (Hydrophobic)

Alanine (Ala or A), Isoleucine (Ile or I), Leucine (Leu or L), Methionine (Met or M), Phenylanine (Phe or F), Proline (Pro or P), Tryptophan (Trp or W), Valine (Val or V)

Polar But Uncharged

Asparagine (Asn or N), Cysteine (Cys or C), Glutamine (Gln or Q), Glycine (Gly or G), Serine (Ser or S), Threonine (Thr or T), Tyrosine (Tyr or Y)

(continued)

Panel 1-1 (continued)

B. Peptide bond formation.

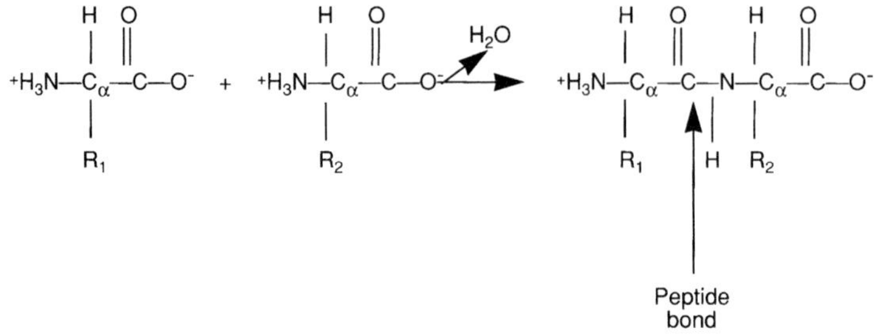

C. The 4 DNA bases.

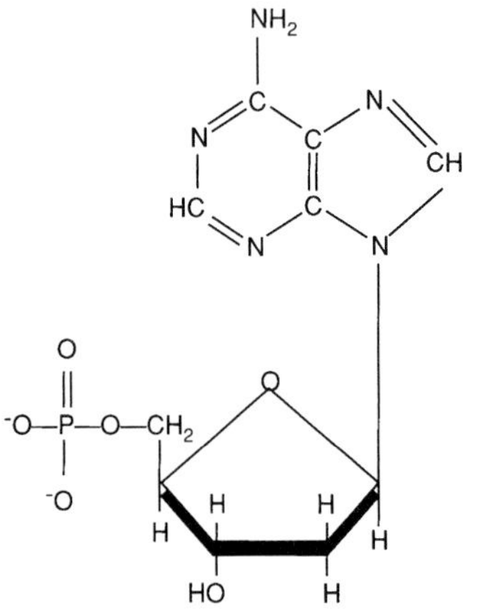

Deoxyadenosine
monophosphate

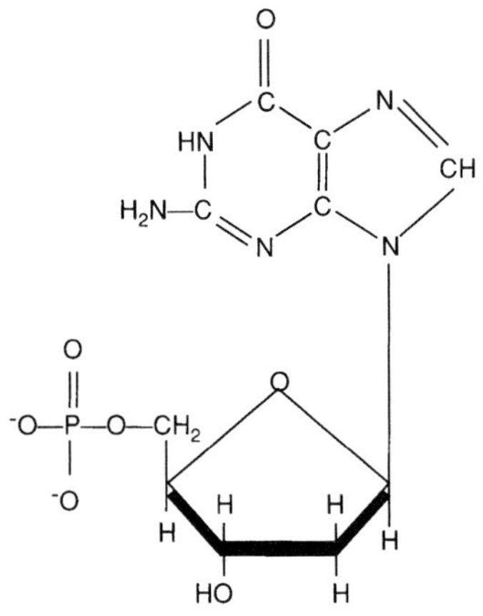

Deoxyguanosine
monophosphate

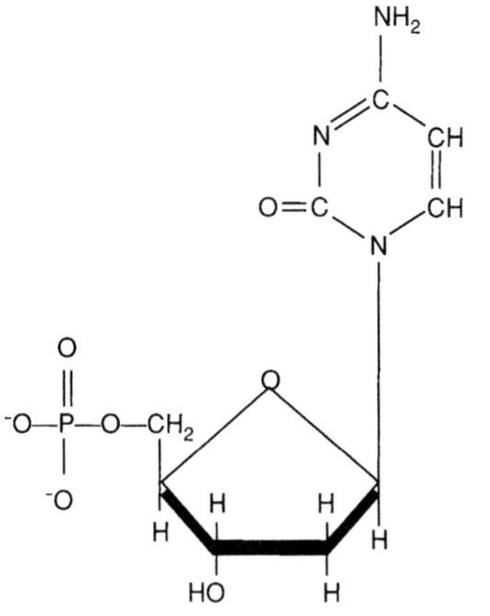

Deoxycytidine
monophosphate

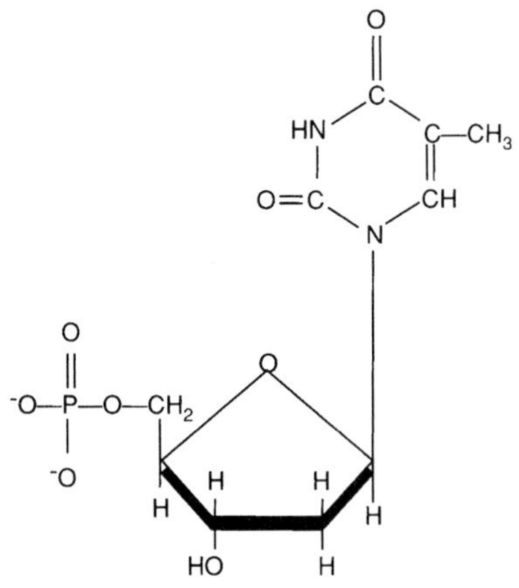

Deoxythymidine
monophosphate

(continued)

Panel 1-1 (continued)

D. *3′–5′ formation in the DNA backbone.*

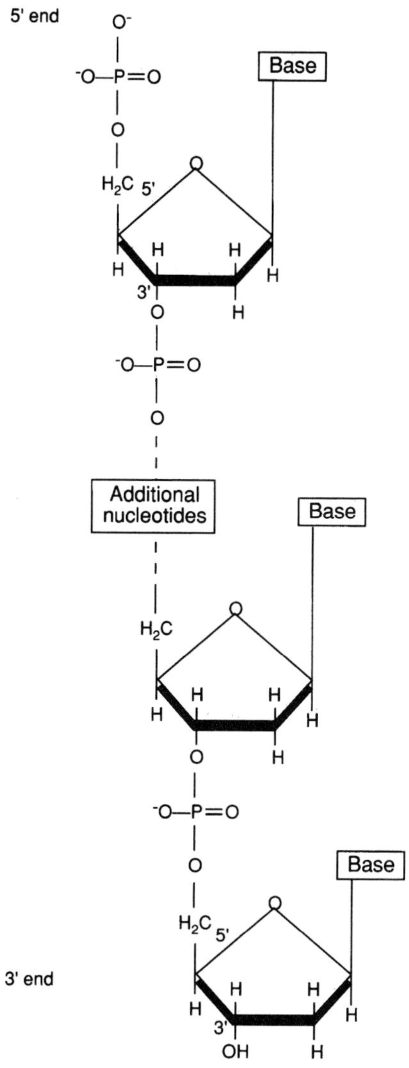

E. Hydrogen bonding during DNA base pair formation.

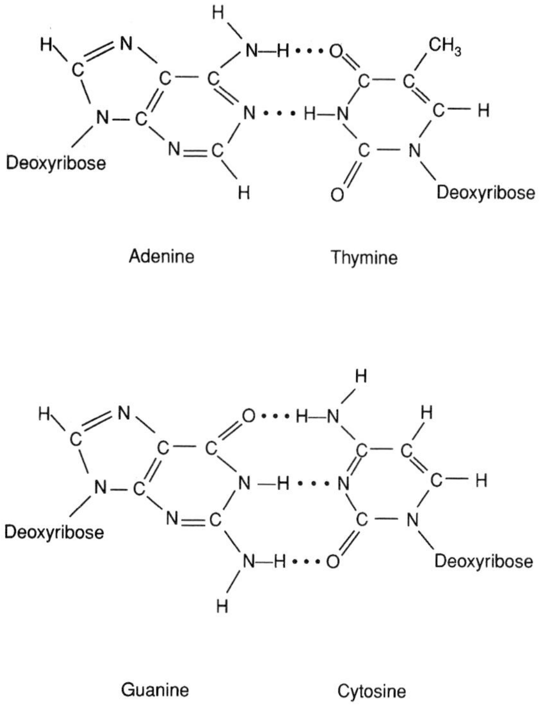

5' end

^-O—P=O

Base

H_2C 5'

3'

^-O—P=O

Additional nucleotides

Base

H_2C

^-O—P=O

Base

H_2C 5'

3' end

3'

OH

Adenine   Thymine

Deoxyribose

Deoxyribose

Guanine   Cytosine

Deoxyribose

Deoxyribose

(continued)

F. The four nucleotide bases of RNA

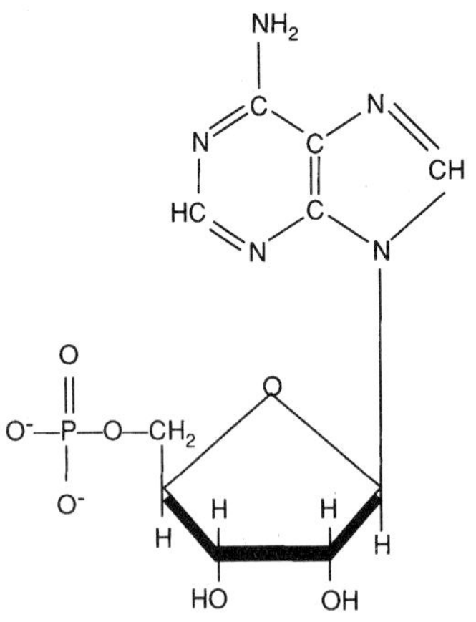

Adenosine
monophosphate

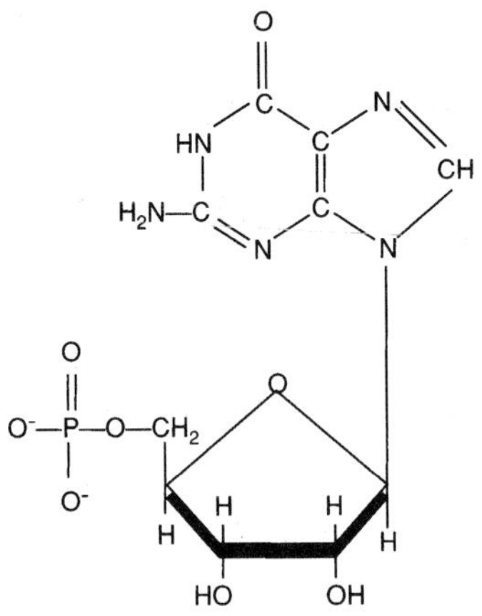

Guanosine
monophosphate

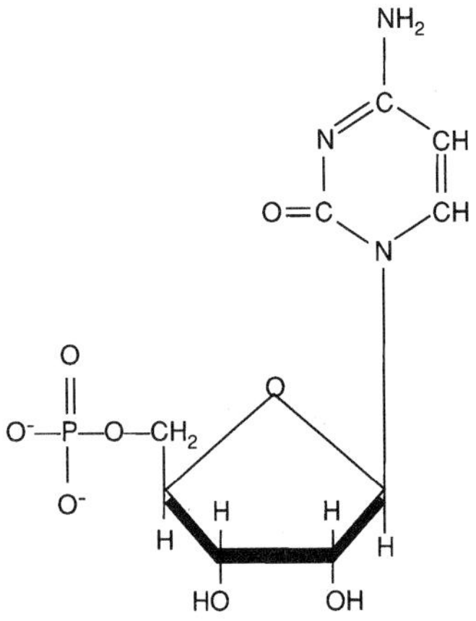

Cytidine
monophosphate

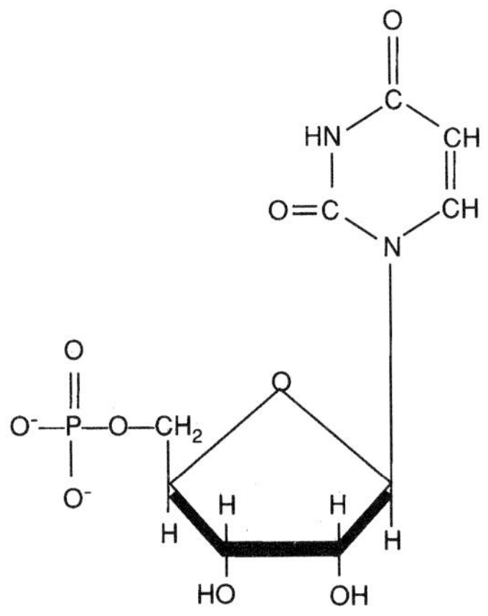

Uridine
monophosphate

(continued)

G. Membrane phospholipids

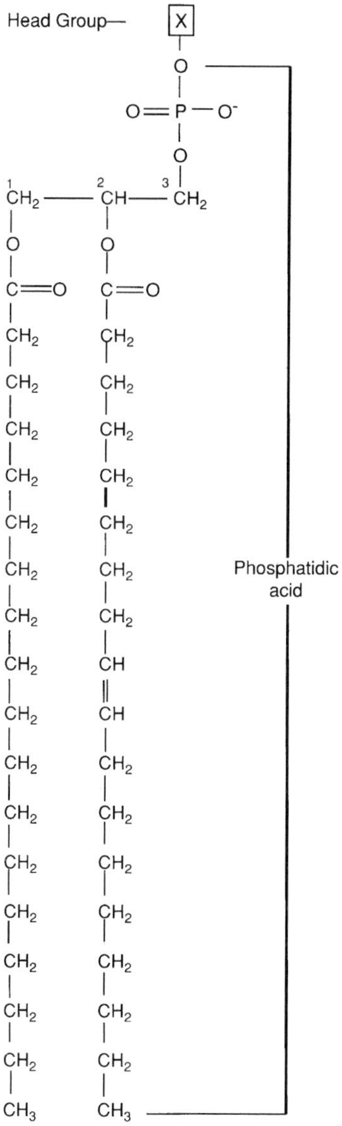

Head Group— [X]

Phosphatidic acid

1. General Structure

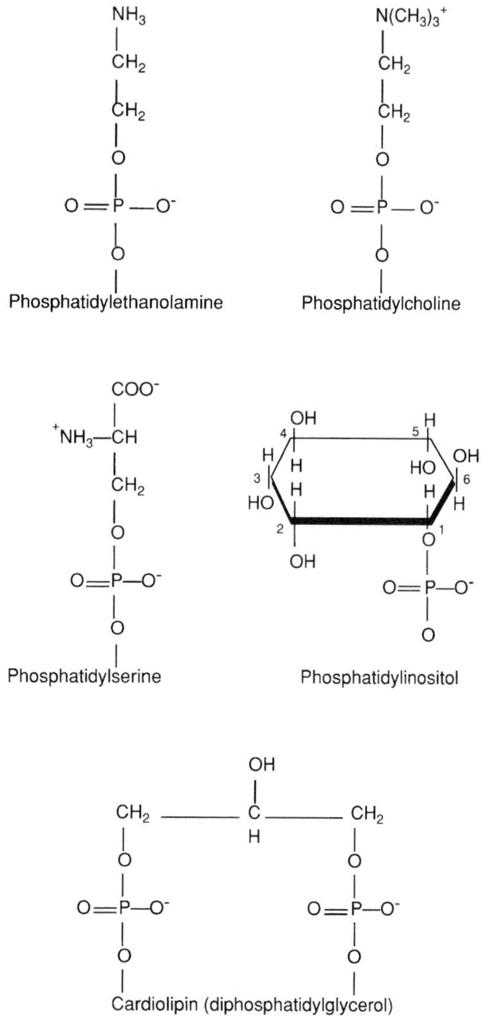

NH₃

Phosphatidylethanolamine

N(CH₃)₃⁺

Phosphatidylcholine

Phosphatidylserine

Phosphatidylinositol

Cardiolipin (diphosphatidylglycerol)

2. Polar Head Groups

(continued)

H. Other important membrane lipids

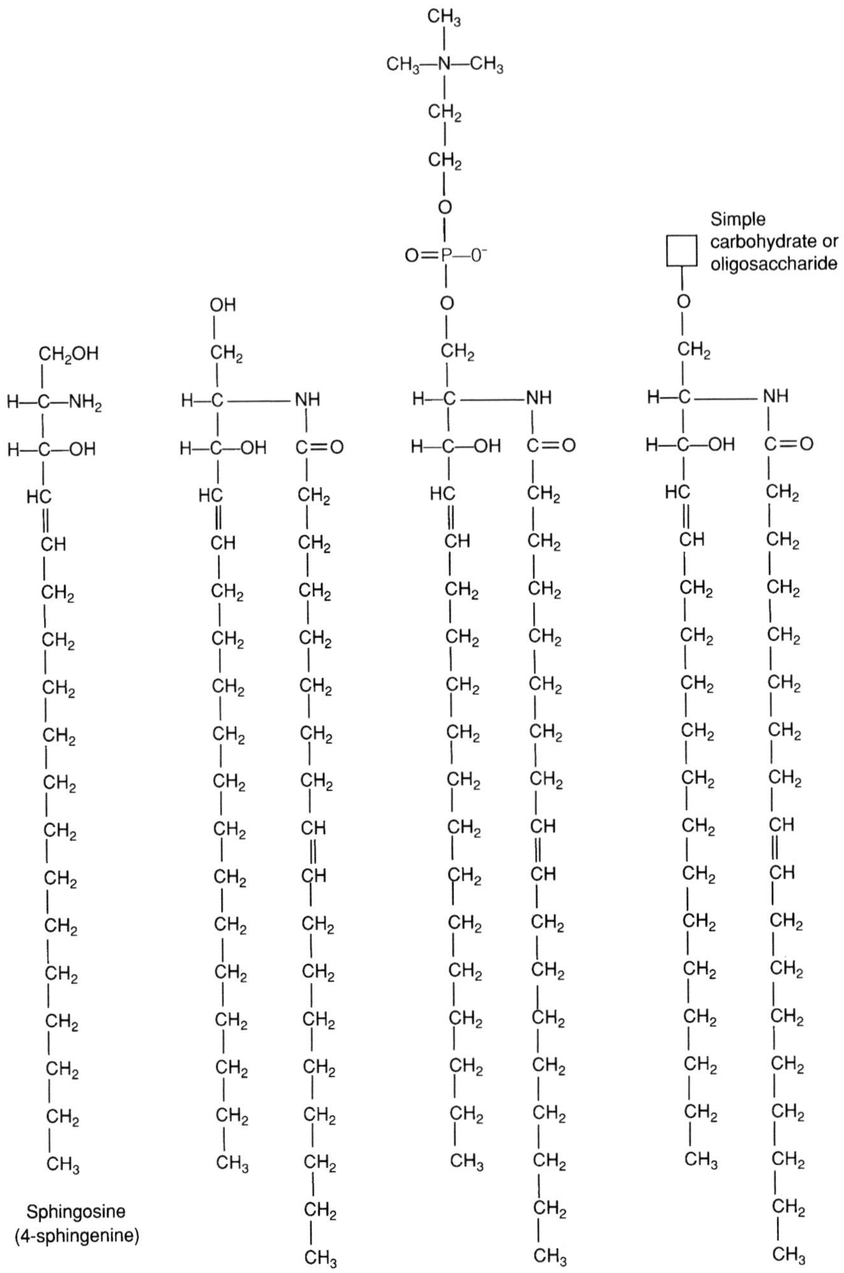

CH_3

CH_3—N—CH_3

CH_2

CH_2

O

O=P—O^-

Simple carbohydrate or oligosaccharide

OH	O	O	
CH_2OH	CH_2	CH_2	CH_2
H—C—NH_2	H—C——NH	H—C——NH	H—C——NH
H—C—OH	H—C—OH C=O	H—C—OH C=O	H—C—OH C=O
HC	HC CH_2	HC CH_2	HC CH_2
‖	‖	‖	‖
CH	CH CH_2	CH CH_2	CH CH_2
CH_2	CH_2 CH_2	CH_2 CH_2	CH_2 CH_2
CH_2	CH_2 CH_2	CH_2 CH_2	CH_2 CH_2
CH_2	CH_2 CH_2	CH_2 CH_2	CH_2 CH_2
CH_2	CH_2 CH_2	CH_2 CH_2	CH_2 CH_2
CH_2	CH_2 CH_2	CH_2 CH_2	CH_2 CH_2
CH_2	CH_2 CH	CH_2 CH	CH_2 CH
CH_2	CH_2 ‖	CH_2 ‖	CH_2 ‖
CH_2	CH_2 CH	CH_2 CH	CH_2 CH
CH_2	CH_2 CH_2	CH_2 CH_2	CH_2 CH_2
CH_2	CH_2 CH_2	CH_2 CH_2	CH_2 CH_2
CH_2	CH_2 CH_2	CH_2 CH_2	CH_2 CH_2
CH_2	CH_2 CH_2	CH_2 CH_2	CH_2 CH_2
CH_3	CH_3 CH_2	CH_3 CH_2	CH_3 CH_2
	CH_2	CH_2	CH_2
	CH_3	CH_3	CH_3

Sphingosine
(4-sphingenine)

Ceramide

Sphingomyelin

Glycolipid

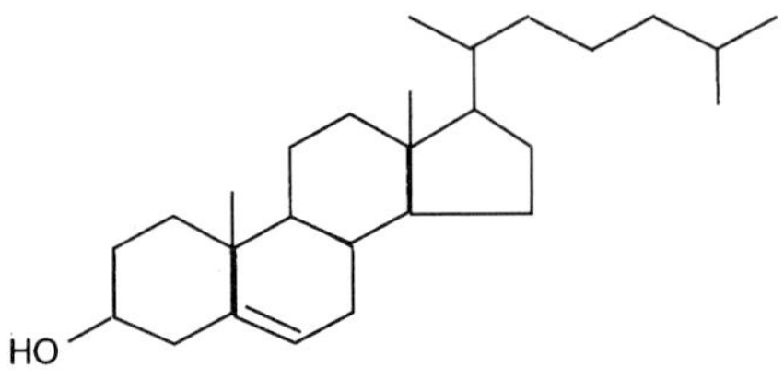

HO

Cholesterol

(continued)

Panel 1-1 (continued)

I. Important sugar residues.

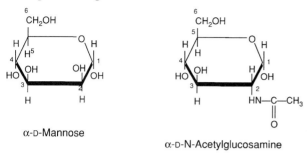

α-D-Mannose

α-D-N-Acetylglucosamine

α-D-Galactose

α-D-N-Acetylgalactosamine

α-L-Fucose

N-Acetylneuraminic acid
(sialic acid)

The Characteristics of Light Waves Can Be Exploited for Visualization of Living Cells

As light waves pass through cell and tissue samples, the light waves are distorted. This bending and retardation of light waves can be exploited to allow detailed observation of living cells and unstained materials. To understand how this can occur, it is necessary to consider some of the physical characteristics of light. Light waves travel through a microscope system by slightly different paths, and occasionally these light paths cross, or interfere with, one another, producing optical diffraction. When two waves intersect, they can either reinforce or cancel each other. For example, if two waves are completely in phase with one another, the crests and troughs of the two waves will reinforce each other to enhance brightness. Alternatively, two waves that are exactly out of phase will completely abolish each other, resulting in darkness.

There are small differences in the index of refraction for various parts of cells, and certain types of microscopes are able to take advantage of this phenomenon to increase the contrast of unstained samples. For example, light that passes through a dense part of a cell, such as the nucleus, will be retarded differently from the light that passes through less dense regions of the cell. This retardation causes a phase shift in the wave form of the light that passed through the nucleus, relative to the light that moved through thinner areas of the cell. The optical system of the phase-contrast microscope contains modifications in the condenser and objective lens that allow the out-of-phase retarded light waves to be paired with other out-of-phase waves, and the net effect is increased contrast. This enables one to visualize cellular components that could not be observed by standard light microscopy. Moreover, fixation and tissue processing are not required; hence, phase-contrast microscopy is a valuable tool for examining unfixed, living cells. A second type of light microscope, the differential interference contrast (DIC) microscope, also exploits the refractive characteristics of cellular components to create superior images of cell structure (Fig. 1–3). The DIC microscope is based on principles similar to those of the phase-contrast microscope, but the DIC microscope incorporates a series of prisms and polarizers to generate contrast in either fixed or living specimens.

Novel light microscope systems also can be developed by altering the illuminating source. For example, in the polarizing microscope, polarized light is passed through a cell or tissue sample, and a second polarizer is used to detect molecular orientation within the sample. Highly ordered structures, such as the spindle fibers of the mitotic apparatus, can be visualized clearly because they are able to alter the behavior of the polarized light (see Fig. 1–3). In another type of system called the *dark-field microscope*, light is aimed at the sample at an angle. Only those light rays deflected by the sample through the objective lens will be observed, and the cell and its components will be visualized as illuminated objects against a dark background.

Light microscopes can be manipulated further by coupling the microscopes to video monitors and computer systems to allow maximal image resolution. This procedure, known as *image processing*, has had a major influence on cell biology in recent years. Image processing incorporates the advantages of video cameras and computer electronics to overcome many of the limitations of a light microscope. The rationale for using these types of systems is quite simple. Because the images that are produced by video systems are electronic signals, the information can be digitized and fed into computers. The computers can then be manipulated to extract as much information as possible from a light microscope image. Moreover, these systems have the capacity to

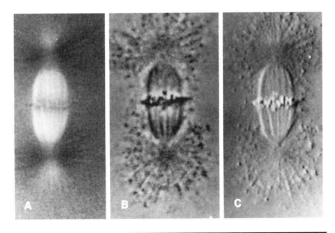

Figure 1–3. An isolated mitotic apparatus viewed by light microscopy. Examples of three different forms of light microscopy. The same isolated metaphase mitotic spindle is viewed with (**A**) polarization microscopy, (**B**) phase-contrast microscopy, and (**C**) DIC microscopy. (Salmon ED, Segall RR. *J Cell Biol* 1980;86:355–365. Reprinted with permission of Rockefeller University Press.)

enhance very slight differences in contrast that exist between objects. By doing so, objects that were indistinguishable from the background can be clearly visualized and resolved. In fact, objects as small as microtubules (24 nm diameter) can be observed routinely with this form of microscopy.

Tissues Usually Are Fixed and Stained for Histologic Examination

In the previous section, several expensive microscope systems were described that permit the investigation of living cells. However, there is little in cells and tissues that allows easy viewing of substructure with a standard light microscope. Therefore, tissues are usually treated with stains and dyes before being viewed microscopically. These dyes interact in a somewhat specific fashion with various cellular and extracellular materials, permitting easy viewing of tissue substructure.

Before staining, tissues generally are treated with a chemical or a mixture of reagents that results in fixation of the tissue. Common histologic fixatives are compounds that covalently cross-link macromolecules together, thereby freezing the tissue in a nearly native conformation. The fixative of choice for most routine histologic studies is usually an aldehyde, such as glutaraldehyde or formaldehyde, that forms covalent bonds between free amino groups on neighboring proteins. Sometimes, chemical fixation can cause deleterious artifacts in the sample. In these instances, tissues can be frozen very rapidly before observation. Rapid freezing of tissues, however, can suffer from artifacts caused by ice crystal formation, and special precautions must be taken to bypass this potential pitfall.

Because tissue pieces are much too thick to be observed at high resolution with the light microscope, fixed tissues must be cut into very thin slices before being stained and observed. These slices, or sections, are produced by using a machine called a *microtome*. A microtome consists of a sharp metal blade and a series of gauges and gears that allows reproducibly thin (1–10 μm thick) sections to be cut. Because the tissues themselves are much too fragile to withstand the shearing forces that are generated by the microtome blade during sectioning, the tissues must be embedded with a supporting medium to avoid sectioning artifacts. For light microscopy, the embedding resin of choice is usually paraffin, although plastic resins also can be used. When tissue pieces are placed into melted paraffin, the paraffin percolates between the cells, as well as into the inside of the individual cells. Upon cooling, the paraffin forms a solid block into which the cells are embedded. The embedded tissue is then ready for sectioning with a microtome.

Tissue sections can be stained by any one of several dyes. Many histologic stains react with specific cellular and extracellular components, whereas others react more generally with biological molecules. For example, silver salts specifically bind to reticular fibers in the extracellular matrix, and periodic acid Schiff (PAS) reagent labels only glycoproteins and other carbohydrate-rich macromolecules. Other reagents stain tissues in a much less specific manner. These types of stains, like hematoxylin and eosin (H&E), bind to tissue substances through simple charge interactions. As a result, many cellular and extracellular materials will be stained to varying degrees by these types of dyes.

Before staining, the paraffin-embedding medium must be removed from a histologic section to allow the stains to interact with the tissue components. This is accomplished by soaking the sectioned material in an organic solvent such as xylene. The extracted tissue is then rehydrated and stained. The most common staining method for histologic preparations is the use of a mixture of H&E. Eosin reacts with positively charged macromolecules (eosin is acidic) through electrostatic interactions. Accordingly, cytoplasm and most of the extracellular filaments will appear pink in tissue sections. Components that are stained by acidic dyes such as eosin are said to be *acidophilic*. Basic dyes carry positive charges and electrostatically couple to negatively charged cellular and extracellular structures. Hematoxylin, although not a true basic dye, stains tissues in much the same way as does a basic dye, owing to the interaction of hematoxylin with a second component that is added to a histologic section before the incubation of the section with H&E. The other component, called a *mordant*, binds to negatively charged cellular structures and then hematoxylin binds specifically to this intermediate between the biological macromolecules and the dye. As a result, DNA, RNA, and areas of dense rough endoplasmic reticulum will stain blue in H&E preparations. Cellular components that interact with basic dyes are said to be *basophilic.*

Individual Proteins Can Be Identified in Cells Using Selective-Staining Procedures

Techniques have been developed that permit specific labeling and localization of individual proteins inside of cells. For example, various enzymes can be identified in cells by incubating the cells in the presence of the appropriate substrates. In certain cases, the catalysis of the substrate can result in the formation of a precipitate in the cell, thereby localizing the enzyme in the cell. A much more common technique for identification and localization of intracellular proteins is to use specific antibody probes. The antibodies recognize their particular cellular antigens and bind tightly to the antigenic molecules. The antigen antibody complexes then can be visualized using either enzymatic or fluorescent secondary antibody probes. The secondary antibodies specifically interact with the primary immunoglobulin molecules that are bound to the cellular antigen of interest (Fig. 1–4A). For example, to study the intracellular distribution of microtubules, antibodies can be generated against the protein tubulin (the protein subunit of microtubules) by injecting purified tubulin into rabbits. The rabbit antiserum can then be added to a fixed

preparation of cells, and the antitubulin immunoglobulins will bind to the microtubules in the cells. After rinsing the preparation, the microtubules can be visualized by adding fluorescently labeled antibodies (prepared in a different species) that specifically recognize rabbit immunoglobulins. Because the antitubulin molecules that are bound to the cellular microtubules are immunoglobulins that were generated in rabbits, the fluorescently tagged antirabbit immunoglobulin antibodies will bind to the first antibodies that were added to the samples and will specifically identify the cellular microtubules when the preparation is visualized using a microscope that is equipped with a fluorescent light source and special filters (see Fig. 1–4B). The use of antibody molecules for the study of cell structure and composition is called *immunocytochemistry.*

For immunocytochemical analysis, pure antibody probes must be generated. Antibodies are produced when a purified protein antigen is injected into an animal. The immune system of the animal then recognizes the injected antigen as being foreign and mounts an immune response against the protein. A blood sample is collected from the immunized animal several days later, and the antibodies of interest are usually purified from the remainder of the serum proteins. The monospecific antibodies are then used for immunocytochemical analysis (see Fig. 1–4). Monoclonal antibodies are produced by a variation of this procedure. For monoclonal

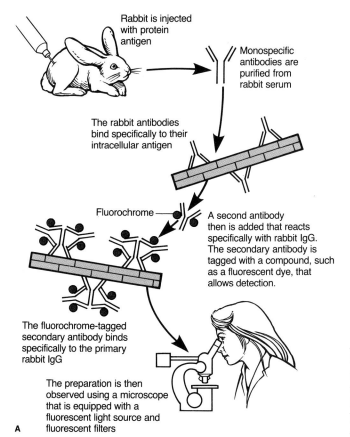

Rabbit is injected with protein antigen

Monospecific antibodies are purified from rabbit serum

The rabbit antibodies bind specifically to their intracellular antigen

Fluorochrome

A second antibody then is added that reacts specifically with rabbit IgG. The secondary antibody is tagged with a compound, such as a fluorescent dye, that allows detection.

The fluorochrome-tagged secondary antibody binds specifically to the primary rabbit IgG

The preparation is then observed using a microscope that is equipped with a fluorescent light source and fluorescent filters

A

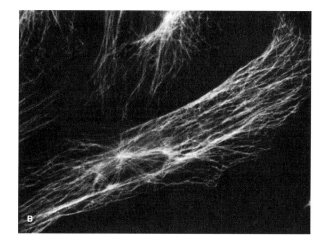

B

Figure 1–4. Immunocytochemical detection and localization of cellular proteins. **A:** A schematic showing how immunocytochemistry is used to localize cellular antigens. **B:** Antitubulin immunofluorescent staining of cultured mammalian cells. Numerous microtubules can be observed coursing throughout the cell. (**B** courtesy of R. Balczon.)

antibody production, the spleen is removed from an immunized animal (usually a mouse) and dissociated into individual cells. The spleen is used for this procedure because numerous plasma cells generally are trapped within the spleen at any instant. The immunoglobulin-secreting lymphocytes are then fused to an immortalized cancerous cell line, producing hybrid cells. The hybridomas that are secreting antibodies specific for the antigen of interest are selected from the population of hybrid cells using any one of several selection procedures (Fig. 1–5). The appropriate antibody-secreting hybridoma cell lines are grown in culture, and the antibodies produced by these cell lines are used for experimental studies. Antibodies produced by this procedure are called *monoclonal* because all of the antibody that is obtained is derived from a single parental plasma cell. Antibodies that are obtained from an animal's serum are said to be *polyclonal* because they were derived from several antibody-secreting plasma cells that were present in the animal's body. For certain experimental procedures (e.g., screening cDNA libraries), polyclonal antibodies are advantageous, whereas

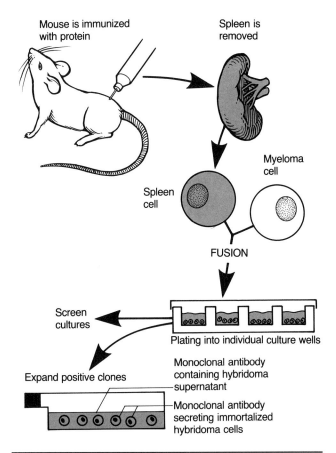

Figure 1–5. Monoclonal antibody production. A diagram demonstrating the procedures used to produce cell lines that generate monoclonal antibodies. Details are provided in the text.

monoclonal antibodies are more appropriate for other procedures (e.g., immunocytochemistry).

Antibodies that are produced by the previously outlined procedure can be used for the localization of tissue antigens by the technique of *immunofluorescence microscopy*. In immunofluorescence microscopy, fluorescent molecules are covalently attached to antibodies, and the antibodies are added to fixed samples. The tagged antibodies then attach to their appropriate cellular antigens and are visualized using a fluorescence microscope. The fluorescence microscope is a form of light microscope that contains a series of filters that permit the excitation and visualization of the fluorescent antibodies. When a fluorescent molecule is illuminated, it is excited and emits light at a second wavelength. The emitted light is visualized as a bright glow against a dark black background.

In addition to being a powerful research practice, immunofluorescence microscopy is also used for several practical pathologic-testing procedures. For example, certain tumor types can be identified because of specific marker antigens (e.g., intermediate filament proteins). In addition, several rheumatoid diseases are characterized by the production of autoantibodies that recognize specific cellular antigens. For example, human patients with lupus erythematosus produce autoantibodies that are reactive with DNA, whereas patients with a certain type of scleroderma generate antibodies that recognize antigens specific for the centromere/kinetochore region of chromosomes. Therefore, a common diagnostic tool when assessing patients exhibiting symptoms of autoimmune disease is to stain cultured cells with an aliquot of a patient's serum to determine whether the person is producing antibodies that recognize a specific cellular component.

The Electron Microscope Permits Analysis of Fine Ultrastructural Detail

Earlier in this section, a mathematical equation demonstrated that the resolving capacity of a microscope is a function of the wavelength of the illuminating source and the numerical aperture of the lens system being used. From the equation

$$d = \frac{0.61\,\lambda}{NA_{objective} + NA_{condenser}}$$

it is clear that one way of increasing the resolving power of a microscope is to use an illuminating radiation of shorter wavelength. Recall that d is the minimum distance between two objects required for those two subjects to be discerned as distinct entities. As d decreases, the resolving power of a microscope becomes greater. This is the rationale behind the electron microscope. In addition to having characteristics of a particle, electrons also exhibit wavelike behavior. Moreover, the wavelength of an electron is very short (0.004 nm). As a result, the theoretical limit of resolution of an electron microscope is approximately 0.002 nm. Because of aberrations in the lenses of an electron microscope, the actual limit of resolution is closer to 1–2 nm.

The design of an electron microscope is similar in principle to that of a light microscope (Fig. 1–6). A heated tungsten filament serves as both the electron source and the cathode of the microscope. An anode is also present and imparts a voltage difference between the two electrodes of up to 100,000 V. This voltage difference accelerates the electrons and drives them through the microscope column. Magnetic coils are placed along the path of the electron beam, and these powerful magnets focus the beam by serving as the condensing, objective, and projection lenses. A specimen holder is located between the condensing lens and the objective lens. As the specimen is bombarded by the electron beam, some of the electrons are deflected by

the sample, whereas others pass directly through the specimen. Those electrons that pass through the sample form an image by being focused onto a phosphorescent screen for viewing. Electron microscopes are also equipped with cameras so that data can be collected and analyzed in detail. A further addition to the complexity of an electron microscope is that electrons are deflected by collisions with air molecules. Therefore, electron microscopes must be equipped with vacuum pumps, and before the formation of the electron beam the air must be pumped completely out of the electron microscope. The electrons are then accelerated down the microscope column.

Samples for Electron Microscopy Require Special Methods of Preparation

To visualize cell or tissue preparations with an electron microscope, certain problems must be overcome. For example, electrons cannot pass through samples that are cut to the thickness of histologic sections. Instead, exceedingly thin sections must be cut. Special methodologies and instruments have been developed that allow this potential pitfall to be overcome. In addition, special stains must be used that allow visualization of the relatively transparent cellular components. These stains are all salts of heavy metals. When these heavy-metal stains bind to cellular constituents, they cause electrons to be scattered, resulting in increased contrast.

Similar to preparation for observation with the light microscope, samples must be fixed before staining and observation with an electron microscope. For electron microscopy, a two-step fixation process with glutaraldehyde and osmium tetroxide is often used. Glutaraldehyde cross-links adjacent protein molecules, whereas osmium tetroxide stabilizes lipid membranes and proteins. After fixation, the cells or tissues are infiltrated and embedded with a tough plastic resin. Plastics are used for electron microscopy, rather than paraffin, because tissues embedded in plastic can be cut into much thinner sections than can paraffin-embedded materials. Material must be cut into extremely thin sections for viewing with the electron microscope because electrons, even when accelerated to 100,000 V, have very limited ability to penetrate biological materials. For electron microscopic observation, a microtome equipped with either a diamond or a glass knife is used to cut thin sections that are approximately 1/100 the thickness of an individual cell.

An image obtained with an electron microscope is as dependent on electrons that are deflected or absorbed by the sample as it is on electrons that pass completely through the section. Consequently, sectioned material is stained with salts of either uranium or lead to increase contrast. Important cellular and extracellular molecules are composed principally of atoms with low atomic

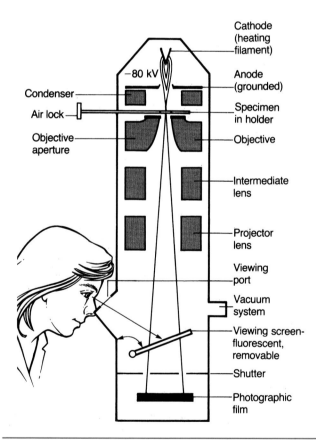

Figure 1–6. Transmission electron microscopy. A schematic showing the components of a transmission microscope. A beam of electrons is emitted from a cathode and focused by a magnetic condenser onto a sample. The scattered electrons are then focused by objective and projection lenses onto a fluorescent screen for viewing. (Modified from Widnell CC, Pfenninger KH. *Essential Cell Biology.* Baltimore: Williams & Wilkins, 1990.)

numbers such as C, H, O, and N. Molecules that are composed of these low molecular mass atoms do not deflect electrons well and, as a result, contrast is very low in unstained material. When sections are stained with uranyl acetate or lead citrate, contrast is increased because individual cellular constituents bind these heavy-metal salts with varying affinities. In addition to lead and uranium, the fixative osmium tetroxide is also an excellent stain for electron microscopy. Specifically, osmium stains the surfaces of membranes very well, giving lipid bilayers their characteristic trilaminar-staining patterns.

The type of electron microscope that develops an image because of electrons that are transmitted through a section is a transmission electron microscope (TEM).

Cell Surface Morphology Can Be Studied with Modified Electron-Microscopic Procedures

A second principal type of electron microscope, the scanning electron microscope (SEM), is used for the study of cell surface topography. In SEM, a fixed sample is coated with a thin film of heavy-metal atoms. The

Figure 1–7. Fertilization of a sea urchin egg visualized by scanning electron microscopy. An example of how scanning electron microscopy can be used to investigate cell surface events. This figure shows a sea urchin sperm binding to the surface of an egg. (Schatten G, Mazia D. *Experimental Cell Research* 1976;98:325–337. Reprinted with permission of Academic Press, Inc.)

sample then is scanned with a beam of electrons, and the electrons that are reflected by the metal-coated surface of the sample are collected by an electron detector and displayed as a three-dimensional image on a television monitor. Scanning electron microscopy provides an accurate presentation of cell surface features, with a resolution of about 10 nm (Fig. 1–7).

A modification in the preparation of a tissue sample provides a novel method for studying membrane morphology using TEM. In this procedure, known as *freeze-fracture electron microscopy*, tissue samples are rapidly frozen to near -180°C in the presence of a compound that prevents ice crystal formation, such as glycerol. The frozen block is then split with a knife blade, and the fracture plane through the frozen sample usually passes directly through hydrophobic lipid bilayers. This exposes the interior leaflets of the lipid bilayer. The fractured surfaces are coated with platinum, and the tissue is dissolved away, leaving a metallic replica of the interior of plasma membranes. Freeze-fracture provides a powerful mechanism for studying the distribution of intramembranous proteins. (This technique is discussed in further detail in Chapter 2.)

Cell Fractionation Allows for the Purification and Analysis of Individual Organelles and Proteins

Microscopy enables detailed examination of cell behavior and cell morphology. However, to elucidate how cells work, it is necessary to purify the individual cellular constituents. This involves lysing cells and using any one of several different purification procedures to isolate fully functional organelles, enzymes, and proteins. The information that is obtained from these *in vitro* analyses provides important clues to how cellular components function.

Cells and tissues can be disrupted into a thick slurry by using either a blender or a homogenizer. If performed properly, the homogenate will contain all of the cellular organelles in an intact state, as well as numerous small membranous vesicles that are fragments of the endoplasmic reticulum. The intact organelles and vesicles can be separated according to size and density by ultracentrifugation. Ultracentrifuges are machines that are capable of generating extremely high centrifugal forces. For example, the best ultracentrifuges available today can rotate at rates of nearly 80,000 rpm and can generate centrifugal forces of approximately $500,000 \times g$. When a tube that contains a cellular homogenate is spun in an ultracentrifuge, the organelles pellet at different rates. For example, at very low centrifugation speeds, large structures such as nonlysed cells and intact nuclei will pellet. At higher speeds, components such as mitochondria and lysosomes will be collected, whereas at very high speeds small membranous structures will pellet. By

Figure 1-8. The purification of cellular components by centrifugation. A schematic demonstrating how centrifugation at progressively higher gravitational forces separates cellular homogenates into enriched organelle fractions.

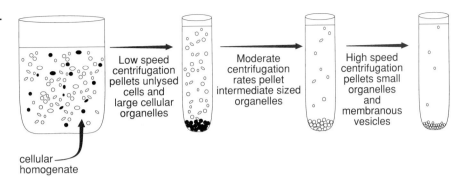

varying the rate and length of centrifugation, a highly enriched fraction can be obtained that contains the organelle of interest (Fig. 1–8). For some procedures, this fraction is suitable for experimental analysis. Alternatively, other methods can be used to purify the organelle of interest further.

A modified form of ultracentrifugation enables one to purify certain cellular constituents. In this procedure, a cellular homogenate or enriched fraction is layered on top of a gradient composed of either sucrose or salt. In this form of centrifugation, called *buoyant density centrifugation*, the cellular constituent of interest will migrate down the gradient during centrifugation until it reaches a region that is the same density as the organelle. At this point, the constituent can move no further and forms a distinct band in the centrifuge tube (Fig. 1–9). The rate at which a cellular organelle or component migrates through a gradient depends on both its shape and its size, and is called the *sedimentation coefficient* of that organelle.

Methods Have Been Developed That Permit the Purification of Individual Proteins

Like organelles, certain large proteins can be enriched considerably by high-speed ultracentrifugation. However, the method of choice for isolating proteins is column chromatography. In column chromatography, a cell or tissue extract is placed on top of a long, cylindrical column that is packed with a chromatographic resin. As the proteins migrate through the column, the individual proteins are retarded to various degrees. As the buffer flows out of the bottom of the column, small fractions are collected and analyzed for the presence of the protein of interest. Proteins that are purified in this fashion usually maintain much or all of their biological activity. As a result, the individual proteins can be studied *in vitro*, and the data that are obtained from these experiments can provide important information regarding the *in vivo* behavior of the protein or enzyme that is being studied.

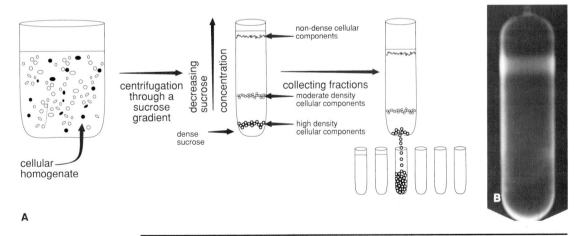

Figure 1-9. Density-gradient centrifugation can be used to purify cellular components. **A:** A schematic showing how samples will fractionate in a gradient during centrifugation in an ultracentrifuge. The individual bands, which contain subcellular components, can be collected and analyzed. **B:** A photograph showing how DNA banded during ultracentrifugation through a gradient of CsCl. (Micrograph courtesy of W. Zimmer and A. Kovacs.)

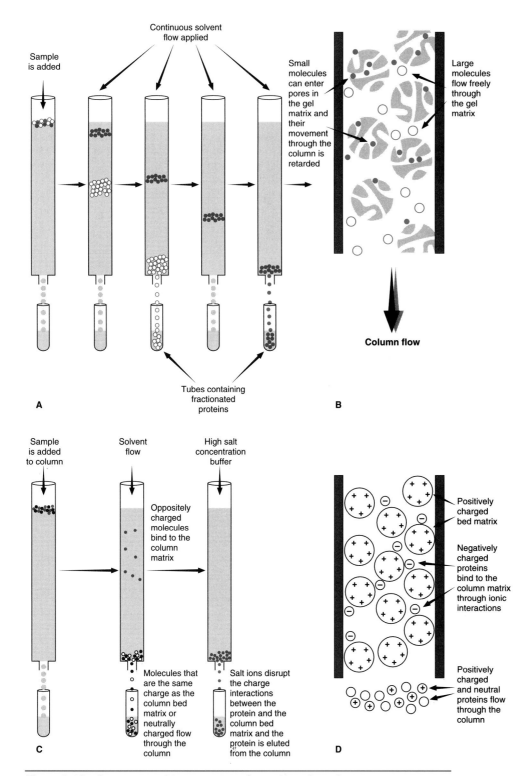

Figure 1–10. Chromatographic separation of proteins. These diagrams represent how proteins can be isolated from a protein mixture by either gel filtration or ion-exchange chromatography. **A:** In gel filtration chromatography, components are separated by size as proteins flow through the column, with larger molecules moving through the column faster. The individual proteins are collected in fractions. **B:** A schematic demonstrating how small molecules are trapped in pores in the gel matrix during gel filtration chromatography, resulting in retardation of the smaller molecules by the column bed. **C:** In ion-exchange chromatography, proteins bind to the charged column matrix because of charge interactions between the column matrix and the protein. The protein can then be eluted from the column by increasing the ionic strength of the buffer that is added to the column. **D:** A schematic demonstrating how charged proteins can bind to an ion-exchange matrix composed of molecules that are oppositely charged.

Different types of chromatographic resins exist, and each form of resin has its own unique protein separation characteristics. Gel filtration columns are packed with a resin that is composed of tiny porous beads. As a cell extract or protein mixture moves through the column, some protein molecules are small enough to enter the pores in the beads, whereas other molecules are so large that they flow through the column in a relatively unretarded fashion. Proteins that are small enough to enter the pores of the beads migrate through the column much slower than do the large proteins. Therefore, gel filtration chromatography allows the separation of proteins by size (Fig. 1–10). Ion-exchange resins, on the other hand, permit the separation of proteins according to charge. Ion-exchange columns are packed with resins that are either positively or negatively charged. As a mixture of proteins migrates through an ion-exchange column, some of the proteins will bind to the column through ionic interactions. The bound protein or proteins can then be eluted by rinsing the column with a high ionic strength buffer (see Fig. 1–10). A slight modification of ion-exchange chromatography is hydrophobic chromatography. In this method of protein separation, the column resin is composed of beads that contain exposed hydrophobic side chains. Proteins that contain exposed hydrophobic domains will be retarded and separated from the other proteins in a preparation. Experimental strategies can be developed using these various forms of chromatography that will permit the purification of a protein to homogeneity while maintaining biological activity in the molecule.

Affinity chromatography is a form of chromatography that permits very rapid purification of biological molecules. Affinity chromatography takes advantage of specific binding interactions that occur between proteins, such as enzymes and immunoglobulins, for their substrates or antigens, respectively. For example, a purified antibody can be coupled covalently to a chromatography resin, then a cellular extract can be passed through the column. All of the cellular proteins will flow through the column except the single protein antigen recognized by the antibody that has been coupled to the inert chromatography resin. The antigen–antibody interaction can be disrupted by running a buffer of either high ionic strength or low pH through the column, and the cellular antigen can be collected. Affinity chromatography allows the rapid purification of proteins to near homogeneity.

Electrophoresis Permits the Size of a Protein To Be Calculated

Because proteins are charged molecules, they will migrate in solution if an electric current is supplied to a protein mixture. This procedure, known as *elec-*

trophoresis, is a very powerful means for determining the size and subunit composition of a protein.

The most common form of electrophoresis used today for the study of proteins is sodium dodecylsulfate–polyacrylamide gel electrophoresis (SDS–PAGE). In this procedure, a protein or protein mixture is treated with a buffer that contains SDS, a negatively charged detergent. The SDS binds tightly to proteins and, in so doing, imparts a negative charge to all of the proteins in the preparation. In addition, a disulfide-reducing agent, such as mercaptoethanol, is usually added to the SDS-containing buffer. By cleaving disulfide bonds, mercaptoethanol allows multisubunit proteins to be broken down into their individual polypeptide subunits. The treated protein sample is then placed atop a polymerized gel matrix composed of polyacrylamide. When an electric current is applied to the preparation, the negatively charged SDS-treated proteins migrate away from the negative electrode and into the polyacrylamide gel matrix located between the sample and the positive electrode. As the proteins migrate through the gel matrix, the polypeptides migrate at different rates, with small peptide chains migrating at much faster rates than large polypeptides. Therefore, SDS–PAGE allows the separation of a mixture of proteins into discrete bands that can be visualized by staining the gel (Fig. 1–11). The relative molecular mass (M_r) of the protein of interest can be calculated by comparing the migration of that protein with the migration rate of known standard proteins.

An extension of SDS–PAGE called *Western blotting* is an important method that is routinely used for standard immunocytochemical studies. In this procedure, a mixture of proteins is separated by SDS–PAGE. The separated proteins are then electrophoretically transferred to a sheet of nitrocellulose paper, and the nitrocellulose sheet is probed with an antibody. The specific protein recognized by the antibody can be visualized by incubating the nitrocellulose with a second antibody that specifically reacts with the primary antibody. The secondary antibody is usually coupled to either a radioactive compound or an enzyme that allows easy detection. Western blotting is important for the characterization of antibodies and for the analysis of the protein composition of cells.

A modified form of standard gel electrophoresis permits resolution of hundreds, and perhaps thousands, of individual cellular proteins. In this procedure, called *two-dimensional gel electrophoresis*, cellular proteins are separated first according to charge and then according to size. The first-dimensional separation, called *isoelectric focusing*, separates polypeptides in a thin cylindrical tube gel on the basis of their isoelectric point (pH at which the protein has no net charge). This occurs because a pH gradient is established in the tube gel during this electrophoresis procedure. Because isoelectric

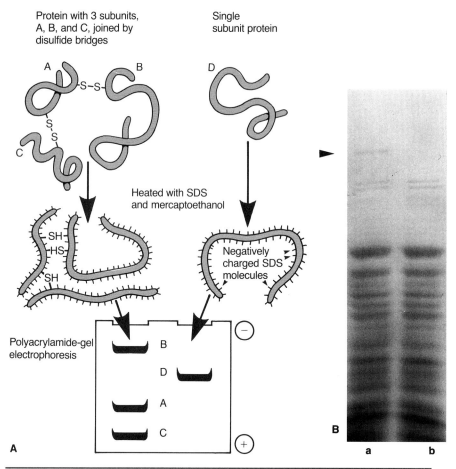

Figure 1–11. Use of SDS–PAGE for characterizing protein preparations. **A:** A diagram demonstrating SDS–PAGE. Proteins are treated with SDS and mercaptoethanol. The SDS binds to proteins, resulting in the denaturing of the protein. It also imparts a negative charge on all proteins. Mercaptoethanol reduces disulfide bonds. The proteins and protein subunits then separate as they migrate through the polyacrylamide gel matrix in an electrical field. **B:** An example of SDS–PAGE. The photograph shows two different samples of *Escherichia coli* cells. In one lane (*lane b*), native *E. coli* extract was run, whereas in the other lane (*lane a*), an extract of a strain of *E. coli* expressing a foreign protein was run. A high molecular mass band can be observed in lane a (*arrowhead*) that is not present in lane b, demonstrating that the treated *E. coli* are expressing the foreign protein. The gel has been stained with Coomassie Blue to visualize the proteins. (**A**, modified from Alberts, et al. *Molecular Biology of the Cell*, 2nd ed. New York: Garland Publishing, 1989; **B**, courtesy of R. Balczon.)

focusing is dependent on the charge of the native protein, protein samples are not treated with SDS for this procedure. For second-dimensional separation, the thin tube gel containing the focused proteins is soaked in a buffer that contains SDS to denature the proteins, after which the entire tube gel is layered on top of a standard polyacrylamide gel. The focused proteins are electrophoresed through the matrix of the polyacrylamide gel, which separates polypeptides according to size. The gel is stained to visualize the protein profile of the sample that is being examined.

Mammalian Cells Can Be Studied by Growing Tissue Preparations in a Culture Flask

An important technique used in most cell biology laboratories is cell culture. Most cells, whether plant or animal, will grow *in vitro* if supplied with a growth medium of composition similar to the environment in which that particular cell type normally flourishes. Because cultured cells maintain many of the characteristics that they exhibit *in vivo*, the ability to study cells in culture makes it much easier to study cell phenomena. For example, cultured fibroblasts usually continue to secrete collagen. For the researcher interested in examining the mechanisms of collagen production and release, it is much simpler to manipulate cultured fibroblasts than to perform experiments on fibroblasts that inhabit an animal's connective tissue.

Cell culture media are designed to mimic the normal physiologic environment. These media are isoosmotically correct mixtures of amino acids, salts, vitamins, and other components such as glucose. In addition, most mammalian cell culture media contain between 5% and 10% horse or calf serum. Serum contains low levels of protein growth factors, such as platelet-derived growth factor, which are responsible for stimulating cell division. The activity of growth factors will be considered in detail in Chapter 9.

Most types of human cells have a limited life span in culture and usually will die out after 50–100 cell divisions. However, several cell lines have been established that are essentially immortal. These cell lines usually have been derived from tumors or by transforming cells with a cancer-causing agent such as an oncogenic virus or chemical. One should remember that transformed cells quite often exhibit characteristics that are considerably different from those of the cells that are present in normal body tissues. Consequently, many researchers prefer to study primary cultures that are prepared by enzymatically disassociating a piece of normal tissue. Cells in a primary culture usually show most of the differentiated properties that they exhibited *in vivo*.

Cultured Cells Can Also Be Used as Systems for the Study of Physiologic Processes, Developmental Events, and Immune System Activity

The boundaries between the various biomedical disciplines are disappearing rapidly. Now, cell biologists routinely use techniques that are associated classically with other fields. Likewise, procedures that were considered to be cell biological are having a distinct effect on other disciplines. Obviously, various forms of microscopy are used by scientists in all fields of endeavor. Cell culture is another technique that is employed for numerous types of studies. Cell culture is used by electrophysiologists to study membrane phenomena, by immunologists and endocrinologists to investigate cellular communication events, and by developmental biologists to study the events of differentiation and morphogenic movement.

Microelectrodes can be inserted into cells to investigate ion concentrations inside cells. For these types of studies, pieces of fine glass tubing are melted and pulled so that the tips are only a fraction of a micrometer in diameter. The micropipet is then filled with an ionic

solution and subsequently poked through the plasma membrane so that the tip is in the cytoplasm. The electrode tip then works in a manner analogous to a common laboratory pH meter and, depending on the composition of the electrode tip, the intracellular concentration of ions such as H^+, Na^+, Cl^-, Ca^{2+}, Mg^{2+}, or K^+ can be measured, and any fluxes that occur in the concentrations of those ions after an experimental manipulation of the cell can be measured.

A modification of this technique is called *patch clamping*. In patch clamp studies, a fine micropipet is placed against the surface of a cell, and a slight vacuum, which causes a small piece of the plasma membrane to be torn away from the surface of the cell, is applied to the tip of the micropipet. This patch of membrane, with its associated ion channels, remains firmly attached to the tip of the pipet (Fig. 1–12), and the movement of ions through the individual channels in the membrane can be measured. This procedure has been used recently to demonstrate that the underlying cellular abnormality

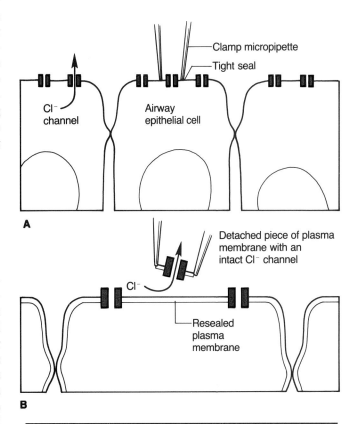

Figure 1–12. Patch clamping is used to study membrane channels. A schematic of how patch clamping works for the study of membrane channel activity. In this example, (**A**) a patch clamp electrode is lowered onto the surface of a human airway epithelial cell. A vacuum is applied to the electrode, causing (**B**) a piece of the membrane, with its associated Cl^- channels, to be torn from the cell surface. The movement of Cl^- ions through the channels can then be measured after various experimental treatments.

that results in the clinical manifestations of cystic fibrosis (CF) is the malfunctioning of chloride channels in the
epithelial cells of CF patients (see Fig. 1–12).

The insertion of micropipets across the plasma
membranes of cells has been modified for additional
types of studies. For example, in microinjection
micropipets are loaded with a protein-containing solution and, after the cell is impaled with the micropipet, a
small volume of the protein is injected into the cytoplasm and the effects on the treated cell are monitored.
A common type of study is to microinject a monospecific antibody into a cell. The antibody then binds to its
intracellular target antigen and inhibits the function of
the target protein. In effect, this specifically removes a
protein from a cell, and the role of this protein in normal cellular processes can be deduced by studying the
cells from which this protein has been removed experimentally. Likewise, microinjection can be used to study
cell coupling and intracellular ion fluxes. For cell-cou-
pling studies, a low molecular mass fluorescent compound can be injected into one cell, and the neighbors
of that cell can be observed with a fluorescence microscope. If the fluorescent dye passes from the injected cell
to a neighboring cell, then those cells must be coupled
by gap junctions (see Chapter 6). For ion flux studies,
an ion-sensitive fluorescent dye (such as the Ca^{2+}-sensi-
tive dye aequorin) is injected into a cell, and when an
intracellular change in the concentration of that ion
occurs the compound fluoresces. The amount of cellular
fluorescence can be quantitated so that the change in
ion concentration can be measured directly.

In recent years, immunologists have made great
strides in understanding the immune response. Many of
the key events that drive the immune response are due
to both direct interactions that occur between cells of
the immune system and to the activity of secreted molecules. These secreted molecules, called *interleukins*, are
proteins that are released from one type of immune cell
that trigger another cell either to proliferate or to
respond in some other fashion to a foreign antigen.
These cell–cell interactions that drive the immune
response were worked out in experimental cell culture
systems. Moreover, many of the key signaling molecules
have been purified from culture supernatants, and these
proteins have been characterized in detail.

Cell culture systems are also being used by developmental biologists to study cell differentiation events.
Certain cell types, such as muscle cells and many cells
found in the nervous system, will carry out their normal differentiation processes *in vitro*. For these types
of studies, undifferentiated tissue is removed from an
embryonic animal and dissociated into single cells. The
cells are plated in a culture vessel, and many of the
cells proceed with their normal differentiation pathways (Fig. 1–13). In many instances, the undifferentiated blast form of a cell will go through all of the nor-

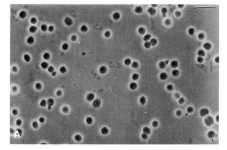
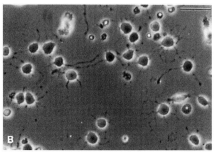
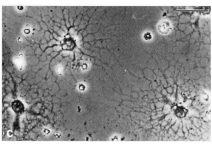

Figure 1–13. Primary culture can test the intrinsic potential of
a cell to undergo differentiation. In this series of phase contrast micrographs, oligodendrocyte progenitor cells have
been removed from the cerebrum of a young rat, the brain of
which has not yet undergone myelination. Despite the
absence of other neural cell types, the progenitors (**A**)
quickly regenerate cytoplasmic processes (**B**) shorn away by
the isolation procedure and evolve into morphologically
complex, mature oligodendrocytes (**C**; bars, 30 μm). (Modified from Gard AL, Pfeiffer SE. *Development* 1993;106:
119–132, reprinted with permission.)

mal events that result in the formation of a terminally
differentiated cell. By manipulating the culture
medium, investigators can ask questions about what
growth factors and hormones drive differentiation
events. In addition, studies can be performed to investigate how specific genes are turned on and off at the
appropriate times during development and terminal
differentiation.

These are but a few examples of how cell culture is
used to study various cellular phenomena. Clearly, cell
culture is a powerful investigative tool that has played a
key role in our elucidation of how cells work and how
cells interact with one another. Recent advances in our
understanding of growth factors, the production of

defined culture media, and novel cell culture techniques will be instrumental in future cell biology studies.

Pulse Chase Experiments Are Used to Study Important Biomolecules in Cells

Radioisotopes are unstable forms of atoms that spontaneously decay. During this disintegration, high-energy particles are given off by the decaying atom. These subatomic particles can be detected easily using either photographic films and emulsions or specialized machines specifically designed to detect emitted radiation. It is this characteristic of radioactive compounds that makes them useful for biological studies.

Pulse chase experiments make it possible to trace chemical pathways in cells. In these types of experiments, a radioactive precursor is added to the culture medium in which cells are being maintained. These radioactive precursors will be actively incorporated into biological molecules as they are being produced (the pulse). The radioactive medium is replaced by culture medium that contains normal nonradioactive constituents (Fig. 1–14). Samples of the cell preparation are collected at regular intervals and analyzed to determine both the intracellular location and chemical form of the biomolecule of interest (the chase). These types of experiments are very useful for determining when and where a molecule is synthesized, transported, and degraded within a cell. For example, pulse chase experiments using radioactively labeled uridine originally demonstrated that mRNA is synthesized in the nucleus and then moves to the cytoplasm to be translated.

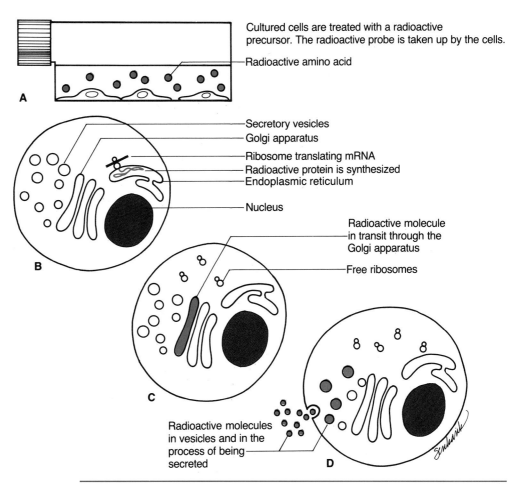

A Cultured cells are treated with a radioactive precursor. The radioactive probe is taken up by the cells.

Radioactive amino acid

B Secretory vesicles
Golgi apparatus
Ribosome translating mRNA
Radioactive protein is synthesized
Endoplasmic reticulum
Nucleus

C Radioactive molecule in transit through the Golgi apparatus
Free ribosomes

D Radioactive molecules in vesicles and in the process of being secreted

Figure 1–14. Pulse chase analysis of protein synthesis. A schematic demonstration of a pulse chase experiment. In this example, a population of cells was treated with a radioactive amino acid (the pulse) by (**A**) adding the probe to culture medium. The radioactive amino acid was taken up by the cells and (**B**) incorporated into a protein being synthesized on the rough endoplasmic reticulum. In addition, the culture medium was changed at this time, and the transit of the protein of interest was followed through the cell (the chase). With time, (**C**) the hypothetical protein moved through the Golgi apparatus into (**D**) secretory vesicles and was secreted by the cell.

Recombinant DNA Technology Is Revolutionizing the Study of Cells

Recombinant DNA methods are having a dramatic influence on how cells are studied. Formerly, the only way of investigating the information that was contained in DNA was through laborious genetic analysis. However, several important discoveries and developments in the past few decades now allow DNA to be investigated with relative ease. Principal among these developments were the discovery and purification of the key enzymes that are necessary for studying DNA and RNA, including various polymerases and restriction enzymes. In addition, techniques have been developed that make possible the determination of DNA sequences. These advances, along with many other important discoveries, enable one to isolate the DNA region that codes for a specific protein and to analyze that protein in ways that were not even dreamed of until recently. For example, recombinant DNA techniques can be used to produce considerable amounts of minor cellular proteins. In this way, proteins that could not be studied previously because they could not be isolated from cells and tissues in sufficient quantities can now be investigated in detail. In addition, once the coding sequence of a protein has been determined, the DNA can be manipulated to produce mutated and truncated portions of that protein. Accordingly, important functional and regulatory domains in proteins can be identified.

Recombinant DNA procedures have had an important effect on medicine. Hormones and other drugs that could be obtained only in small quantities by laborious purification schemes now can be produced in bulk by using the techniques of molecular biology. In addition, cloned genes are powerful diagnostic tools in many instances. For example, fetuses can be examined for the presence of serious mutations. In the future, one hopes that genetic engineering may permit damaged genes to be replaced by their normal copies in patients who are afflicted with certain genetic diseases such as CF, muscular dystrophy, and sickle cell anemia. The techniques and procedures that are commonly used for recombinant DNA studies are described in more detail in Chapter 5.

Clinical Case Discussion

As explained in this chapter, immunofluorescence can be used for pathological testing procedures. The example given was of a woman with lupus erythematosus, a disease in which patients produce autoantibodies against their own DNA. Systemic lupus erythematosus (SLE) is the prototype of immune complex disease. This disorder is more common in the African-American population, with an 8:1 female-to-male ratio. The pathogenesis of SLE is related to deposition in the tissue of soluble immune complexes that circulate in the bloodstream. The trigger for the formation of immune complexes in SLE has not been identified. Multiple systems can be involved in the clinical manifestations. The most common presenting symptoms include generalized weakness, anorexia, fever, fatigue, and weight loss. A nondeforming arthritis and myositis also are commonly present. Skin lesions include a butterfly erythematous rash over the cheeks and purpura. Alopecia and Raynaud's phenomenon may also occur. Other signs include hepatosplenomegaly, lymphadenopathy, pleurisy, pericarditis, nephritis, and behavior disturbances. Laboratory findings include: pancytopenia, a Coomb's positive hemolytic anemia, elevated sedimentation rate, and hypergammaglobulinemia. Renal involvement is indicated by the presence in the urine of red cells, white cells, red cell casts, and proteinuria. The differential diagnosis for SLE include rheumatic fever, rheumatoid arthritis, and viral infection.

The antinuclear antibody (ANA) test is the most sensitive diagnostic test for lupis and is invariably positive in patients with active disease. An autoantibody profile should be obtained for patients with a positive ANA screen, which identifies disease-specific antibodies. The most common autoantibodies present at the time of diagnosis and correlated with disease activity are DNA-anti-DNA complexes. Positive anticardiolipin antibody and the lupus anticoagulant identify patients at risk of developing thrombosis. Treatment consists of immunosuppression with prednisone in cases with renal, cardiac, or central nervous system involvement. If disease control is inadequate, agents such as azathioprine (Imuran) or cyclophosphamide (Cytoxan, NEOSAR) should be added. The skin lesions are treated with hydrochloroquine.

Suggested Readings

Cell Structure

Alberts B, Bray D, Lewis M, et al. *Molecular Biology of the Cell*, 3rd ed. New York: Garland Publishing, 1994.

Widnell C, Pfenninger K. *Essential Cell Biology*. Baltimore: Williams & Wilkins, 1990.

Tools of the Cell Biologist

Freshney RI. *Culture of Animal Cells: A Manual of Basic Techniques*, 3rd ed. New York: Wiley-Liss, 1994.

Inoue S. *Video Microscopy*. New York: Plenum Press, 1986.

Walker J. *Methods in Molecular Biology, vol 1, Proteins*. Clifton, NJ: Humana Press, 1984.

Review Questions

1. Cellular components that are stained by acidic dyes such as eosin are referred to as being:
   a. Monoclonal
   b. Acidophilic
   c. Lipophilic
   d. Histologic
   e. Basophilic

2. The principal reason for using a microscope for the observation of cells and tissues is:
   a. Microscopes allow the resolution of structural detail
   b. Microscopes magnify objects
   c. Cells and tissues need never undergo fixation prior to observation, allowing the material to be viewed in its natural state
   d. All of the above
   e. None of the above

3. The use of antibodies to study cell structure and composition is called:
   a. Histology
   b. Hybridoma technology
   c. Patch clamp analysis
   d. Immunocytochemistry
   e. Polyclonal antibody production

4. Gel filtration chromatography allows cellular components to be purified based on:
   a. Migration rate through an electric field
   b. Charge
   c. Lipid composition
   d. Migration rate through a sucrose gradient
   e. Size

5. The autoantibody best correlated with active disease in patients with systemic lupus erythematosus is:
   a. Antierythrocyte
   b. Antithyroid
   c. Anti-DNA
   d. Antithrombin
   e. Anticardiolipin

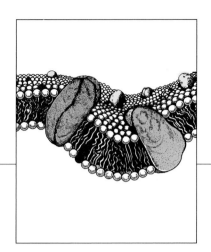

Chapter 2
Cell Membranes

Clinical Case

A 9-month-old Caucasian infant boy, Melvin Hochberg, was referred for evaluation of failure to thrive and recurrent infections. The baby is a poor feeder, passes bulky, greasy stools, and has increased flatulence. Mom states he tastes salty when she kisses him. Birth history reveals an intestinal obstruction due to a meconium ileus. The baby subsequently developed two episodes of pneumonia and one episode of wheezing, which responded to bronchodilator treatment. Physical examination revealed a malnourished infant with pale mucous membranes, diffuse respiratory wheezes and rales, hepatomegaly and edema in the lower extremities. Laboratory tests showed decreased hemoglobin, total protein, and albumin levels. Serum immunoglobulin levels were normal. Diffuse perihilar infiltrates were seen on chest x-ray. Elevated concentrations of sodium and chloride three times the normal level were demonstrated by a sweat chloride test.

Membrane Structure

The Basic Structure of Biological Membranes Is a Lipid Bilayer

In 1925, Gorder and Grendel, two Dutch biochemists, extracted the phospholipids from human erythrocyte membranes and placed them in a water trough. These phospholipids formed a monolayer at the air–water interface, with their hydrocarbon tails facing the air and polar head groups in the water. When the phospholipids were compressed with a movable barrier, the surface area covered by the phospholipids was twice the surface area of the erythrocyte membranes from which they were extracted. Gorder and Grendel correctly predicted from this simple experiment that biological membranes were bilayers, in which the polar head groups of the phospholipids face the water and the hydrocarbon fatty acid tails face the hydrophobic core.

If phospholipids are placed in water, they will spontaneously form micelles or bilayers (Fig. 2–1). The driving force for the formation of micelles and bilayers is the hydrophobic effect. If you have ever seen gasoline or oil spilled on a rainy day and witnessed the beading of the oil and its separation from the water, you have witnessed the hydrophobic effect. Water is a polar molecule in which the hydrogen is partially electropositive and oxygen partially electronegative. Because of their polar nature, water molecules tend to form clusters by hydrogen-bonding to each other (see Fig. 2–1).

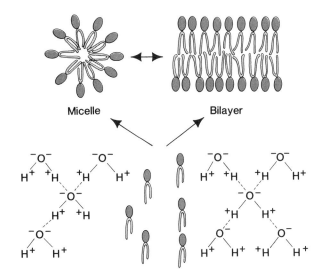

Figure 2–1. The hydrophobic effect drives bilayer formation. Phospholipids placed in water disrupt the hydrogen bonding of water clusters. Therefore, the hydrophobic effect that causes the phospholipids to bury their nonpolar tails by forming micelles or bilayers is really based on the driving force of the water molecules' maximal hydrogen bonding. The bilayer is a more stable conformation than the lipid micelle.

When hydrocarbons are dropped into an aqueous solution, they disrupt the hydrogen bonding of the water molecules. The coalescing of the hydrocarbons is driven by reestablishment of the water molecules' hydrogen-bonding pattern. This burying of hydrocarbons to maximize water's hydrogen-bonding is referred to as the *hydrophobic effect*. Now, how does the hydrophobic effect operate in the formation of phospholipid micelles and bilayers? Phospholipid molecules are amphipathic, meaning they have hydrophilic (water-loving) polar head groups and hydrophobic (water-fearing) fatty acyl chains. When placed in water, these amphipathic phospholipids try to bury their hydrophobic fatty acyl chains away from water by forming either spherical micelles in which the fatty acyl chains face the center of the sphere and the polar head groups are at its surface or bilayers that are sheets of two phospholipid monolayer or leaflets (see Fig. 2–1). The bilayers eventually must form a spherical vesicle so that there is no hydrophobic edge facing the water. With phospholipids that have two fatty acid chains, bilayers are a more stable conformation than are micelles.

In the 1930s, with the realization that biological membranes also contain protein, Danielli and Davson incorporated protein into the model of membrane. However, they visualized the protein as being attached to only the periphery of the membrane by association with the polar head groups of the phospholipids.

With the realization in the 1950s that proteins were involved in the transport of polar molecules across bio-

logical membranes, Danielli altered the model so that "unrolled" membrane proteins covered the membrane surface and wound their way through the bilayer to form a "pore." In 1972, Singer and Nicholson proposed the fluid mosaic model of membranes (Fig. 2–2). The basic principles of this model were that membrane proteins can be globular, just as can water-soluble proteins. The globular membrane proteins are embedded within the bilayer, with hydrophobic portions of the proteins buried within the hydrophobic core of the lipid bilayer and hydrophilic portions of the protein exposed to the aqueous environment. It was suggested that many of these proteins could be tripartite; that is, they have an embedded hydrophobic domain and two hydrophilic domains exposed on the inside and outside of the cell. Finally, these embedded membrane proteins were believed to be mobile, likened to "icebergs within a lipid sea." Most of the points of this model remain intact today, except that we now know that proteins found only on the periphery of the membrane also play important functions, and not all membrane proteins are freely mobile.

All biological membranes, both plasma membranes and organelle membranes, have the same basic structure and function. All biological membranes contain the lipid bilayers and serve as a selective permeability barrier. Consequently, the ionic composition of the extracellular fluid differs from that of the cytoplasm, and the polar molecules within the cytoplasm differ from those within the lumen of the endoplasmic reticulum or Golgi complex. Here, we will focus primarily on the plasma membrane.

The Lipid Composition of Human and Animal Biological Membranes Includes Phospholipids, Cholesterol, and Glycolipids

The cellular plasma and organelle membranes within the body contain 40–80% lipid. Among these lipids, the phospholipids are the most prevalent. In a cell mem-

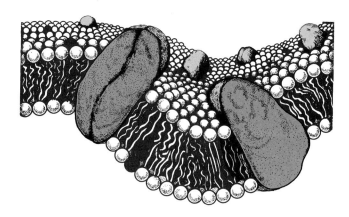

Figure 2–2. The fluid mosaic model of biological membranes. In 1972, Singer and Nicolson proposed a model in which globular integral membrane proteins (*red*) are freely mobile within a sea of phospholipids (*white*) and cholesterol (*black*).

brane that contains 50% phospholipid by weight, there are approximately 50 molecules of phospholipid per protein molecule. Four of the major phospholipids found in human and animal membranes are phosphatidylcholine (PC), phosphatidylserine (PS), phosphatidylethanolamine (PE), and phosphatidylinositol (PI). Figure 2–3 gives the structure of phosphatidylcholine (as an example of the phospholipids) and sphingomyelin. The polar head group for the phospholipids can be either choline, serine, ethanolamine, or inositol, linked by a phosphate ester bond to carbon 3 of the glycerol backbone, whereas the hydrophobic portion of the phospholipids contain two hydrocarbon fatty acyl chains linked to carbons 1 and 2 of the glycerol backbone. In human and other animal cellular plasma membranes, the phospholipids contain fatty acids that typically contain an even number of carbon atoms (16, 18, or 20), and one of the fatty acids is unsaturated (contains at least one double bond). The saturated fatty acid is straight and flexible, whereas the unsaturated fatty acid (which most typically contains one cis double bond) has a rigid kink at the site of the double bond. Sphingomyelin is also an amphipathic membrane lipid (see Fig. 2–3). It contains sphingosine, which is an amino alcohol with an unsaturated fatty acid chain, linked by an amide bond to another fatty acid to form ceramide. The hydroxyl group of ceramide forms an ester linkage with phosphocholine in sphingomyelin. Thus, similar to phospholipids, sphingomyelin has a polar head group and a hydrophobic portion that includes two fatty acyl

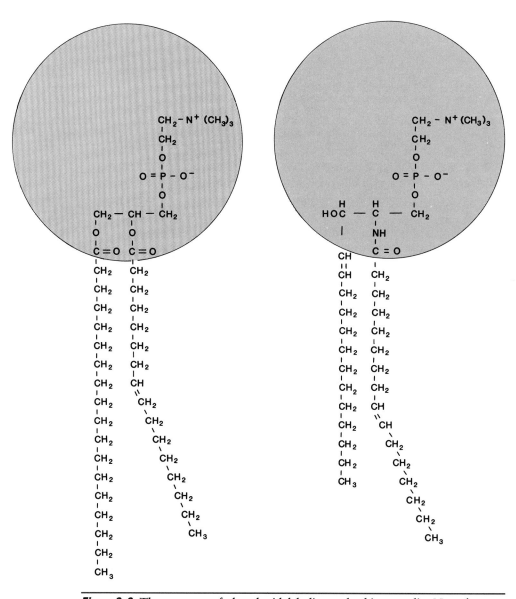

Figure 2–3. The structure of phosphatidylcholine and sphingomyelin. Note that both phosphatidylcholine and sphingomyelin have a choline-containing polar head group and nonpolar hydrocarbon tails.

chains. The hydrophobic effect acting on phosphatidyl-choline, phosphatidylserine, phosphatidylethanolamine, phosphatidylinositol, and sphingomyelin, when in an aqueous solution, will result in a bilayer with polar head groups facing the water and fatty acyl tails forming a hydrophobic core. These fatty acyl tails also interact with each other by weak van der Waals contacts. At neutral pH, only phosphatidylserine contains a net negative charge. Although most human and other animal membranes contain these four major membrane lipids (as well as other minor phospholipids; Table 2–1), the percentage composition of phospholipids varies significantly from one membrane to the next; that is, various biological membranes differ in their absolute phospholipid composition, but all form the lipid bilayer.

Cholesterol is also a major component of biological membranes (see Table 2–1). The structure of cholesterol indicates that it is amphipathic, with a polar hydroxyl group, hydrophobic planar steroid ring, and attached hydrocarbon. Cholesterol intercalates between the phospholipids, with its hydroxyl group near the polar head groups and its steroid ring and hydrocarbon tail parallel to the fatty acid chains of the phospholipids and perpendicular to the membrane surfaces.

Glycolipids, which are lipids with attached sugar residues, are more minor components of human and other animal membranes. The sugar residues of plasma membrane glycolipids almost always face the outside of the cell; that is, they have an asymmetric distribution, being found only in the outer leaflet of the bilayer. Human and animal cell glycolipids are produced primarily from ceramide and are referred to as *glycosphingolipids*. Such cell membranes contain neutral glycolipids with 1–15 uncharged sugar residues, and gangliosides with 1 or more negatively charged sialic acid sugars. The glycolipids are probably important in cell–cell and cell–interstitial matrix interactions, contribute to the negative charge of the cell surface, and play a role in immune reactions.

Whereas Membrane Lipids Form the Foundation of the Bilayer, Membrane Proteins Are Primarily Responsible for Function

Even though the lipids form the bilayer structure, the proteins are primarily responsible for membrane functions, including transport of ions and polar molecules, binding of hormones, signal transduction across the membrane, and structural stabilization of the bilayer. It is no surprise, therefore, that the protein:lipid ratio varies substantially in biological membranes and correlates with the degree of membrane function. The membrane protein:lipid ratio varies from 0.23 in myelin, which serves as an ensheathment or insulator of axons, to a value of 3.2 for the biologically active mitochondrial inner membrane in which one finds the proteins responsible for electron transport and oxidative phosphorylation (Fig. 2–4).

There are two basic types of membrane proteins: integral and peripheral. *Integral membrane proteins* are embedded in the lipid bilayer and cannot be removed without disrupting the bilayer. Most frequently, the agent used to disrupt the membrane is a detergent that, like the phospholipids, is amphipathic. Detergents have a charged or polar group attached to a single hydrocarbon chain and form micelles when added in sufficient concentration to water. When added to membranes, detergents disrupt the bilayer by forming mixed phospholipid-detergent micelles and remove the integral membrane proteins by coating their hydrophobic domains with the hydrocarbon portion of the detergent.

The second major classification of membrane protein is *peripheral membrane proteins*. Peripheral membrane proteins are operationally defined as those proteins that can be removed from membrane without dissolving the bilayer. Most frequently, these peripheral proteins are removed by shifting the ionic strength or pH of the aqueous solution, thereby causing dissociation of the ionic interactions of the peripheral protein

Table 2–1
Lipid Composition of Biologic Membranes: Percentage of Total Lipid

Membrane	PE	PS	PC	SM	GL	CH	Other
RBC plasma membrane	18	7	17	18	3	23	13
Liver plasma membrane	7	4	24	19	7	17	22
Mitochondrial membrane	35	2	39	0	trace	3	21
Endoplasmic reticulum	17	3	40	5	trace	6	27

From Alberts B, Bray D, Lewis J, et al. *Molecular Biology of the Cell*, 2nd ed. New York: Garland Publishing, 1989.
GL, glycolipids; CH, cholesterol.

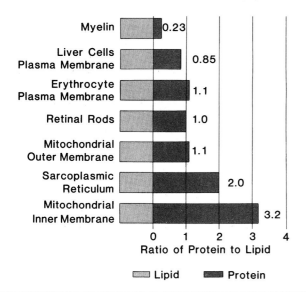

Figure 2–4. The protein:lipid ratio of biological membranes. There is a correlation between the function of a biological membrane and its protein:lipid ratio. Myelin, which functions as an insulator of axons, has a very low protein:lipid ratio. The mitochondrial inner membrane, which is very active and includes proteins for electron transport and oxidative phosphorylation, has a very high protein:lipid ratio. (Modified from Weissman G, Claiborne R. *Cell Membranes: Biochemistry, Cell Biology and Pathology.* New York: HP Publishing, 1975.)

with either phospholipid polar head groups or other membrane proteins.

There are various subtypes of integral and peripheral membrane proteins (Fig. 2–5). Some integral membrane proteins are transmembrane and make a single

pass through the membrane (see Fig. 2–5, example 1). This type of protein usually has a hydrophilic section containing charged and polar amino acids in the aqueous environment outside the cell, a hydrophobic stretch of 20–25 nonpolar amino acids forming an α-helix within the hydrophobic core of the bilayer and a hydrophilic portion within the aqueous interior of the cell. This type of integral protein (with three parts) is sometimes referred to as a *tripartite single-pass transmembrane protein.* A well-studied example of a single-pass transmembrane protein is glycophorin A, the major sialoglycoprotein in the erythrocyte plasma membrane. Glycophorin A is a membrane glycoprotein containing 131 amino acids, with most of its mass and all of its carbohydrate (~100-sugar residues) on the external surface of the membrane. The NH_2-terminal segment is highly hydrophilic; the hydrophobic segment, of 23 nonpolar amino acids (amino acids 73–95), makes a single pass through the membrane as an α-helix, and a COOH-terminal hydrophilic segment is exposed at the cytoplasmic surface. The function of a glycophorin A is unknown.

Not all transmembrane proteins are of the single-pass type. In Fig. 2–5, example 2, we see a protein that makes multiple α-helical passes through the membrane. All transporters and ion channel yet studied are multipass transmembrane proteins. Unlike single-pass transmembrane proteins, multipass transmembrane proteins can contain polar, and even charged, amino acids within the bilayer core. These polar amino acids, when facing one side of the α-helix, contribute to the formation of aqueous pores. An example of an important multipass transmembrane protein is the erythrocyte anion-exchange protein called *band 3.* Band 3 is a

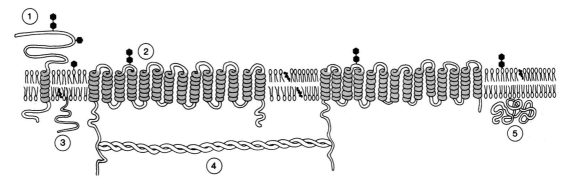

Figure 2–5. Integral and peripheral membrane proteins. Integral and peripheral membrane proteins can interact with the lipid bilayer in many different ways. The following situations are presented: (**1**) a single-pass glycosylated integral membrane protein. Note that a single α-helical segment of the protein crosses the bilayer; (**2**) a multipass glycosylated integral membrane protein, this structure is found in transporters and membrane channels; (**3**) an integral membrane protein for which the protein itself does not enter the bilayer—instead, it is covalently linked to a fatty acid chain or by sugars to phosphatidylinositol; (**4**) a peripheral membrane protein associated by ionic interactions to an integral (or another peripheral) membrane protein; and (**5**) a peripheral membrane protein associated with the polar head groups of phospholipids by an ionic interaction.

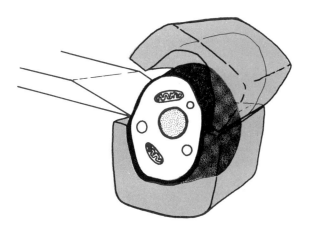

Figure 2–6. Freeze fracture electron microscopy. **A:** A cell membrane frozen rapidly to –180°C is fractured with a cold knife, splitting the bilayer into its two leaflets. Integral membrane proteins are pulled to one leaflet or the other, leaving behind a pit. **B:** Freeze fracture electron micrograph of the red blood cell membrane. The extracellular leaflet is called the *E face*, and the protoplasmic leaflet is called the *P face*. Intramembranous particles (*IMPs*) are found on both leaflet faces, but are more densely distributed on the convex P face. (**B**, courtesy of D. Branton.)

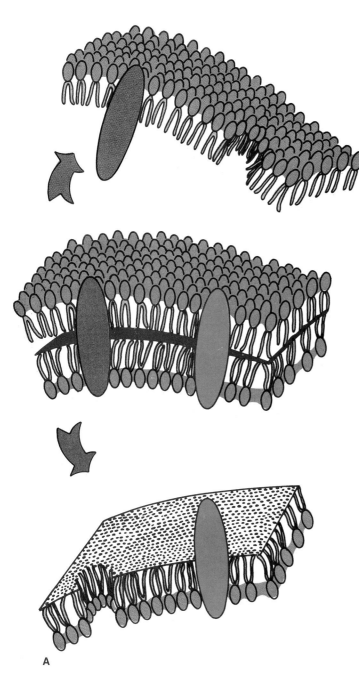

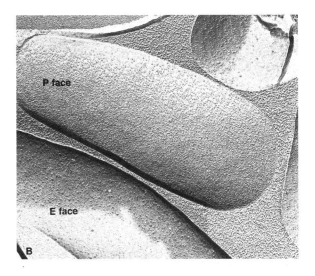

homodimer or homotetramer in which each chain contains 929 amino acids. Band 3 is responsible for the one-for-one exchange of HCO_3^- for Cl^- across the erythrocyte membrane that allows the release of CO_2 in the lungs. Unlike glycophorin A, band 3 makes 12–14 α-helical passes through the membrane. It contains a small hydrophilic COOH-terminus and a longer hydrophilic NH_2-terminal domain extending into the cytoplasm. The NH_2-terminal domain has binding sites for glycolytic enzymes, hemoglobin, and the cytoskeletal-linking protein, termed *ankyrin*. The carbohydrate moieties associated with band 3 are on the outside surface of the red blood cell membrane. Indeed, most transmembrane proteins are glycoproteins, and the sugar residues are almost always found on the noncytoplasmic side of the membrane.

Not all integral proteins are transmembrane proteins. Example 3 (see Fig. 2–5) is an integral membrane protein, but the protein itself is not embedded in the membrane; it is covalently linked to a fatty acid chain (most frequently myristic acid) or by sugars to phosphatidylinositol, that are embedded in the hydrophobic core. To be sure we understand this concept, all transmembrane proteins are integral membrane proteins, but not every integral membrane protein is a transmembrane protein.

The peripheral proteins can attach to the membrane surface by ionic interactions with an integral membrane protein (or another peripheral membrane protein) (see Fig. 2–5, example 4), or by interaction with the polar head groups of the phospholipids (see Fig. 2–5, example 5). A very well-studied peripheral membrane protein is red blood cell spectrin. Spectrin is a structural protein that is found on the cytoplasmic surface of the erythrocyte membrane and constitutes 20–25% of the total membrane protein. Spectrin is attached to the membrane by an ionic interaction with the peripheral membrane protein ankyrin which, in turn, associates with the NH_2-terminal hydrophilic domain of band 3. (We will discuss the functions of spectrin in Chapter 3.)

An interesting technique that supports the bilayer concept of biological membranes and allows the visualization of the hydrophobic domains of multipass integral proteins is freeze-fracture electron microscopy (Fig. 2–6). In freeze-fracture electron microscopy, a biological membrane is rapidly frozen (to $-180°C$) and fractured with a cold knife. Because the plane between the bilayer leaflets is held together by weak van der Waals forces, it tends to be the fracture plane. The fractured membranes are shadowed with heavy metals, and this replica of the inner surfaces of the bilayer leaflets is viewed in the electron microscope. In pictures from the red blood cell membrane (see Fig. 2–6), we see a concave leaflet that is called the *E face* (because it represents the extracellular leaflet face) and a convex leaflet that is called the *P face*

(because it is the protoplasmic leaflet face). Several facts are demonstrated by these electron micrographs. First, they are studded with 7.5-nm-diameter intramembranous particles (IMPs). In the red cell membrane, these IMPs represent the now-exposed hydrophobic domains of the multipass transmembrane protein band 3. Also, there are more IMPs on the P face than on the E face. As we will learn in the section on the spectrin membrane skeleton, band 3 is associated with the spectrin membrane skeleton on the cytoplasmic surface of the membrane. Therefore, on fracturing, most of the band 3 molecules are pulled by the spectrin skeleton into the P face.

Protein and Lipids Are Asymmetrically Distributed Across Biological Membranes

The carbohydrate portion of glycoproteins and glycolipid have an asymmetric distribution across biological membranes. The sugar residues are almost always found on the noncytoplasmic side of the membrane. For the plasma membrane of human cells, this means that the carbohydrate will be outside the cell and, for organelle membranes, the carbohydrate will be found within the lumen. Oligosaccharides are attached to glycoproteins by N-linkage to asparagine residues or O-linkage to serine or threonine. In both cases, the sugar would be asymmetrically distributed.

The carbohydrate moieties of glycolipids and glycoproteins, as well as glycosaminoglycans, which are oligosaccharides bound together by small protein cores, make up a fuzzy coat observed on electron microscopy of the outer surface of the plasma membrane. This fuzzy coat is often referred to as a *glycocalyx*, meaning "sugar chalice."

Proteins also have an asymmetric distribution. Spectrin is always associated with the inner cytoplasmic leaflet of the erythrocyte membrane. Glycophorin A always has its NH_2-terminal domain outside the red cell and its COOH-terminal domain within the cytoplasm. Proteins cannot flip-flop from one leaflet of the bilayer to the other, so their asymmetry is absolute.

The question then arises as to whether phospholipids are asymmetrically distributed across the plasma membrane of human cells. This was first studied in the erythrocyte membrane, because the human erythrocyte has no organelles; therefore, the plasma membrane can be isolated simply by lysing the cell, releasing the hemoglobin and other cytoplasmic proteins, then washing the membranes.

In the human erythrocyte membrane, the choline-containing lipids—phosphatidylcholine and sphingomyelin—are primarily in the outer leaflet, whereas the amine-containing phospholipids, phosphatidylserine and phosphatidylethanolamine, are in the inner leaflet (Fig. 2–7). A similar asymmetry was subsequently found in kidney cell plasma membranes. Phospholipids within

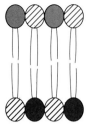

Figure 2–7. Asymmetry of red blood cell membrane phospholipids. The major phospholipids in the rbc membrane are distributed so that the choline containing phospholipids [phosphatidylcholine (*red*) and sphingomyelin (*red stipple*)] are found in the E leaflet and the amine containing phospholipid [phosphatidylethanolamine (*black*) and phosphatidylserine (*black stipple*) are found in the P leaflet].

human and animal cell plasma membranes rarely flip-flop from one leaflet to the other. The phospholipids are synthesized in the endoplasmic reticulum (ER), and their asymmetry is established by specific translocating enzymes within this membrane. Once the newly synthesized membrane reaches the plasma membrane, asymmetry has already been established.

Cholesterol has a property different from phospholipids in its distribution across the bilayer. Because the polar head group of cholesterol is a small hydroxyl group, cholesterol (unlike phospholipids) can readily flip-flop from one leaflet of the bilayer to the adjacent leaflet. Therefore, cholesterol is distributed on both sides of the bilayer and can move across the bilayer in response to shape changes within the plasma membrane.

What is the significance of the asymmetric distribution of proteins and phospholipids across the membrane? The answer is simple: It allows the two sides of the membrane to be functionally distinct.

Biological Membranes Are Fluid, But That Does Not Mean That Every Membrane Macromolecule Is Mobile

Membrane phospholipids are capable of several types of motion within the biological membrane (Fig. 2–8). The phospholipids can rotate very rapidly around a central long axis. The fatty acid chains of phospholipids are flexible, with greatest flexion toward the center of the hydrophobic bilayer core. We learned in the last section that the flip-flop of phospholipids from one leaflet of the plasma membrane to the other is a rare event but must occur to establish asymmetry after synthesis in the endoplasmic reticulum. Phospholipids can move laterally across a biological membrane at a rate of approximately 1×10^{-8} cm^2/sec at 37°C, which means that they exchange places with their nearest neighbor about 10^7 times per second. Cholesterol, which increases the packing of the phospholipids, tends to slow their lateral mobility.

Integral membrane proteins can also rotate along their axis within the membrane, but they do not flip-flop from one leaflet to the other. Several lines of evidence have indicated that membrane proteins are also capable of lateral movement within the plane of the membrane. When bivalent antibodies against cell surface antigens are added to a motile cell, the surface proteins tend to cross-link into large clusters called *patches* and, subsequently, cap at the trailing edge of the motile cell. This process, which requires energy in the form of adenosine triphosphate (ATP) and is blocked by chemicals that cause depolymerization of actin, demonstrates that these surface antigens are capable of lateral movement within the plane of the plasma membrane. Furthermore, it suggests the involvement of an energy-dependent cytoskeletal-mediated event.

The classic experiment demonstrating lateral mobility of integral membrane proteins within the plasma membrane was performed by Frye and Edidin (Fig. 2–9). These scientists fused human cells and mouse cells, forming a hybrid cell or heterokaryon (because they contained one nucleus from the mouse and one from the human cell). Specific human surface antigens were then labeled with an antibody coupled to a fluo-

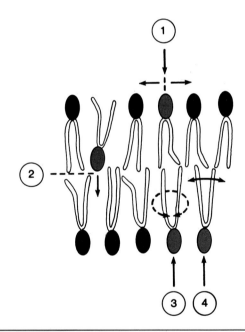

Figure 2–8. The motion of phospholipids within the lipid bilayer. Phospholipids are capable of varying types of movement within lipid bilayers: (**1**) Phospholipids are able to move laterally across the membrane at very fast rates; (**2**) they are capable of transbilayer movement (flip-flop) in the endoplasmic reticulum, but flip-flop is a rare event in the plasma membrane; (**3**) they can rotate rapidly around a central axis; and (**4**) the fatty acyl tails undergo constant flexion. (Modified from Alberts B, Bray D, Lewis J, et al. *Molecular Biology of the Cell*, 2nd ed. New York: Garland Publishing, 1989.)

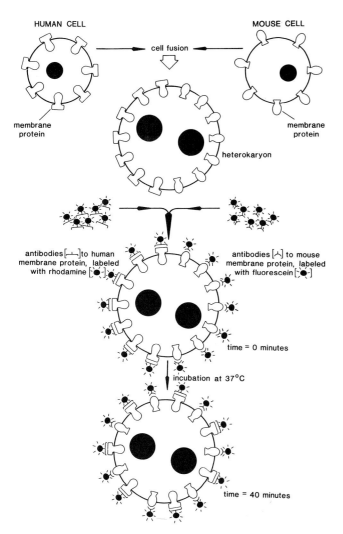

Figure 2–9. **Proteins are capable of lateral mobility across membranes.** The classic experiment of Frye and Edidin demonstrated that surface antigens of mouse and human cells were capable of free diffusion, after cell fusion to form a human mouse heterokaryon. (Modified from Alberts B, Bray D, Lewis J, et al. *Molecular Biology of the Cell*, 2nd ed. New York: Garland Publishing, 1989.)

rescent dye of one color and specific mouse surface antigens labeled with an antibody coupled to a fluorescent dye of a different color. Immediately after fusion, the human and mouse antigens were found in opposite hemispheres of the heterokaryon membrane, but less than an hour later, at 37°C, they were totally intermixed when viewed under the fluorescence microscope. The intermixing was not affected by ATP inhibitors or inhibitors of protein synthesis, but was slowed by decreasing the temperature. This experiment clearly demonstrated that the human and mouse surface antigen proteins had the ability to move laterally within the plasma membrane.

A more quantitative measurement of protein lateral mobility can be obtained by fluorescence recovery after

photobleaching (FRAP). In these studies, a surface protein is labeled with a fluorescently coupled monovalent antibody. A focused laser beam bleaches a small selected region of the plasma membrane, decreasing fluorescence in that spot (Fig. 2–10). The fluorescence of the bleached area returns with time because unbleached, labeled surface molecules diffuse into it. The percentage recovery in the bleached spot is proportional to the fraction of that specific integral membrane protein that is mobile. The rate of recovery of fluorescence allows a diffusion coefficient to be calculated. From this approach, we know that integral proteins within artificial lipid vesicles diffuse at a rate of 10^{-9}–10^{-10} cm^2/sec, whereas in a biological membrane they diffuse at a rate of 10^{-10}–10^{-12} cm^2/sec. Hence, what restricts the movement of proteins within a biological membrane? If we return to the band 3 molecules within the erythrocyte membrane, a major fraction of these molecules cannot move laterally within this membrane but are rapidly mobile when purified and placed within an artificial lipid vesicle. The reason is that the erythrocyte band 3 molecules are restricted by their interaction with spectrin on the cytoplasmic membrane surface (via ankyrin). So, although band 3 is capable of movement, its lateral

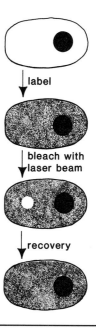

Figure 2–10. **Fluorescence recovery after photobleaching (FRAP) allows measurement of diffusional rates within biological membranes.** In the FRAP techniques, a fluorescent label is specifically associated with the membrane protein or lipid under study. A portion of the plasma membrane fluorescence is bleached with a focused laser beam. If the fluorescent molecule is capable of lateral mobility, the bleached spot will eventually recover its fluorescence. The rate of recovery is a measure of diffusional rate, and the amount of recovery determines the percentage of labeled molecules capable of lateral movement.

mobility is restricted by its interaction with the membrane skeleton. The lateral mobility of integral membrane proteins can be restricted by their interaction with each other, with the membrane skeleton or cytoskeletal components, or by interaction outside the cell with other cell surfaces.

Diffusion

The Cell Membrane Is a Selective Permeability Barrier That Maintains Distinct Internal and External Cellular Environments

Small, uncharged molecules can pass through a lipid bilayer by simple diffusion. For example, small gaseous molecules such as O_2, CO_2, and N_2, and small uncharged polar molecules such as ethanol, glycerol, and urea can simply diffuse down their concentration gradient across the lipid bilayer. To understand how this works, we must first describe simple diffusion. The particles of a substance dissolved in liquid or gas solvent are in continuous random (Brownian) movement. They tend to spread from areas of high concentration to areas of low concentration until the concentration is uniform throughout the solution. If areas of high and low particle concentration are separated by a permeable membrane, it is equally probable that a particle on the high- or low-concentration side will move to its opposite side in a given time. However, because there are many more particles on the high-concentration side, the net flux will be from high to low concentration. This will continue until the concentrations on both sides of the membrane are equal, at which time no further net diffusion will occur.

Given the situation in Figure 2–11, in which a small nonpolar molecule is present in two solutions at concentrations C_1 and C_2 ($C_1 > C_2$), and these solutions are separated by a membrane permeable to the nonpolar molecule, but not to solvent, Fick's first law of diffusion states that:

$$\text{The net rate of diffusion} = J = -DA \frac{dc}{dx}$$

This equation tells us that the rate of diffusion will be proportional to the surface area of the membrane (A), the concentration gradient across the membrane dc/dx, and a factor called the *diffusion coefficient* (D), which is a measure of the rate at which the molecule can permeate the membrane. The minus sign in this equation is because diffusion is positive in the direction of higher to lower concentration. D is proportional to the speed that the diffusing molecule can move in the surrounding medium; for spherical molecules, this is described by Einstein's equation: $D = KT/6\pi r\eta$ where K is equal to the Boltzmann constant, T is the absolute

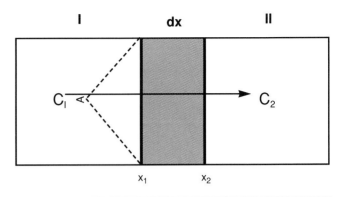

Figure 2–11. Rate of diffusion. Given the situation in which a small nonpolar molecule is present at concentrations C_1 and C_2 on both sides of a semipermeable membrane (*red*) that is permeable to the nonpolar molecule—but not to solvent—the nonpolar molecule will move down its concentration gradient. In the situation given in this diagram, $C_1 > C_2$. Hence, movement will be from chambers I to II. The rate of diffusion will be proportional to the concentration gradient ($C_1 - C_2$) and the surface area of the membrane (A), and inversely proportional to the distance across the membrane (dx).

temperature in degrees Kelvin, r is equal to the molecular radius, and η is the viscosity of the medium. Therefore, diffusion is inversely proportional to the radius of the diffusing molecule and to the viscosity of the surrounding medium. As the molecular mass of large molecules is proportional to r^3, D will be inversely proportional to (molecular mass)$^{1/3}$, so that a molecule that is 1/64 the molecular mass of another will have a diffusion constant four times as large.

When considering the diffusion of a molecule through a biological membrane, one must also consider the lipid solubility of the diffusing molecule. The permeability of the plasma membrane to a particular molecule increases with its partition coefficient (as measured by ability to partition in olive oil versus water; Fig. 2–12). For molecules with the same partition coefficient, there is decreasing permeability, with increasing molecular mass. Molecules that are small and soluble in the lipid core, such as O_2, CO_2, N_2, ethanol, glycerol, and urea, will diffuse down their concentration gradient through the bilayer. Diethylurea, which has a 50-fold greater partition coefficient than has urea, diffuses through the membrane at approximately 50 times the rate. For a substance diffusing through a biological membrane, its concentration at the outer face of the bilayer will be βC_O (β = partition coefficient), whereas its concentration at the inner face will equal βC_I. The concentration gradient within the membrane will be

$$\frac{dc}{dx} = \beta \frac{C_O - C_I}{dx}$$

Figure 2–12. Permeability through biological membranes as a function of lipid solubility. The permeability of the alga *Chara ceratophylla* plasma membrane to various nonelectrolytes is plotted versus the olive oil/water partition coefficient of the molecule. (Redrawn from Christenson HW. *Biological Transport*, 2nd ed. Menlo Park, CA: WA Benjamin, 1975. Data from Collander R. Trans Faraday Soc 1937;33:985.)

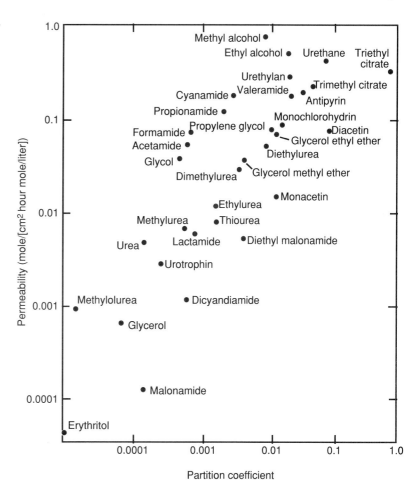

The more lipid-soluble the substance, the larger the partition coefficient (Fig. 2–13); therefore, the larger the effective concentration gradient within the membrane. If we substitute the concentration gradient across a biological membrane into Fick's equation:

$$J = -DA \frac{dc}{dx}$$

$$= -DA\beta \frac{C_O - C_I}{dx} = -\frac{D\beta}{dx} A(C_O - C_I)$$

The term $D\beta/dx$ is referred to as the *permeability coefficient* or P. The permeability coefficient for a diffusing molecule is proportional to the partition coefficient β and to the diffusion coefficient within the membrane D, but is inversely proportional to the distance across the membrane dx. Therefore Fick's equation applied to biological membranes simplifies to

$$J = -PA(C_O - C_I)$$

Fick's equation applies to small uncharged molecules. The diffusion of charged molecules across biological membranes is determined not only by the concentration gradient but also by the membrane potential; that is, the

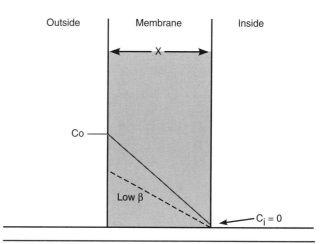

Figure 2–13. Role of partition coefficient (β) in the rate of diffusion across membranes. This situation is one in which an uncharged molecule is present at concentration C_O outside a plasma membrane, but its concentration inside the cell (C_I) is equal to zero. If there were no contribution of the partition coefficient, the rate of diffusion would be simply proportional to the concentration gradient (C_O-C_I), as shown (*solid red line*). However, the rate of diffusion is also proportional to the partition coefficient (β); therefore, the rate of diffusion is proportional to β (C_O-C_I).

movement of small charged molecules across the membrane is related to the electrochemical gradient, not simply to the chemical gradient.

Osmosis

Water Can Freely Diffuse Across Membranes in the Direction of Lower to Higher Osmolarity

Osmosis is the flow of water across a membrane from a compartment where solute concentration is lower to one in which the solute concentration is higher. In Fig. 2–14, a semipermeable membrane (permeable to water but not solute) separates a solution (A) from pure water (B). Water flow will occur from side B to A. The amount of pressure that would have to be applied to a piston on side A to keep water from entering is called the *osmotic pressure*. The osmotic pressure of a solution depends on the number of particles in solution; therefore, it is referred to as a *colligative property*. The degree of ionization must be taken into account when discussing colligative properties. In calculating osmotic pressure, one molecule of NaCl yields two particles, whereas Na_2SO_4 yields three particles. By Van't Hoff's law, osmotic pressure can be calculated as follows:

$$\text{Osmotic pressure } (\pi) = iRTc$$

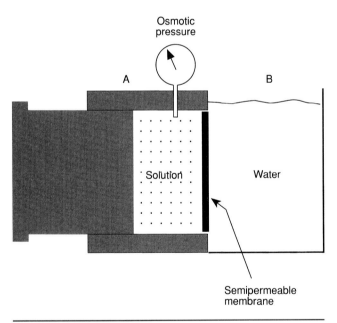

Figure 2–14. Osmotic pressure. Two chambers are separated by a membrane that allows water to diffuse, but not the solutes dissolved in water. Chamber *A* contains solute in water, whereas chamber *B* contains only water. Water will flow from chamber *B* to chamber *A* in an attempt to equalize the osmotic concentration on each side. The amount of pressure that would have to be placed on the piston in chamber *A* to stop the flow of water from *B* to *A* is equal to the osmotic pressure of the solution in chamber *A*.

where

i = number of ions formed by dissociation of a solute molecule
R = ideal gas constant
T = absolute temperature, °K
c = molal or molar concentration

At physiologic concentrations of solutes, such as NaCl, the values obtained for osmotic pressure differ from theoretical values based on Van't Hoff's law. Therefore, a correction factor called the *osmotic coefficient* (Φ) is inserted into the Van't Hoff formula as follows:

$$\pi = \Phi iRTc$$

The factor Φ approaches a value of 1 as the solution becomes increasingly dilute. The term in the equation Φic is referred to as the *osmolar concentration* with units of osmoles per liter (osm/L).

To calculate the osmotic pressure (at 37°) of an isotonic 154-mM NaCl ($\Phi = 0.93$) solution:

$$\pi = \Phi iRTc = (0.93)(2)$$
$$\times(8.2\times10^{-2})$$
$$\times(310°K)(0.154)$$
$$\therefore \pi = 7.28 \text{ atm}$$

To calculate the osmolarity of this solution,

$$\text{Osmolarity} = \Phi ic = 0.93\times2\times0.154$$
$$= 0.286 \text{ osm/L}$$

The osmotic pressure of a multicomponent system can be easily measured from another colligative property such as freezing point depression. Freezing point depression in degrees centigrade $\Delta T_F = 1.86 \, \Phi ic$. If an unknown mixture of solutes lowers the freezing point of H_2O by 0.186°C, the osmolarity of that solution would equal 0.1 osm/L. The osmotic pressure of this solution (at 37°C) would be equal to $\Phi icRT = (0.1)(8.2 \times10^{-2})(310) = 2.54$ atm. As this solution has a lower osmotic pressure than 154-mM NaCl at 37°C, it is hypoosmotic compared with the NaCl solution. The 154-mM NaCl solution is hyperosmotic compared with the multicomponent system. If the osmotic pressures were equal, they would be called *isoosmotic solutions*.

Cellular membranes are highly permeable to water. Indeed, the rate of diffusion of water across membranes is much higher than would be predicted based on its partition coefficient β in the lipid core versus aqueous solutions. Therefore, it is believed that water moves rapidly through protein channels in direct response to differences in osmotic pressure on two sides of the biological membrane. Erythrocytes are very useful when discussing the osmotic properties of cells because they

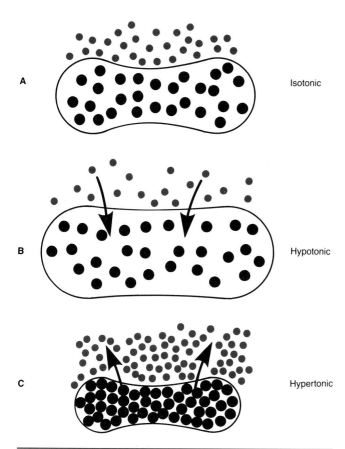

Figure 2–15. The red cell as an osmometer. **A:** When red cells are placed in a solution of 154 mM NaCl *(isotonic)*, they maintain their volume. **B:** If the solution concentration is less than 154 mM NaCl *(hypotonic)*, they swell. **C:** If it is greater than 154 mM NaCl *(hypertonic)*, they shrink.

behave as almost perfect osmometers. At 154 mM NaCl, erythrocytes have their normal plasma volume (Fig. 2–15). Therefore, this concentration of NaCl is termed *isotonic*. The erythrocyte shrinks in more concentrated solutions *(hypertonic)*, and swells in more dilute solutions *(hypotonic)*. When the erythrocyte reaches 1.4 times its original volume, it lyses, releasing hemoglobin and other cytosolic components. The intracellular substances within the erythrocyte that produce its cytosolic osmotic pressure include the O_2-carrying protein hemoglobin, K^+, and organic phosphates; in total it must have an osmolar concentration of 286 milliosmolar and an osmotic pressure of 7.28 atm at 37°C to maintain its normal volume. The major osmotic particle in the extracellular fluid or plasma that balances the osmotic pressure within cells is the Na^+ ion. Sodium is maintained at high concentrations in the extracellular fluid by Na/K-ATPase, as will be discussed later. If Na/K-ATPase is inhibited by ouabain such that it can no longer pump Na^+ out into the extracellular fluid, cells swell and then lyse.

Donnan Effect and Its Relation To Water Flow

Cells contain many negatively charged ions in the cytoplasm, such as protein and RNA, that cannot diffuse through the membrane. These impermeable anions cause a redistribution of permeable anions in a manner predicted by Donnan and Gibbs. In Figure 2–16, there are two compartments, *A* and *B*, separated by a semipermeable membrane. In compartment A, we have a protein with a net negative charge (protein⁻) that cannot diffuse through the membrane. On both sides of the membrane we have sufficient Na^+ and Cl^- that the total concentration of cations and anions are equal in compartments A and B. The Cl^- initially will be at higher concentration in compartment B than in A. Because the membrane is permeable to Cl^- (and Na^+), it will move down its chemical gradient toward compartment A. The Na^+ will follow Cl^- into compartment A to maintain electroneutrality on side A. After these events, the concentration of protein⁻, Cl^-, and K^+ will be higher in compartment A, compared with compartment B. Here, for single cations and anions with identical valence, their distribution across the semipermeable membrane at equilibrium can be expressed by the Donnan equation:

$$[Na]_A[Cl]_A = [Na]_B[Cl]_B$$

We have not yet discussed the movement of water. In the example given, the total number of particles in compartment A will be greater at equilibrium than in compartment B; therefore, water will flow toward compartment A. For this same reason, the Donnan effect tends to cause water to flow into cells but is balanced by the active outward pumping of ions by Na/K-ATPase.

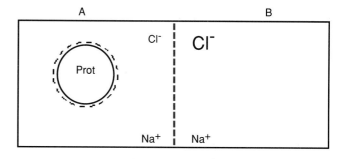

Figure 2–16. The Donnan effect. A semipermeable membrane will allow the diffusion of Na^+ and Cl^- ions, but not the negatively charged protein, in chamber A. Initially, there are an equal number of cations and anions in chambers A and B, which means that the Cl^- is at much higher concentration in B than in A to balance the anionic charge of the impermeable protein. With time, Cl^- will flow down its concentration gradient from B to A. To maintain electroneutrality, Na^+ would also move from B to A. However, this would cause the osmotic concentration to be higher in chamber A than in chamber B; therefore, water would move from B to A.

Transport

Multipass Transmembrane Proteins Facilitate the Diffusion of Specific Molecules Across Biological Membranes

Some molecules, because of their size or polar nature, have very low permeability coefficients (P); therefore, they diffuse very slowly through artificial lipid bilayers. These same molecules cross biological membranes far more rapidly than would be predicted by Fick's equation because of a process called *facilitated diffusion*. The uptake of glucose into the erythrocyte will be used to make this point. The concentration of glucose in blood plasma (approximately 5 mM) is much higher than the concentration of glucose within the erythrocyte. However, because of its large size, glucose cannot simply diffuse through the lipid bilayer (Fig. 2–17). It is rapidly moved across the erythrocyte membrane, down its concentration gradient, by associating with a glucose transporter or permease that changes conformation and allows passage of glucose. Several points are made in Fig. 2–17. First, the velocity of uptake of D-glucose is saturable. At high concentrations of D-glucose, all of the erythrocyte glucose transporters are occupied so

that a maximal velocity is reached (V_{max}). The K_m represents the concentration of D-glucose at which the velocity of uptake is one-half the V_{max}. It is a measure of how tightly D-glucose binds to its transporter. For example, whereas the K_m for D-glucose is 1.5 mM, the K_m for its stereoisomer L-glucose is >3,000 mM. Facilitated diffusion, unlike simple diffusion, is stereospecific. The glucose transporter will bind D-glucose but not L-glucose. Other six-carbon sugar molecules that are similar in structure to D-glucose can be transported by the glucose transporter, but they have a higher K_m value. For example, the K_m values for D-mannose and D-galactose are 20 and 30 mM, respectively. These sugars can competitively inhibit the uptake of D-glucose into erythrocytes. The curve in Fig. 2–17 nicely fits the Michaelis-Menton equation:

$$V = \frac{V_{max}}{1 + K_m/C}$$

where C is the concentration of substrate.

This is because a transporter carries out a function very similar to that of an enzyme. Instead of chemically converting a substrate, as an enzyme does, transporters move a substrate across a biological membrane in a saturable manner.

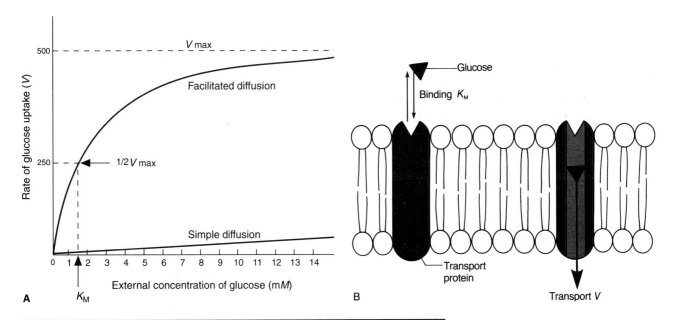

Figure 2–17. **Glucose transport into red blood cells: facilitated diffusion. A:** Glucose moves down its concentration gradient into red blood cells at a rate much faster than would be predicted by simple diffusion through the lipid bilayer *(black line)*. The plot for rate of glucose uptake versus external glucose concentration is hyperbolic *(red line)*. The rate increases with external glucose concentration until it reaches a maximal velocity (V_{max}). The concentration of external glucose at which half maximal velocity is reached is called the K_M. **B:** K_M is a measure of the affinity of the glucose transport protein (permease) for glucose. (Modified from Darnell J, Lodish H, Baltimore D. *Molecular Cell Biology*, 2nd ed. New York: Scientific American Books, 1990.)

Extracellular

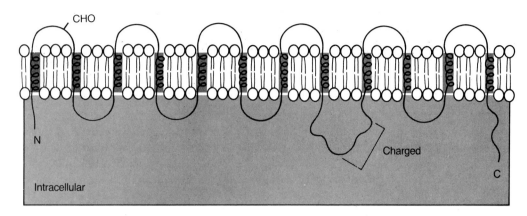

Figure 2–18. Predicted structure of the erythrocyte glucose transporter. The glucose transporter contains 12 membrane-spanning domains. *CHO*, glycosylation site.

All transport and channel proteins yet studied are multipass transmembrane proteins. For example, the entire amino acid sequence of the erythrocyte glucose transporter has been determined by cDNA cloning and sequencing (Fig. 2–18). The glucose transporter has 12 α-helical transmembrane segments. Multipass transmembrane proteins have a higher proportion of polar amino acids within the bilayer core than do the single-pass ones. The binding of D-glucose to an extracellular domain of the glucose transporter is thought to cause a conformational change in the protein, which allows the polar amino acids within the bilayer core to hydrogen-bond with the hydroxyl groups of glucose, thereby facilitating its movement down its concentration gradient. Facilitated diffusion is sometimes called *passive transport* because it requires no external source of energy. D-glucose simply moves down its concentration gradient.

Active Transport Requires ATP Hydrolysis, Either Directly or Indirectly

Active transport and selective permeability of the plasma membrane for ions creates very large differences in the ionic composition of the cytosol and extracellular fluid of human and animal cells (Table 2–2). The Na^+ and Cl^- are maintained at approximately 10- to 20-fold higher concentrations outside than inside cells, whereas K^+ is at a 20- to 40-fold higher concentration inside the cells. Much of the remainder of this chapter will deal with how this distinction in ionic composition is maintained. The enzyme Na/K-ATPase uses the energy of ATP hydrolysis to pump three Na^+ ions out of the cell against their electrochemical gradient and two K^+ ions into the cytosol against their electrochemical gradient (Fig. 2–19). This enzyme is an important example of *primary active transport*, for which energy derived from ATP

Table 2–2
Typical Ion Concentrations Within Mammalian Tissue

Ion	Cytoplasm (mM)	Extracellular Fluid (mM)
K^+	139	4
Na^+	12	145
Cl^-	4	116
HCO_3^-	12	29
Protein	138	9
Mg^{2+}	0.8	1.5
Ca^{2+}	<0.0005	1.8

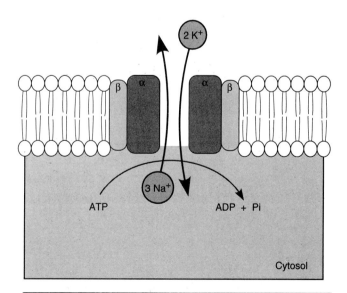

Figure 2–19. The Na/K-ATPase is an electrogenic pump. It moves $3Na^+$ out of the cell and $2 K^+$ into the cytoplasm at the expense of ATP hydrolysis to ADP and P_i.

hydrolysis directly moves molecules across membranes against their electrochemical gradient.

The Na/K-ATPase has been isolated from several mammalian tissues and appears in the membrane as an $(\alpha\beta)_2$ tetramer, with a catalytic 120-kDa α-subunit and a glycosylated 50-kDa β-subunit of unknown function. The complete sequence of the α- and β-subunits are known from cDNA cloning and sequencing; their membrane topography is shown in Fig. 2–20. The extracellular domains of the α-subunit contain binding sites for two K^+ ions and the inhibitor ouabain, a compound related to the cardiac glycoside digitalis. The intracellular domains contain binding sites for three Na^+ ions, ATP, and a phosphorylation site. This Na/K pump is linked directly to ATP hydrolysis, as the α-subunit is autophosphorylated in a single aspartate residue, conformationally converting the ATPase from a form that transports K^+ to a form that transports Na^+. At the outer surface of the membrane, binding of K^+ promotes hydrolysis of the phosphate group from the α-subunit. Cleavage of the phosphate converts the carrier back to a form that preferentially transports K^+.

That three Na^+ ions are pumped out and two K^+ ions enter the cell with hydrolysis of one ATP molecule is important in two ways. First, Na/K-ATPase is said to be electrogenic because each cycle leads to one net positive charge finding its way to the outside surface of the membrane and thereby contributing in a small way to the development of the membrane potential. Second, the large impermeable anions within cells would normally cause cells to swell if Na^+ was not pumped to the outside of the cell by Na/K-ATPase. Remember that three osmotic particles are pumped out, whereas only two osmotic particles are pumped in for each cycle of ATP hydrolysis.

Other similar ATPases pump Ca^{2+} out of cells or back into the sarcoplasmic reticulum of muscle, or pump protons into the lumen of the stomach. These are called the *P class of ATPases* because they contain nearly identical sequences surrounding the phosphorylated aspartate.

The lumen of the lysosome and vacuoles is highly acidic (pH 5.0–5.5) (see Chapter 4). It is maintained at this pH by a V-type ATP-dependent proton pump. This proton pump is composed of several subunits, with relative molecular mass (M_r) of 70, 60, and 17 kDa, that are structurally distinct from the subunits of the P-type ATPases. The V-type ATPase pumps protons into the lysosome at the expense of ATP hydrolysis and is not autophosphorylated in the process.

Secondary Active Transport Involves the Cotransport of a Molecule with Sodium as It Moves Down Its Electrochemical Gradient

The movement of Na^+ down its electrochemical gradient can be coupled to the movement of another molecule against its gradient. This is referred to as *secondary active transport* because the Na^+ electrochemical gradient is maintained by Na/K-ATPase. The transported molecule and cotransported molecule can move in the same direction, which is called *symport*, or they can move in opposite directions across the membrane, which is referred to as *antiport* (Fig. 2–21). Glucose is transported against its concentration gradient from the intestinal lumen into the intestinal epithelial cell across the apical membrane (Fig. 2–22). The glucose can then be transported down its concentration gradient at the basolateral membrane into the blood by facilitated diffusion. The process at the apical membrane is carried

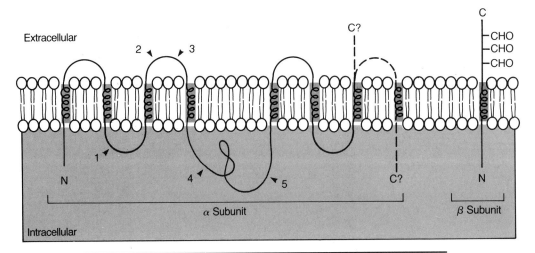

Figure 2–20. Predicted structure of the α and β subunits of Na/K-ATPase. The following binding sites on the α-subunit are identified: (**1**) Na^+ binding, (**2**) K^+ binding, (**3**) ouabain binding, (**4**) phosphorylation site, and (**5**) ATP binding. The function of the β-subunit is unknown.

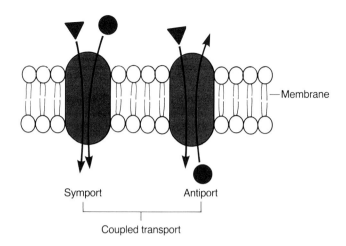

Symport   Antiport

Coupled transport

Figure 2–21. Coupled transport. Sometimes the transport of one molecule across a biological membrane is coupled by the transport protein to the movement of another molecule. If the movement of both molecules is in the same direction, the cotransport is referred to as *symport*. If the molecules are being moved in opposite directions, the cotransport is referred to as *antiport*. (Modified from Darnell J, Lodish H, Baltimore D. *Molecular Cell Biology*, 2nd ed. New York: Scientific American Books, 1990.)

out by a glucose Na^+ symport protein (Fig. 2–23). The binding of Na^+ and glucose to the symport protein causes a conformational change that allows both Na^+ and glucose to enter the cell together. The energy that moves glucose against its chemical gradient is derived from Na^+ moving down its electrochemical gradient. This process is indirectly linked to the ATP hydrolysis,

which fuels the pumping of Na^+ out of the epithelial cell. The Na^+ glucose symport protein is a multipass transmembrane protein with 11 membrane-spanning domains. Similar symport proteins are used to transport amino acids against their concentration gradient into the intestinal epithelial cell at the apical membrane surface. The Na/K-ATPase that pumps Na^+ out of the cell is located at the basolateral surface, constrained by interactions with ankyrin and nonerythroid spectrin (protein components of the membrane skeleton discussed in Chapter 3). Thus, glucose, amino acids, and Na^+ enter at the apical surface and exit the epithelial cell into the blood at the basolateral surface.

Antiport of Sodium and Calcium Ions Plays an Important Role in Cardiac Muscle Contraction

The uptake of Ca^{2+} into cardiac muscle triggers contraction (see Chapter 3). The accumulated Ca^{2+} is then moved out of the cardiac muscle cell by an antiport protein, which is powered by the simultaneous movement of Na^+ down its electrochemical gradient into the heart muscle cells. Again, the Na^+ electrochemical gradient is maintained by Na/K-ATPase; therefore, this is another form of secondary active transport.

The drugs digoxin (Crystodigin, Lanoxin) and ouabain are used clinically because they increase the force of heart muscle contraction and, therefore, are of great value in treating congestive heart failure. These drugs inhibit Na/K-ATPase, causing an increase in intracellular Na^+, thereby dissipating the Na^+ electrochemical gradient and inhibiting the Na^+-Ca^{2+} antiport. The

Figure 2–22. Transport of glucose into and out of intestinal epithelial cells. Glucose enters from the intestinal lumen through the apical membrane of the epithelial cells by a glucose Na^+ symport transporter. The Na^+ ions are pumped back out of the cell by Na/K-ATPase, located on the basolateral membrane. Glucose then exits the cell by facilitated diffusion, by permeases located in the basolateral membrane. (Modified from Darnell J, Lodish H, Baltimore D. *Molecular Cell Biology*, 2nd ed. New York: Scientific American Books, 1990.)

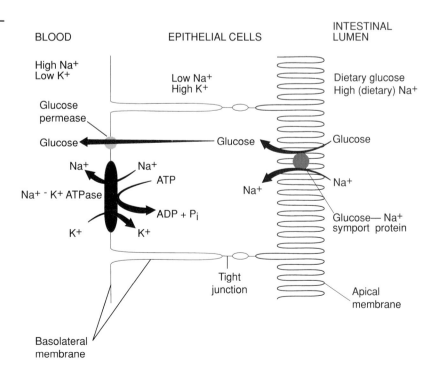

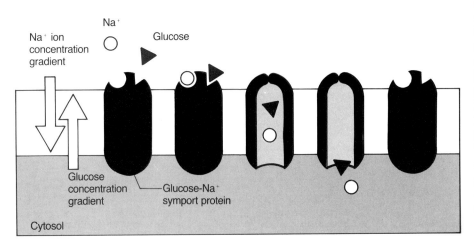

Exterior

Figure 2–23. Mechanism of glucose Na⁺ symport. The proposed mechanism for glucose movement against its concentration gradient, empowered by Na⁺ movement down its concentration gradient, is presented. The binding of one Na⁺ molecule and one glucose molecule to separate sites on the extracellular domain of a glucose Na⁺ symport protein causes the transporter to change conformation. The change of conformation creates a channel through which Na⁺ and glucose can be transported into the cytosol. The symport protein then returns to its original conformation.

final result is increased intracellular Ca^{2+} and, consequently, stronger and more frequent heart contraction.

ABC Transporters Are a Distinct Class of ATPase

Earlier we discussed the P-class of ATPases (such as the Na/K-ATPase) and the V-class of ATPase (such as the lysosomal proton pump). A third class of transmembrane ATPase is called the *ABC transporter superfamily* because, as shown in Fig. 2–24, this family of transporters contains an ATP-binding-cassette (ABC). Two important members of the ABC transporter superfamily are the multidrug resistance (MDR) ATPase and the cystic fibrosis (CF) transmembrane regulator.

The MDR ATPase was initially discovered when oncologists realized that tumors were often resistant to a broad range of distinct anticancer drugs. The reason for this resistance was that tumor cells overexpress the 170,000 M_r MDR ATPase (also called *P170*). The MDR ATPase utilizes energy of ATP hydrolysis to pump hydrophobic molecules (such as a broad range of anticancer drugs) out of cells. Normally, the MDR ATPase is expressed in liver, kidneys, and intestines; where it is thought to pump toxic substances into the bile, urine, and intestinal lumen, respectively. However, in tumor cells, the protein is overexpressed and also pumps out, of cells, various drugs that would be medicinally efficacious. Therefore, a liver cancer (hepatoma) is difficult to

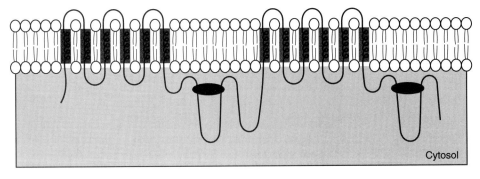

Figure 2–24. Predicted structure of the ABC transporters. The ABC transporters such as the multidrug resistance (MDR) ATPase and the cystic fibrosis transmembrane regulator (CFTR) protein contain hydrophobic segments consisting of six membrane spanning α-helices that serve as translocation pathways. The proteins also contain two large hydrophilic domains that face the cytosol and contain ATP-binding domains (*red*). The ABC transporters hydrolyze ATP, which causes a change in protein structure, facilitating the movement of hydrophobic molecules (MDR) or Cl⁻ (CFTR).

treat because the affected cells are resistant to a wide range of chemotherapeutic drugs. The MDR-ATPase (P170) also appears to function as a volume-sensitive Cl^- channel, which requires ATP for activity.

The cystic fibrosis transmembrane regulator (CFTR) protein has a structure that is closely related to the MDR-ATPase. Genetic defects in CFTR lead to CF. Cystic fibrosis is due to a dysfunctional CFTR in epithelial cells of the lung and other tissues. The CFTR is an ATP- and cyclic adenosine monophosphate (cAMP)-sensitive Cl^- channel. Whereas epithelial cells from unaffected individuals carry out normal cAMP-dependent Cl^- flux through CFTR; in CF patients, the CFTR often becomes insensitive to cAMP. How ATP hydrolysis and cAMP affect Cl^- flux through the normal CFTR is currently an area of active study. However, it is clear that the severity of an individual CF case is linked to the extent that Cl^- flux is defective through that individual's epithelial CFTR. Therefore, cystic fibrosis is receiving a great deal of attention by many researchers who are trying to replace the defective CFTR gene by gene therapy approaches.

Ion Channels and Membrane Potentials

The hydrophobic environment of the phospholipid membrane bilayer is virtually impermeable to ions in aqueous solution. To appreciate why this occurs, recall that ions interact electrostatically with water molecules, which behave as dipoles (Fig. 2–25). Strong bonds form between a single ion, such as K^+, and the electronegative oxygens of surrounding water molecules. Whether hydrated or in an energetically unfavorable dehydrated state, ions are immiscible with the nonpolar hydrocarbon fatty acid chains of membrane phospholipids and, hence, are unable to cross the lipid bilayer by simple diffusion.

Protein Channels Permit the Rapid Flux of Ions Across Membranes

Earlier in this chapter, we learned how the problem of transporting ions and other hydrophilic substances across membranes is surmounted by permease proteins. The other major class of transport proteins works very differently, by forming aqueous ion channels that traverse the lipid bilayer (Fig. 2–26). Ion channels are distinguished physiologically from carrier proteins by the high velocity of ion flux they allow across the membrane, without expending energy. Whereas permease-mediated transport operates at a maximum of 10^5 ions/sec, ions flow at a rate 100–1,000 times faster through channels that mediate not only rapid electrical signaling of nerve impulses and muscle contraction but also many biologic responses common to nonexcitable cells.

An important principle is that ion channels influence the rate, but not the direction, of ion flow across cell membranes. Movement of a nonionic solute

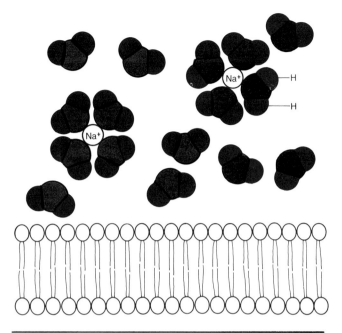

Figure 2–25. Hydrated cations in solution. Water molecules attracted by their electronegative oxygens to a cation (e.g., Na^+) form an aqueous shell surrounding the ion. These waters of hydration increase the effective size of the ion. It is energetically unfeasible for a diffusible ion to trade its polar environment of water and attempt to enter the nonpolar lipid bilayer.

through a channel is strictly passive, determined only by its concentration gradient across the membrane. For an ion, however, the direction of flow depends on both the chemical concentration gradient and the electrical potential across the membrane (discussed later).

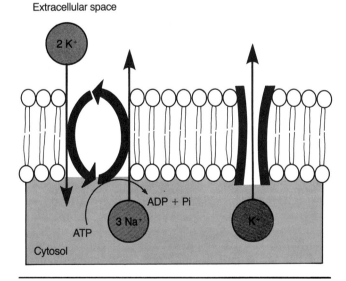

Figure 2–26. Two classes of membrane transport proteins for K^+. The influx of K^+ across the plasma membrane is mediated actively by Na/K-ATPase and passively through K^+ channel-forming proteins.

Ion Channels Have a Common Structural Motif: The Transmembrane α-Helix

Although the concept of the ion channel is nearly 150 years old, insight concerning the basic properties of their construction and function was not attained until the 1960s. A major breakthrough came with the discovery that ion channels could be artificially introduced into cell membranes treated with small hydrophobic proteins, termed *ionophores*. Made primarily by different fungi, some ionophores have been used as antibiotics, but many more have become important tools for the cell biologist. Gramicidin A, for example, is a linear polypeptide with

15 amino acids that, like all ionophores, is readily miscible with the phospholipid bilayer. This ionophore is distinct because, when inserted into the membrane, it forms an unusual β-helix (Fig. 2–27). Alternating *d*- and *l*-amino acid residues in the molecule orient the polar (hydrophilic) carbonyl oxygen atoms and amide nitrogens of the peptide bonds toward the hollow center of the helix, where they form the wall of the channel pore. Hydrophobic amino acid side chains radiate outwardly from the helix to anchor the ionophore in surrounding lipid. Two helical molecules of gramicidin A are thought to align end to end, to completely span the lipid bilayer. Given the different flow rates of various ion species across membranes punctuated with these simple channels, it is apparent that gramicidin pores prefer small cations to anions and that partial dehydration is necessary for ions to traverse the channel in single file, as they do through many animal membrane channels.

Channel-forming proteins expressed by mammalian cells also form helices in membranes, but they differ fundamentally from gramicidin A in several ways. Membrane-spanning proteins in eukaryotes consist solely of *l*-amino acids and thus adopt an α-helix that lacks a central pore. In this configuration, carbonyl oxygens and amide nitrogens of the polypeptide backbone do not behave as hydrophilic. Instead, they hydrogen-bond to each other and are thus prevented from interacting with water. As a result, α-helices are naturally hydrophobic and highly suited for association with the lipid bilayer.

What makes some transmembrane α-helical proteins prone to forming aqueous channels? A simple

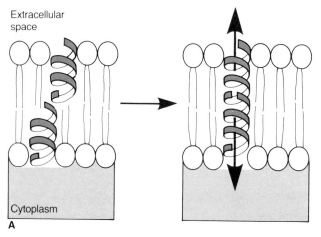

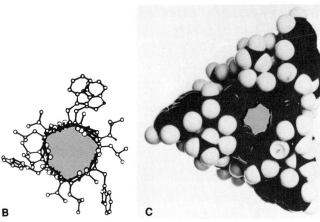

Figure 2–27. Structure of the β-helix and transmembrane channel formed by gramicidin A. A: Two gramicidin A peptides dimerize head to head to span the lipid bilayer. **B:** In this stereoperspective through one-half of a gramicidin A transmembrane, channel hydrophilic carbonyl and amide groups of the peptide backbone line the central pore of the β-helix, whereas hydrophobic amino acid side chains project outward. **C:** The pore size (4Å) of the gramicidin A channel is appreciated in this space-filling model of B. (Modified from: **A**, Sawyer DB, Koeppe R, Andersen O. *Biochemistry* 1989;28:6571; **B**, Venkatachalam CM, Urry WJ, *J Comput Chem* 1983;4:461; **C**, Urry DW, Long MM, Jacobs M, Harris RD. *Ann NY Acad Sci* 1975;264:203.)

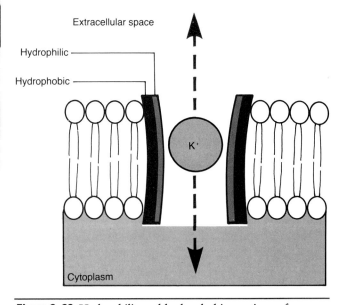

Figure 2–28. Hydrophilic and hydrophobic portions of an aqueous ion channel. A simple schematic model of an aqueous channel-forming protein predicts that a hydrophilic portion of the molecule lines the pore to interact with solute, and a hydrophobic portion adjoins the surrounding membrane lipid.

Figure 2–29. Prototypical ion channel constructed from transmembrane α-helical domains. **A:** A designer peptide containing only serine and leucine residues can form an α-helix with the two amino acid species largely segregated on opposite sides. **B:** When added to membrane, six of these identical α-helices coassemble to form an ion channel lined by the hydrophilic serine residues; the hydrophobic leucines face the surrounding membrane phospholipids. **C:** A view through the channel shows the pore in relation to the surrounding α-helices. (Modified from Hall Z. *An Introduction to Molecular Neurobiology.* Sunderland, MA: Sinauer, 1992.)

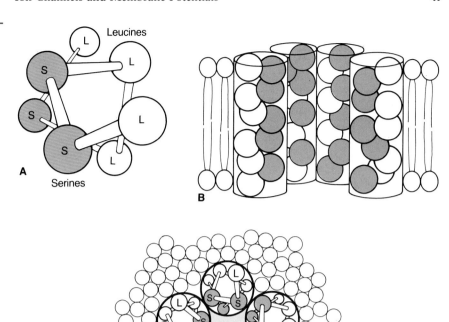

Figure 2–30. Model for the acetylcholine receptor ion channel at the neuromuscular junction. Five homologous subunits, with an M_r totaling about 300 kDa, surround a central pore to form the pentagonal configuration of the receptor. The polypeptide chain of each subunit crosses the lipid bilayer four times as an α-helix. As modeled in Fig. 2–29, one of the helices of each subunit (*red*) has more polar amino acid residues than do the others and forms part of the hydrophilic channel wall. (Modified from Alberts B, Bray D, Lewis J, et al. *Molecular Biology of the Cell*, 2nd ed. New York: Garland Publishing, 1989.)

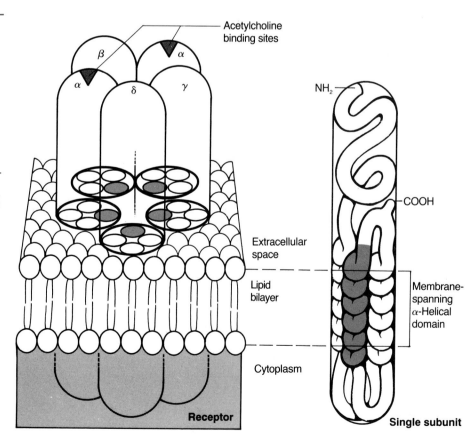

schematic model of a membrane channel shows a pore lined by transmembrane protein (Fig. 2–28). We can deduce that in its most energetically favorable form, the protein must be configured such that its most hydrophilic residues point toward the aqueous channel, and its hydrophobic residues face the lipid bilayer. In fact, a subunit of the simplest ion pore, constructed by a multimer of a synthetic peptide, demonstrates this organization. Consisting only of serine (hydrophilic) and leucine (hydrophobic) residues, the amino acid sequence of a 21-amino acid polypeptide can be designed to form an α-helix that is internally polarized (i.e., most of the serines align along one side of the helix, whereas most of the leucines segregate on the other side; Fig. 2–29). This peptide will spontaneously insert itself into a lipid bilayer so that six copies coassemble as subunits of a hexagonal structure. They form a central pore lined by the serine side of each component helix. These channels bear a striking resemblance in dimension and electrophysiologic properties to those formed at the neuromuscular junction by the acetylcholine receptor, the channel of which is circumscribed by five subunits arranged in a pentagonal configuration (Fig. 2–30). Thus, in principle, the ion channel can be constructed by a very simple molecule.

Ion Channels Are Allosteric Proteins, the Conformation of Which Is Regulated by Different Types of Stimuli

In nature, more than 75 different ion channels have been reported, and the list continues to grow. All channels are glycoproteins that contain several α-helical membrane-spanning regions, flanked by hydrophilic portions protruding into the extracellular space and cytoplasm. Functioning as an allosteric protein, the ion channel exists in two or more conformations, including a brief but stable open state and a stable closed state. Different types of stimuli that control the opening and closing, or *gating,* of ion channels, distinguish three major varieties of rapidly gated ion channels and a fourth, slowly gated type—the gap junction (Fig. 2–31). Voltage-gated channels are responsible for the propagation of electrical impulses over long distances in nerve and muscle, and they open specifically in response to a change in the electric field that normally exists across the plasma membrane of cells at rest. Ligand-gated ion channels are insensitive to voltage change but opened by the noncovalent, reversible binding of a chemical ligand. These substances include neurotransmitters or drugs that bind to the extracellular portion of the receptor. Alternatively, an intracellular second messenger or enzyme interacting with the cytoplasmic face of the channel can also influence its conformational state. Ligand-gated ion channels enable rapid communication between different neurons and between neurons and muscle or glandular cells across synapses. A few cell

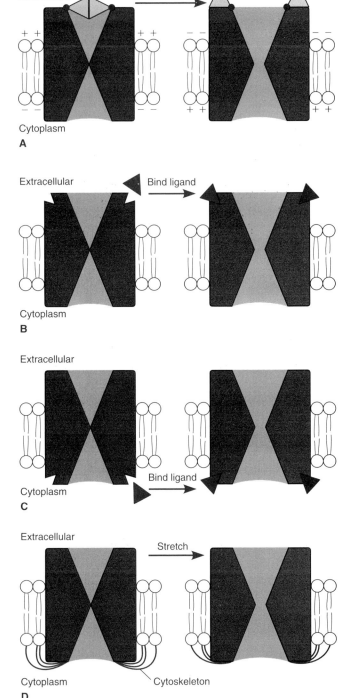

Figure 2–31. **Classes of ion channels stimulated by different gating mechanisms.** Ion channels are distinguished according to the signal that opens them. **A:** Voltage-gated channels require a deviation of the transmembrane potential. Ligand-gated receptors respond to the binding of a specific ligand, either (**B**) an external neurotransmitter molecule or (**C**) an internal mediator such as a nucleotide or ion. **D:** Mechanically gated channels can sense movement of the cell membrane linked by cytoskeletal filaments to the channel protein. Each effector causes an allosteric change that opens the channel, thereby causing an ion flux across the membrane.

types have mechanically gated ion channels for which the opening is controlled by cellular deformation. A fourth class of ion channel, the gap junction, enables ions to flow between adjacent cells without traversing the extracellular space. Gap junctions are not rapidly gated, but open and close in response to changes in the intracellular concentration of Ca^{2+} and protons.

Structural Studies Provide Insight Concerning Channel-Gating Mechanisms

Knowledge of the various ion channel-gating mechanisms has emerged with the purification of the channel proteins and some understanding of how individual subunits are structured and arranged. The best understood is the ligand-gated class, in particular, the channel gated by acetylcholine on skeletal muscle cells. These channels are restricted to special junctions, termed *synapses*, between nerve terminals and apposed stretches of the muscle cell membrane. Acetylcholine released from a stimulated nerve terminal diffuses across a cleft in the synapse and functions at the muscle cell as a neurotransmitter molecule that binds to external sites on the acetylcholine receptor channel protein. Ligand binding causes a momentary conformational change that opens the channel to small cations, but not anions. Many different neurotransmitters other than acetylcholine operate elsewhere, primarily by binding to transmitter-gated ion channels (see Chapter 9). As a rule, each of these ligands has its own specific ion channel receptor which, when opened, is selectively permeable to a certain ion.

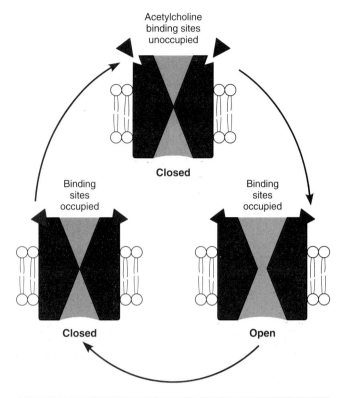

Figure 2–32. Three conformational states of the acetylcholine-gated ion channel. The binding of two acetylcholine molecules alters the protein conformation to open the channel pore. However, the effect is only transient; the pore soon closes with acetylcholine still bound to the receptor sites. Once the ligand dissociates from the receptor, the channel can return to a closed but receptive conformation.

Figure 2–33. Schematic view of the channel-forming polypeptides for voltage-sensitive Na^+ and Ca^{2+} channels. Four interconnected domains of a single large polypeptide subunit are arranged as a tetramer, lining the ion pore. Additional accessory proteins do not form part of the channels per se. (Modified from Hall Z. *An Introduction to Molecular Neurobiology.* Sunderland, MA: Sinauer, 1992.)

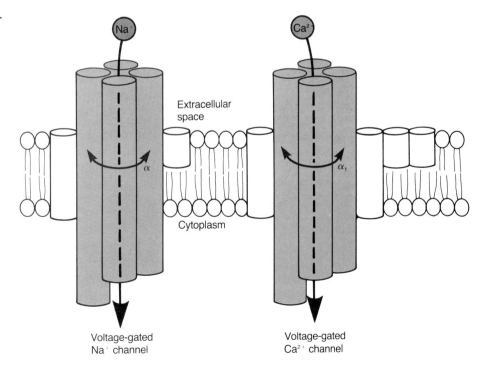

A model of the acetylcholine receptor, based on x-ray diffraction and electron-microscopic studies, is shown in Fig. 2–30. The pentameric configuration of the receptor is typical of many other members of the ligand-gated ion channel family (see Chapter 8). Two of the subunits are identical and three are different; each of the two α-subunits contain an extracellular binding site

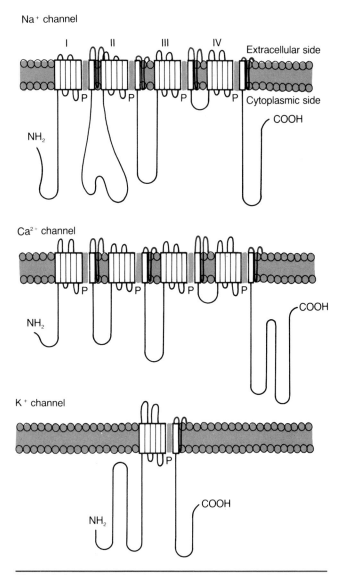

Figure 2–34. Models for the secondary structure of voltage-sensitive Na⁺, Ca²⁺, and K⁺ channels in the lipid bilayer. For Na⁺ and Ca²⁺ channels, the channel-forming α-polypeptide (see Fig. 2–31) is unraveled and schematically modeled in the membrane to emphasize the homologous construction of the four (I–IV) interconnected pseudosubunits, each containing six transmembrane α-helical regions. Four identical subunits of the K⁺ channel (only one shown) demonstrate a highly similar segmental construction. The S4 helices (*red*) are thought to act as a voltage sensor for all three channel types. The nonhelical P sequence between S5 and S6 dips into the channel pore and may confer ion selectivity to the pore. (Modified from Catterall WA. *Science* 1988;242:50; Stevens CF. *Nature* 1991;349:657.)

for acetylcholine. The binding of two acetylcholine molecules to the α-subunits evokes a conformational change that briefly opens the channel for about 1 msec before the channel recloses, still remaining bound to acetylcholine (Fig. 2–32). Once the channel is closed, acetylcholine dissociates from the channel, which returns to an unbound conformation. Lining the circumference of the channel pore is one of four α-helical-spanning regions in each subunit that contains more hydrophilic amino acid residues than do the others.

Unlike the multimeric construction of ligand-gated channels, voltage-gated ion channels for Na⁺ and Ca²⁺ are constructed from a single large polypeptide chain. In both cases, the polypeptide includes four internally homologous domains (I–IV), each containing six hydrophobic membrane-spanning regions that probably correspond to α-helices. These four interconnected domains are thought to arrange as a tetramer surrounding the voltage-gated ion channel (Fig. 2–33) in a manner analogous to the construction of the ligand-gated channel by the five subunits of the acetylcholine receptor. Differing somewhat from Na⁺ and Ca²⁺ channels is the formation of the voltage-gated K⁺ channel, which consists of four identical subunits. Nevertheless, the proposed secondary structures of K⁺, Ca²⁺, and Na⁺ channels in the membrane are strikingly similar (Fig. 2–34), suggesting that all three channels have evolved from a common ancestral gene.

In voltage-gated receptors, one of the proposed membrane-spanning helices, S4, is thought to function as the actual voltage sensor of depolarization that causes a conformational change to open the channel pore. The amino acid sequence of the S4 helix is highly conserved for all three of the voltage-sensitive channels and contains an unusually large number of basic amino acid residues that may be instrumental to the sensing mechanism (Fig. 2–35). Another recurrent sequence between helices S5 and S6 exists in domains I–IV of Na⁺ and Ca²⁺ channels and in all four K⁺ channel subunits. This nonhelical sequence, termed *P3*, is believed to dip into and line the channel pore with hydrophilic amino acid residues that influence ion selectivity.

Pore Size and Charge Interactions Determine the Ion Selectivity of Channels

Ion channels in mammalian cell membranes can be remarkably selective for the type of ions that flow through them. Conversely, gap junctions form nonselective large pores, permitting the flow of ions as well as second-messenger molecules and other metabolites with an M_r of up to about 1.5 kDa. Some gated channels with larger pores, such as the acetylcholine receptor at the neuromuscular junction, are permeable to only small cations (K⁺, Na⁺, Ca²⁺). Others are very discriminate; channels permeable to a single cation, either K⁺, Na⁺, or Ca²⁺, have been identified, as have anion-selec-

Figure 2–35. Model for the action of the S4 region voltage sensor. At rest (*left*), the interaction of basic (positively charged) amino acids in the S4 helix with fixed negative charges (e.g., organic fatty acids) in the membrane stabilize the channel in a closed conformation. On depolarization (*right*), the introduction of positive charge to the cytoplasmic side of the membrane is thought to repel the positively charged S4 region and cause it to rotate axially and outwardly toward the extracellular surface. (Modified from Hall Z. *An Introduction to Molecular Neurobiology.* Sunderland, MA: Sinauer, 1992.)

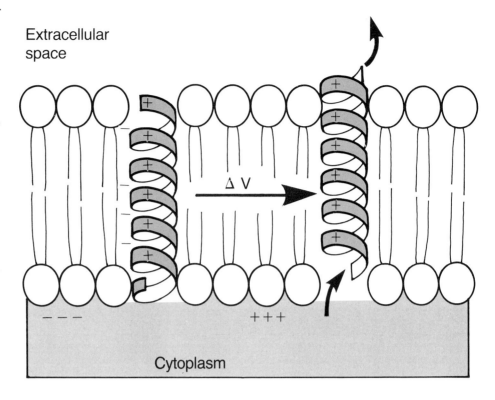

tive channels permeable only to Cl⁻. The voltage-gated K⁺ channel, for example, is too small to pass Ca^{2+} and shows a 100-fold greater selectivity for K⁺ than for Na⁺, even though Na⁺ is the smaller ion (Fig. 2–36). Thus, although varying the pore size would seem to be an easy way for channels to restrict the flow of certain ions, this alone does not control ion specificity.

Another determining factor can be the difference in hydration of specific cations in solution. Because Na⁺ is smaller than K⁺ in ionic diameter, its charge density and electric field is stronger. As a result, Na⁺ interacts strongly with more surrounding water molecules than does K⁺ does. The pull on a moving Na⁺ ion created by the larger shell of electrostatically bound water molecules reduces its mobility relative to K⁺, as though Na⁺, in effect, had the larger diameter. Hydrated Na⁺ cannot permeate the smaller-diameter K⁺ channels (Fig. 2–37).

Although the waters of hydration account for the selectivity of K⁺ channels, this does not explain why Na⁺ channels, which exceed the diameter of K⁺ as well as that of Na⁺, accept only Na⁺. Current thinking predicts that Na⁺ channels narrow at a region where polar residues lining the pore interact electrostatically for an instant (about 1 msec) with Na⁺ in passage. Importantly, the strength of this interaction is thought to be sufficient to transiently weaken the attraction of Na⁺ for its waters of hydration, thereby allowing free passage of Na⁺ through this narrow "molecular sieve" within the channel (Fig. 2–38). Although Na⁺ channels can admit K⁺, these cations are thought to be too large to closely approach the polar residues in the lining. Consequently, the interaction between the Na⁺-binding site and K⁺ is too weak to sufficiently dehydrate K⁺ for further passage through the narrow channel inlet. These examples emphasize that a small pore size and the biochemistry of

Figure 2–36. Dimensions of ion channels and solute molecules. This schematic representation relates the size of a molecule of water, Na⁺, and K⁺ to the pore size of the acetylcholine receptor at the neuromuscular junction and that of the voltage-sensitive Na⁺ and K⁺ channels. (Modified from Hall Z. *An Introduction to Molecular Neurobiology.* Sunderland, MA: Sinauer 1992.)

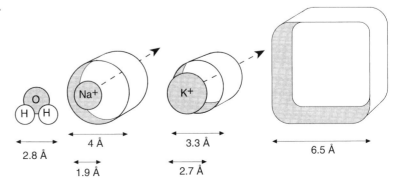

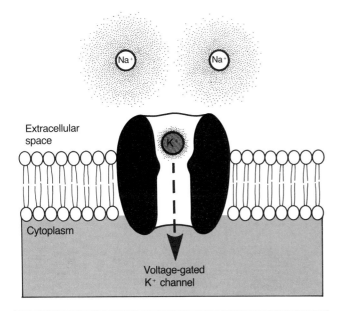

Figure 2–37. Ion selectivity of the K⁺ channel, based on waters of hydration. The high electric field strength of the smaller Na⁺ ion attracts a larger shell of surrounding water molecules than does K⁺. The diameter of the hydrated Na⁺ ion is larger than is the pore of the K⁺ channel.

the pore lining conspire to determine the ion selectivity of a particular type of channel.

Although ion channels are most often recognized in the role of electrical signaling in nerve and muscle cells, they exist—to some degree—in all cells to mediate other functions. The selectivity of a cell membrane for permeant ions depends on the relative proportions of various types of ion channels. A single neuron can have as many as four different ion-selective channels. The most common one among neural and nonneural cell types is termed the *K⁺ leak channel* because its opening does not require a specific gating stimulus. These channels enable **all** cells in the body to maintain a voltage difference across their plasma membrane, the *membrane potential*.

Membrane Potential Is Caused by a Difference of Electric Charge on the Two Sides of the Plasma Membrane

Cells maintain slightly more negative than positive ions in the cytosol and, conversely, more positive than negative ions in the extracellular fluid (Fig. 2–39). This causes a transmembrane voltage difference owing to the resemblance of the plasma membrane to an electrical capacitor. Membranes comprised of a lipid bilayer are hydrophobic and, therefore, poor conductors of ionic current. As a result, an accumulation of negative charges along the cytosolic side of the membrane attracts positively charged ions on the extracellular side of the bilayer. The potential difference, or voltage gra-

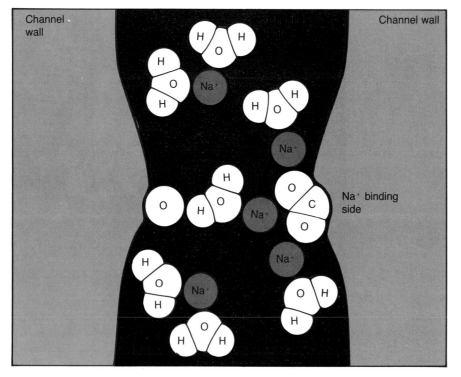

Figure 2–38. Model of a selectivity filter in the Na⁺ channel. Along the channel wall is a postulated narrowing at which an Na⁺-binding site faces another polar amino acid residue lining the opposite side of the channel. At this inlet, an incoming Na⁺ loses at least some of its waters of hydration, allowing further passage, and transiently binds to the polar site with sufficient energy to "stabilize" the unbalanced positive charge of the cation. Steric hindrance is thought to prevent the larger K⁺ ions from closely approaching the Na⁺-binding site, thereby precluding a similar exchange. (Modified from Hille B. *Ionic Channels of Excitable Membranes.* Sunderland, MA: Sinauer, 1992.)

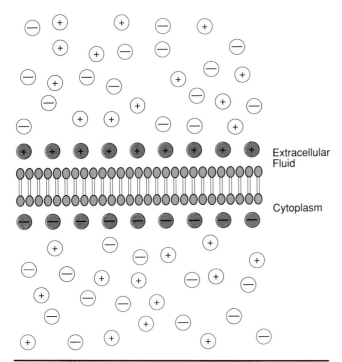

Figure 2–39. The separation of charge across the cell membrane forms a membrane potential. A net excess negative charge inside the membrane and a matching net excess positive charge outside the membrane form a transmembrane potential difference that is maintained across an impermeable lipid bilayer. Charge on either side of the membrane is concentrated in a thin layer (<1 nm thick) and formed by an extremely small percentage of the total ions in the cell.

dient, that arises across the membrane thickness (about 5.0 nm) is enormous, nearly 200,000 V/cm. As a consequence, membrane potential and transmembrane ionic gradients provide a driving electrical force for many biologic processes.

Any membrane potential V_m is defined as $V_m = V_i - V_o$, where V_i is the voltage inside the cell and V_o is the voltage outside the cell. Because V_o is arbitrarily set to zero, V_m in the undisturbed, or *resting*, cell becomes negative solely because of the slightly negative net ionic charge of the cytoplasm. To understand how the membrane potential is generated requires consideration of the major contributing ions and their distribution across the plasma membrane.

Membrane potential is based primarily on four ion species: K^+, Na^+, Cl^-, and organic anions (A^-), such as amino acids and other metabolites. Of these, Na^+ and Cl^- are concentrated in the extracellular fluid, whereas K^+ and A^- are preponderant inside the cell (see Table 2–2). Recall that the active extrusion of Na^+ from the cell by Na/K-ATPase maintains the osmotic balance of the cytosol by preventing the influx of water that would otherwise occur. In exchange for pumping three Na^+ ions out of the cell, two K^+ ions enter the cytosol to

counterbalance organic anions that do not permeate the plasma membrane. For a cell with many Na^+ channels in its membrane, such as a neuron or muscle cell, almost all of them remain closed when the cell is at rest. Therefore, Na^+ extruded by Na/K-ATPase cannot readily reenter the cell down its steep concentration gradient. Only nongated K^+ leak channels remain open. As a result, K^+ will tend to passively leak out of the cell through the K^+ leak channels down its steep concentration gradient until the force of outward diffusion is counterbalanced by a second and opposing inward electrical force created by the attraction that organic anions in the cytosol have for K^+ (Fig. 2–40). Collectively, these two influences—the concentration gradient and the voltage gradient—for a particular ion across a membrane determine the net electrochemical gradient that drives the flow of that ion species through a membrane channel. When the electrical and chemical forces counterbalance one another for an ion species, the electrochemical gradient is zero, and no net flow of this ion occurs across the membrane. Given the concentration of K^+ inside and outside the cell, the voltage necessary to achieve this equilibrium, termed the *equilibrium potential*, can be calculated from the Nernst equation:

$$E_K = \frac{RT}{ZF} \ln \frac{[K^+]_o}{[K^+]_i}$$

where E_K is the value of the equilibrium potential of K^+, R is the gas constant (2 cal mol^{-1} °K^{-1}), T is the absolute temperature in degrees Kelvin (°K), F is the Faraday constant (2.3×10^4 cal V^{-1} mol^{-1}), and $[K^+]_o$ and $[K^+]_i$ are the concentrations of K^+ outside and inside of the cell. For a monovalent cation, such as K^+, the charge (Z) = $^+1$, and at 37°C, RT/ZF is 27 mV. By using typical values for $[K^+]_o/[K^+]_i$ in mammalian tissue (Table 2–2) and converting natural logarithms to base 10 logarithms, the equation simplifies to

$$E_K = 27 \text{ mV} \times 2.3 \log_{10} \frac{4}{139} = -96 \text{ mV}$$

Given the concentration of any ion species across the membrane, the Nernst equation can be used to determine the equilibrium potential for that ion.

Membrane Potential for Any Cell at Rest Depends Primarily on the Transmembrane Potassium Ion Gradient Through Nongated Potassium Ion Leak Channels

The resting membrane potential (V_R) for a cell is determined by permeability of the membrane to specific ions and their concentrations inside and outside the cell. For a glial cell, having only nongated K^+ leak channels in its membrane, V_R is essentially equal to E_K. However, consider the artificial scenario resulting from the insertion

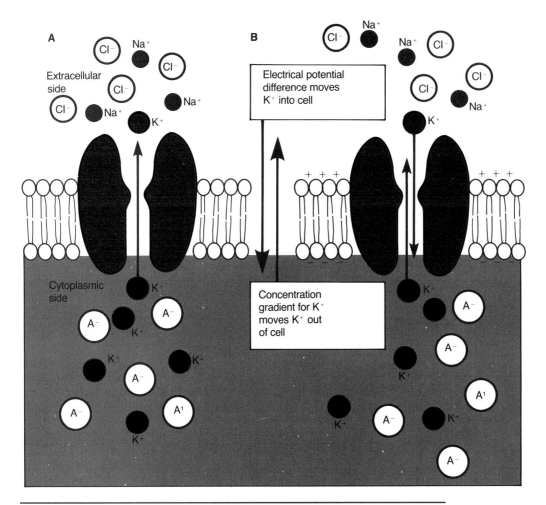

Figure 2–40. Opposing forces regulate K⁺ flux across the plasma membrane. **A:** The resting membrane potential of a cell permeable only to K⁺ depends on the passive diffusion of K⁺ out of the cell down its concentration. **B:** If left unchecked, K⁺ efflux would eventually create an excess negative charge in the cell (an overbalance of organic anions) and a buildup of [K⁺]ₒ, were it not for an electrical driving force moving K⁺ in the opposite direction. An equilibrium results when these two opposing forces counterbalance each other. (Modified from Koester J. In: Kandel ER, Schwartz JH, Jessell TM, eds. *Principles of Neural Science*, 3rd ed. New York: Elsevier, 1991.)

of a few open Na⁺ channels to the glial cell membrane. In this event, a strong electrochemical gradient for Na⁺ would be realized and tend to push Na⁺ into the cell. Why? First, because Na⁺ is more concentrated outside than inside the cell, it will passively flow into the cell through the opened donor Na⁺ channels. Second, the electrical force of attraction of a slightly electronegative membrane potential generated by the Na/K-ATPase and E_K (−96 mV) will also attract Na⁺ into the cell. The magnitude of the latter force can be appreciated by calculating the value of the equilibrium potential for Na⁺ (E_{Na}) from the Nernst equation and typical values for Na⁺ concentration (Table 2–2), where

$$E_{Na} = \frac{RT}{ZF} \ln \frac{[Na^+]_o}{[Na^+]_i}$$

simplifies to

$$E_{Na} = 27 \text{ mV} \times 2.3 \log_{10} \frac{145}{12} = +67 \text{ mV}$$

This tells us that E_{Na} is 163 mV away from the V_R established by E_K alone (−96 mV). As a result, both the electrical and chemical forces work in the same inward direction to form a strong electrochemical gradient driving Na⁺ into the cell.

In principle, Na⁺ influx should then depolarize the cell; that is, it should reduce the charge separation across the membrane by making the interior less electronegative relative to outside the cell. In a neuron that normally makes its own Na⁺ channels in addition to K⁺ channels, this effect is minimal, because when the cell is at rest, the ratio of open K⁺ leak channels to Na⁺ chan-

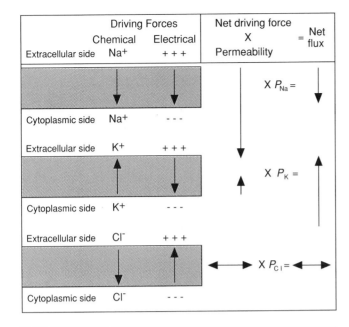

Figure 2–41. Electrochemical driving force and ion permeability codetermine net ion flux across the plasma membrane. The direction of net flux for Na+, K+, and Cl- across the plasma membrane is a function of the net electrochemical driving force and the permeability of the membrane for each ion species. (Modified from Koester J. In: Kandel ER, Schwartz JH, Jessell TM, eds. *Principles of Neural Science*, 3rd ed. New York: Elsevier, 1991.)

nels is very high. Thus, the cell is much more permeable to K+ than to Na+, so that the contribution of Na+ influx to V_R is minimal. In reality, a balance of −80 to −90 mV for V_R is attained by mammalian cells, a value *near* E_K and far from E_{Na}. For a nerve or muscle cell with multiple ion-selective channels in its membrane, V_R is influenced by the permeability of the membrane for each diffusible ion species (K+, Na+, Cl−; Ca^{2+} is too scarce to make a difference) and the concentration of each inside and outside the cell. These influences on V_R are related by the Goldman equation, where

$$V_R = \frac{RT}{F} \ln \frac{P_K[K^+]_o + P_{Na}[Na^+]_o + P_{Cl}[Cl^-]_i}{P_K[K^+]_i + P_{Na}[Na^+]_i + P_{Cl}[Cl^-]_o}$$

and P is permeability. Ultimately Cl− ions play a negligible role in determining V_R because the value of V_R is chiefly set by Na+ and K+ fluxes, lying between E_K and E_{Na} (Fig. 2–41). Chloride equilibrates across the membrane through nongated Cl− channels, but most remains in the extracellular fluid to counterbalance nonpermeable intracellular anions. If the permeability to one ion far supersedes that of others ($P_K \gg P_{Cl}, P_{Na}$), as seen for glial cells, the Goldman equation is reduced to the Nernst equation for that ion, where

$$V_R = \frac{RT}{F} \ln \frac{[K^+]_o}{[K^+]_i}$$

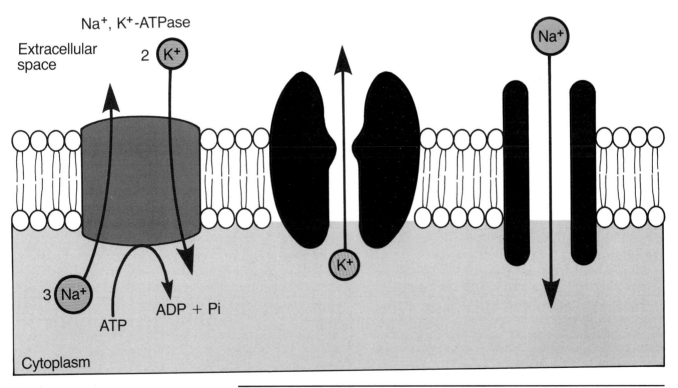

Figure 2–42. Passive and active fluxes maintain the resting membrane potential. The cell at rest maintains a steady state whereby Na+ influx and K+ efflux occurring by passive diffusion is balanced by active transport of these ions in the opposite direction by Na/K-ATPase. (Modified from Koester J. In: Kandel ER, Schwartz JH, Jessell TM eds. *Principles of Neural Science*, 3rd ed. New York: Elsevier, 1991.)

Resting Membrane Potential Is Maintained by Sodium/Potassium ATPase

If K$^+$ efflux and the smaller Na$^+$ influx were allowed to continue unchecked across the plasma membrane of a neuron, transmembrane gradients for both ions would eventually dissipate, because [K$^+$]$_i$ would plummet and [Na$^+$]$_i$ would gradually increase, thereby reducing the V_R. Opposing this effect, to keep V_R constant, is Na/K-ATPase, which continues to pump three Na$^+$ ions out of the cell for every two K$^+$ ions pumped in. Because there is a net transfer of positive charge out of the cell, the pump is said to be *electrogenic*, setting up the slight excess of negative charge inside the cell membrane. Thus, passive channel-mediated ion fluxes occurring by simple diffusion are balanced by active fluxes that require energy (Fig. 2–42). A steady state between the two processes is reached where the *net* ion flux across the membrane is zero to define the resting membrane potential. Importantly, any disturbance to a cell that increases the membrane permeability for an ion will drive the membrane potential away from V_R in the direction of the equilibrium potential for that ion. These transient deviations from V_R and the opening of gated ion channels that cause them are the basis of electrical signals that convey information not only along and between nerve cells, but also from nerve to muscle.

Action Potentials

Information conveyed over long distances in a nerve or muscle cell requires a special signal, the *action potential*. An action potential is a fleeting, self-renewing wave of membrane depolarization that propagates without

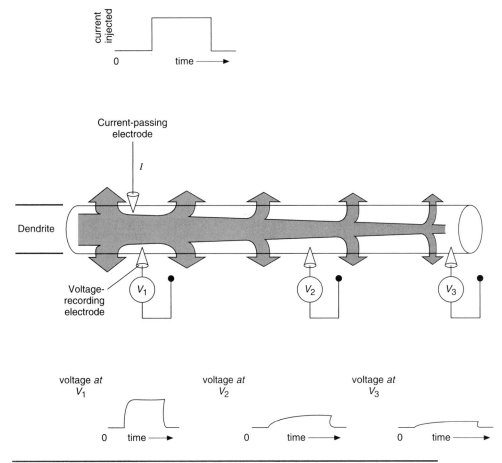

Figure 2–43. Attenuation of a local potential over distance. When a brief pulse of current (*upper tracing*) is injected into a dendrite through a microelectrode, most of it leaks out through channels (K$^+$) near the injection site. As a result, the disturbance in membrane potential measured with recording electrodes (V_1–V_3) decreases exponentially (*lower tracing*) with increasing distance from the original site of depolarization. (Modified from Alberts B, Bray D, Lewis J, et al. *Molecular Biology of the Cell*, 2nd ed. New York: Garland Publishing, 1989.)

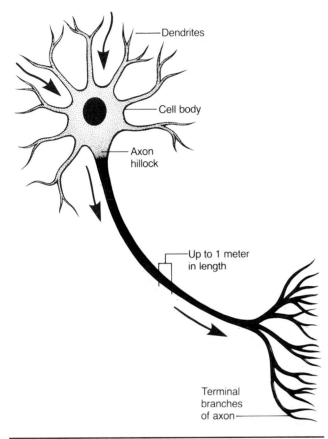

Figure 2–44. **Prototypical neuron and the direction of neural transmission.** Incoming local potentials from dendrites converge at the cell body and reach the axon at its origin. The action potential, typically triggered at a region known as the *axon hillock*, is propagated down the length of the axon (*red*), where it will initiate a chain of events leading to neurosecretion from the terminal branches.

decrement along the length of a nerve axon at high speed (up to 120 m/sec). The gating of two types of ion channels in the membrane, the voltage-dependent Na^+ and K^+ channels, coordinately forms this moving signal by a mechanism that we will now examine in greater detail.

Transmembrane Electrical Signals Not Conveyed as Action Potentials Are Limited by the Intracellular Distance They Can Travel

Neurons convey signals along their length by forming local potentials, which spread passively, and self-propagated action potentials. A local potential arises when the cell membrane is stimulated to become more permeable to certain ions than it is at rest. If Na^+ influx results, the voltage drop across the membrane is reduced, and the signal is said to be *depolarizing*. Alternatively, enhanced Cl^- influx will increase the voltage drop in some neurons and *hyperpolarize* the cell relative to its resting state.

When current enters the neuron through open membrane channels, it diffuses in all directions for a distance that depends on the intrinsic properties of the neuron. For example, much of the current leaks out of the axon near the site of entry (by nongated K^+ leak channels) in lieu of the resistance to longitudinal flow through the cytosol (Fig. 2–43). As a consequence, current flows, and the spreading wave of depolarization decays exponentially with increasing distance from its origin in the fiber. For most neurons, the spread of local potentials over distances up to about 1 mm is sufficient to convey electrical signals toward the cell body along relatively short input fibers called *dendrites*. However, such signals would dissipate well before leaving the cell along its axon, which is typically a much longer structure (Fig. 2–44). For axonal transmission, neurons must enlist another electrochemical mechanism, the action potential.

Figure 2–45. Threshold depolarization triggers the action potential. **A:** An excitable cell is impaled with two microelectrode pipets—one used to deliver a brief pulse of current and the other for recording membrane potential. **B:** Increasing the amount of current causes correspondingly large depolarization until a threshold pulse triggers an action potential. (Modified from Hall Z. *An Introduction to Molecular Neurobiology.* Sunderland, MA: Sinauer, 1992.)

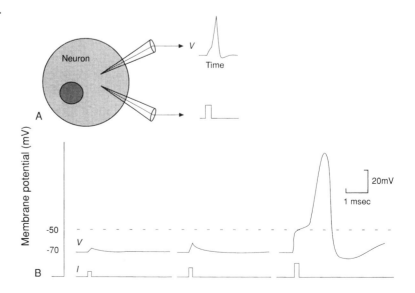

Voltage-Gated Sodium and Potassium Ion Channels Make Cells Electrically Excitable

An action potential in a neuron or myocyte is produced when a momentary electrical disturbance raises the resting membrane potential across a small patch of the cell to a less negative, critical threshold value; subthreshold depolarization produces only a local potential. (This will be discussed more fully in later chapters.) The same effect can be produced experimentally by introducing a brief pulse of current into a neuron impaled with a microelectrode. If the current load is sufficient to raise the resting membrane potential to threshold, this partial depolarization of the cell will trigger the opening of a few local voltage-gated Na^+ channels in the axon, allowing a small amount of Na^+ to enter the axon down its steep electrochemical gradient. Increased intracellular Na^+ further depolarizes the cell and recruits more voltage-sensitive channels to open, resulting in more Na^+ entry. This progressive, self-amplifying mechanism for depolarization reaches a limit when the membrane potential has shifted from −90 mV or −80 mV (V_R) for mammalian myocytes and neurons, respectively, to about −50 mV, having nearly attained the equilibrium potential for Na^+ (Fig. 2–45). Two mechanisms next engage to return the same membrane patch to its original resting potential.

First, opened voltage-gated Na^+ channels rapidly close in the depolarized membrane because they are conformationally unstable in the open state. Once reclosed, the channels remain in an inactivatable state until after the membrane is repolarized. Thus, voltage-gated Na^+ channels can exist in three different conformations: (1) closed, but activatable; (2) open; and (3) closed, but inactive (Fig. 2–46). The transition of recruited channels through each of these states contributes to the shape of the action potential (Fig. 2–47), in particular, the rising (depolarizing) phase.

The second mechanism contributing to the decay of the action potential is the opening of voltage-sensitive K^+ channels. These channels respond more slowly to depolarization than do voltage-gated Na^+ channels and are not opened until the action potential is near its peak. With a membrane potential of −50 mV, a strong electrochemical gradient for K^+ efflux builds inside the cell and is released when the voltage-sensitive K^+ channels transiently open. As a result, the sudden loss of positive charge repolarizes the cell more quickly (Fig. 2–48) than if Na^+-channel inactivation and nongated K^+ leak channels merely acted alone (see Fig. 2–47). The return of the membrane potential toward the K^+ equilibrium potential closes voltage-gated K^+ channels and allows inactivated Na^+ channels to regain their activatable state. Thus, by accelerating the decay of the action potential and membrane repolarization, voltage-gated K^+ channels reduce the length of a refractory period (<1 msec) before another action potential can be triggered at the same site.

Action Potentials Travel Long Distances Without Decrement

"Nerve impulses" are transmitted along most neurons as local potentials that travel toward the cell body through dendritic processes and propagate away from the cell body along the axon as action potentials. The action potential is typically triggered at a membrane segment where the axon emanates from the cell body, a region called the *axon hillock* (see Fig. 2–44). Only there do electrical signals finally encounter voltage-gated Na^+ channels at sufficient density in the membrane to trigger the action potential. Current carried by the same local influx of Na^+ ions that automatically triggers an action potential will also diffuse longitudinally along the axon in both directions. In this way local depolarization reaches the threshold to initiate a new action potential at a neighboring membrane segment as the signal decays at the original initiation site. Thus, the action potential self-propagates as a transient depolarizing wave form racing toward the axon terminal at speeds in vertebrates ranging from 1 to 100 m/sec, depending on the type of axon. Although Na^+ diffusion spreads bidirectionally after the initial influx, note that channels adjacent and immediately proximal (in the direction of the cell body) to the traveling action potential are always transiently inactivated (Fig. 2–49). As a result, the action potential moves in the direction toward voltage-gated Na^+ channels that are in a closed but activatable state, i.e., toward the axon terminal.

Two Factors Accelerate the Velocity of an Action Potential: Axonal Caliber and Myelin

The all-or-none nature of the action potential overcomes the problem of signal attenuation over long distances. Nevertheless, additional properties of the axon determine the speed at which these nerve impulses travel. As the action potential propagates from one patch of axolemma to the next, local current diffuses within the axon to depolarize neighboring segments to the threshold value. The effective distance over which this occurs is determined by the internal resistance (r_i) of the axonal cytosol to current flow, and the resistance of the membrane (r_m) to current (K^+) escape through open channels. A decrease in r_i and an increase in r_m will allow larger currents to flow longer distances inside the axon, thereby charging the capacitance of neighboring membranes at a faster rate. Consequently, a mere doubling of the axonal diameter can significantly speed the conduction rate of nerve impulses. Although r_m would actually decrease as a result of doubling the axonal cir-

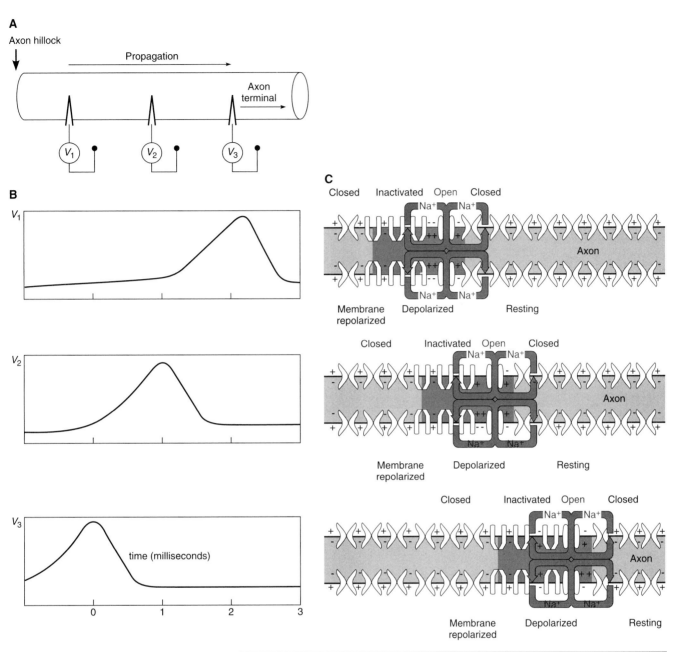

Figure 2–46. The traveling action potential. **A:** An axon impaled with recording electrodes at three progressively more distal sites shows that (**B**) an action potential propagates toward the terminal without decrement as an invariant wave of depolarization. **C:** Unidirectional propagation of the action potential toward the axon terminal does not relate to the direction of ion flow following the initial Na^+ influx; Na^+ ions entering the axon diffuse in all directions. However, movement of the action potential toward the cell body is blocked because voltage-gated Na^+ channels are expressed selectively in the axolemma, beginning at the axon hillock. At any instant, Na^+ channels are left transiently inactivated toward the cell body, relative to the position of the action potential. Therefore, depolarization can spread only toward the axon terminal. (Modified from Alberts B, Bray D, Lewis J, et al. *Molecular Biology of the Cell*, 3rd ed. New York: Garland Publishers, 1994.)

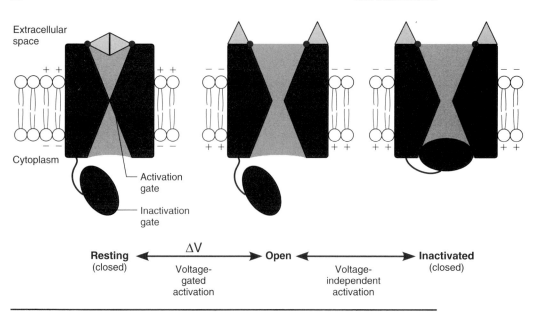

Figure 2–47. Three conformation states of the voltage-gated Na⁺ channel. Membrane depolarization mediates the opening of the voltage-gated Na⁺ channel by opening an activation gate formed by the channel protein lining. The open state is metastable and rapidly inactivated by a separate inactivation-gating mechanism provided by the cytosolic portion of the channel protein. Repolarization of the cell returns the inactivated channel to the closed state. Each of the four subunits of the receptor has an inactivation gate; only one is shown.

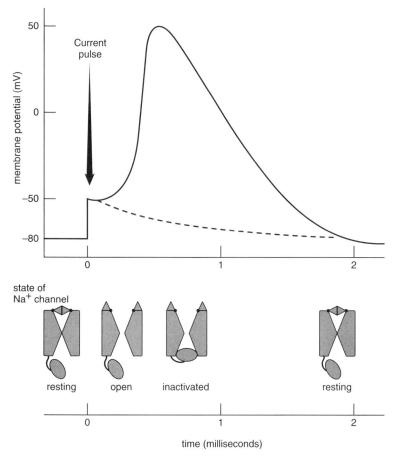

Figure 2–48. Time course of an action potential shaped by the changing conformational state of voltage-gated Na⁺ channels. In response to a brief pulse of current, depolarization of a mammalian neuron to a threshold of about −50 mV triggers an action potential (*upper graph*) by opening of voltage-sensitive Na⁺ channels in a cascade of activity (*lower graph*). After reaching a maximum of +50 mV, the membrane potential declines to its resting state because Na⁺ channels inactivate, and the efflux of K⁺ ions through nongated leak channels continues unabated. Another action potential is not possible until the Na⁺ channels have returned to the closed but activatable state. In the absence of voltage-gated Na⁺ channels, the modest depolarization evoked by the current stimulus would have immediately begun to decay (*dashed line*). (Modified from Alberts B, Bray D, Lewis J, et al. *Molecular Biology of the Cell,* 3rd ed. New York: Garland Publishing, 1994.)

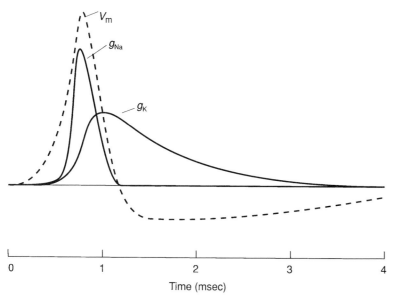

0   1   2   3   4

Time (msec)

Figure 2–49. Ion conductance changes producing the action potential. Voltage-gated Na⁺ channels open, leading to increased Na⁺ conductance (gNa) into the cell and the upward phase of the action potential (V_m). As gNa rapidly falls, gK⁺ slowly rises to accelerate the falling phase of the action potential. [Modified from Hodgkin A, Huxley A. *J Physiol (Lond)* 1952;117:500.]

cumference and, thereby, the number of open ion channels per segment, this is compensated by a fourfold increase in cross-sectional area of the axon, which reduces r_i by a factor of four. The net effect of enlarging the axonal diameter is to increase the propagation velocity of the action potential. If axonal diameter were the sole determinant of the speed of nerve impulses, the human body would require a spinal cord the size of a tree trunk. Rapid signaling along giant axons evolved by the squid is achieved in vertebrates with considerable savings of energy and space by insulating many axons with a myelin sheath.

The formation and maintenance of myelin is the task of two glial cell types. Oligodendrocytes in the central nervous system, and their counterparts in peripheral nerves, the Schwann cells, make enormous quantities of flattened plasma membrane and wrap it around axons in a concentric fashion to form a biochemically specialized sheath up to 200 layers thick (Fig. 2–50). Every myelinating Schwann cell invests a single axon to form a segment of sheath, termed an *internode*, that occupies approximately 1 mm of axon length. By contrast, a single oligodendrocyte extends branches from its cell body that flatten and expand to ensheathe as many as 40 different axons (Fig. 2–51). Between one internode of myelin and the next are regions of axon, varying in length from 0.5 to 20 μm, termed *nodes of Ranvier* (Fig. 2–52). Along a myelinated axon, almost all membrane channels, including voltage-gated Na⁺ channels, are confined to nodes of Ranvier. Owing to the extremely high lipid:protein ratio and the stacking of the individual myelin membranes, transmembrane resistance (r_m) across the axon is greatly increased along internodes. As a result, current entering the axon through Na⁺ channels at nodes of Ranvier spreads locally with greater

efficiency than would otherwise occur in a nonmyelinated axon, because the increased r_m minimizes current leakage from the axon (Fig. 2–53). Myelination also decreases the capacitance of the axolemma. In a nonmyelinated axon, the buildup of opposite charges on either side of the membrane occurs along its entire length. For a myelinated axon, this is confined to the nodes of Ranvier. Having the lower capacitance, myeli-

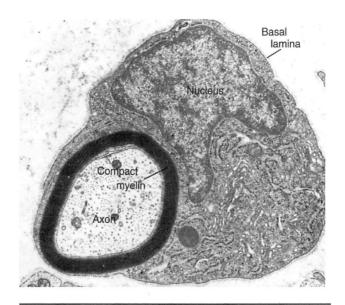

Figure 2–50. Electron micrograph of a myelin sheath surrounding an axon in peripheral nerve. The Schwann cell cytoplasm is electron-dense, owing to an abundance of free and bound ribosomes necessary for membrane biogenesis. Well-defined Golgi apparatus is also prominent. Magnification × 32,000. (Raine C. In: Morell P, ed. *Myelin*, 2nd ed. New York: Plenum Press, 1984, with permission.)

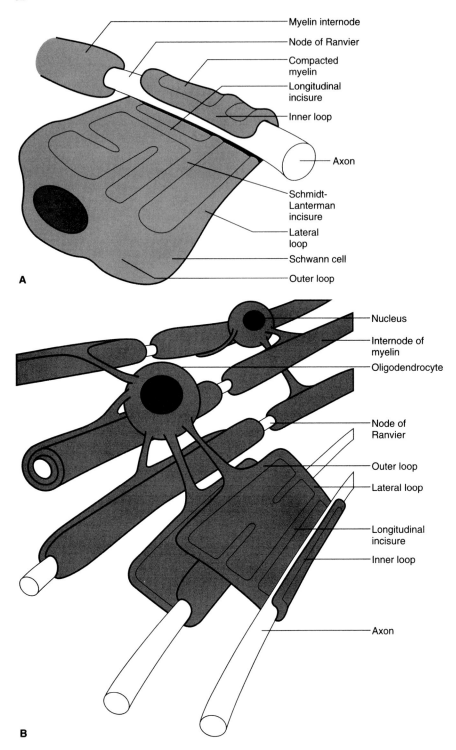

A

B

Figure 2–51. Unrolled myelin made by (A) the Schwann cell and (B) oligodendrocyte. **A:** The Schwann cell makes a single internode of flattened myelin that encircles one segment of one axon. **B:** An oligodendrocyte extends from its soma multiple cytoplasmic processes, each expanding to form a myelin internode around a different axon. During myelination, cytoplasm is largely extruded from the myelin membrane of both cell types. This results in compaction of the myelin, and the flattened plasmalemma collapses against itself. The effect is analogous to removing the air from a balloon, except that compaction is incomplete. Where channels of cytoplasm, termed *loops* or *incisures*, persist, they keep the cell body and myelin sheath in communication with each other.

nated axons require the influx of fewer positive charges to reduce the transmembrane potential to the threshold for an action potential. As a result of the increased r_m and decreased capacitance signal conduction along a myelinated axon jumps from node to node and is said to be saltatory. In addition to speeding the nerve impulse, myelination drastically reduces the amount of energy needed for axonal conduction. In the unmyeli-

nated axon, Na/K-ATPase is needed to restore the membrane potential along the entirety of the fiber. In the myelinated axon, this occurs only at the depolarized nodes of Ranvier.

Because myelin speeds the conduction of nerve impulses, diseases or toxins that injure the myelin sheath or myelin-producing cells can cause significant neurologic problems. Axons stripped of their myelin

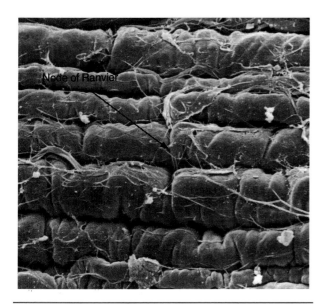

Figure 2–52. Nodes of Ranvier visualized by scanning electron microscopy. Bundled axons coursing in the horizontal plane of the micrograph are surrounded by internodes of myelin. A Node of Ranvier (*arrow*) appears as a deep furrow between internodes. Thin strands are external collagen fibers of the nerve sheath. Magnification 1885. (Kessel RG, Kardon RH. *Tissues and Organs: A Text Atlas of Scanning Electron Microscopy.* New York: WH Freeman & Co, 1979.)

sheath conduct impulses slowly or not at all. Conduction deficits resulting from damage to the brain and spinal cord are particularly severe because oligodendrocytes, unlike Schwann cells, fail to regenerate sufficiently in most circumstances. Moreover, damage to a single oligodendrocyte is of greater physiologic consequence because it produces internodes of myelin surrounding segments of several different axons. Multiple sclerosis is the most prevalent, and is the prototype, of demyelinating diseases affecting the central nervous system. The symptoms of this disease that are usually manifested clinically are the result of dysfunctional sensory and motor neuron systems that are most dependent on rapid neurotransmission.

Clinical Case Discussion

Cystic fibrosis is an autosomal recessive disorder affecting approximately 1 in 2,000 children, with a carrier rate of 1:22 persons in the Caucasian population. The main characteristics of CF are malabsorption due to exocrine pancreatic insufficiency, recurrent bacterial infections in the lower respiratory tract, increased salt loss in sweat, and male infertility due to absence or stenosis of the vas deferens. The diagnosis is made with

Figure 2–53. Saltatory condition along a myelinated axon. The multilaminar myelin membrane largely restricts current flux across the axolemma and transmembrane separation of charge (capacitance) to nodes of Ranvier. (Modified from Hall Z. *An Introduction to Molecular Neurobiology.* Sunderland, MA: Sinauer, 1992.)

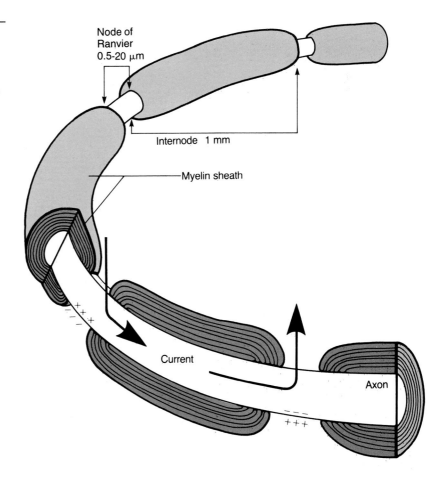

a positive sweat chloride test. The sweat of individuals with CF contains elevated concentrations of chloride and sodium. The gene responsible for CF, located on chromosome 7, encodes for the cystic fibrosis transmembrane conductance regulator (CFTR). The most common mutation, F508, involves a deletion of three adjacent base pairs leading to the absence of one amino acid: phenylalanine at position 508. This deletion accounts for 70% of the CF mutations. In addition to the F508, more than 500 mutations have been identified in different populations. The normal CFTR gene product functions as a cAMP-regulated chloride channel on the apical membrane of epithelial cells. The net effect of the abnormal CFTR protein is both decreased secretion and increased absorption of electrolytes through chloride channels, leading to dehydration of secretions covering the respiratory epithelium. A similar mechanism is operational in the epithelial cells in the biliary tract, gut, and pancreas. Pancreatic insufficiency is present in 85% of patients, due to obstruction of the pancreatic duct.

Standard therapies for CF include antibiotics, chest physiotherapy, and supplemental pancreatic enzymes. Chronic obstructive pulmonary disease and respiratory failure due to chronic *Pseudomonas aeruginosa* endobronchial infections, acquired in more than 90% of CF patients, is the most common cause of death. Recent clinical trials have shown a beneficial effect for recombinant human DNase I (Dornase alpha), which depolymerizes high-molecular-weight DNA released from neutrophils in the respiratory tract. The fatal respiratory manifestations of CF are amenable to gene therapy where the normal CFTR gene would be transferred directly to the airway epithelium.

Suggested Readings

Membrane Structure

Singer SJ, Nicholson GL. The fluid mosaic model of the structure of cell membranes. *Science* 1972;175:720.

Singer SJ. The structure and insertion of integral proteins in membranes. *Annu Rev Cell Biol* 1990;6:247.

Tanford C. *The Hydrophobic Effect*, 2nd ed. New York: John Wiley & Sons, 1980.

Diffusion

Stein WD. *Transport and Diffusion Across Cell Membranes.* Orlando: Academic Press, 1986.

Osmosis

Eveloff JL, Warnock DG. Activation of ion transport systems during cell volume regulation. *Am J Physiol* 1987;252:F1.

Kregenow FM. Osmoregulatory salt transporting mechanisms: control of cell volume in anisotonic media. *Annu Rev Physiol* 1981;43:493.

Transport

Fambrough DM. The sodium pump becomes a family. *TINS* 1988;11:325.

Scott DM. Sodium cotransport systems: cellular, molecular and regulatory aspects. *Bioessays* 1987;7:71.

Walmsley AR. The dynamics of the glucose transporter. *TIBS* 1988;13:226.

Ion Channels and Membrane Potentials

Catterall WA. Molecular properties of voltage-sensitive sodium channels. *Annu Rev Biochem* 1986;55:953.

Hille B. *Ionic Channels of Excitable Membranes*, 2nd ed., Sunderland, MA: Sinauer, 1992.

Action Potentials

Keynes RD. Ion channels in the nerve-cell membrane. *Sci Am* 1979;240:126.

Waxman SG, Ritchie JM. Organization of ion channels in the myelinated nerve fiber. *Science* 1985;228:1502.

Review Questions

1. The intramembranous particles observed on freeze fracture of biological membranes represent exposed:
   a. Multipass transmembrane proteins
   b. Single-pass transmembrane proteins
   c. Phospholipid clusters
   d. Cholesterol clusters
   e. All of the above
2. The Frye and Edidin experiment utilized human-mouse heterokaryons to demonstrate:
   a. The flip-flop of phospholipids
   b. The rotation of phospholipids
   c. The lateral mobility of phospholipids
   d. The lateral mobility of transmembrane proteins
   e. The rotational movement of proteins
3. The movement of Na$^+$ and glucose through the apical membrane of an intestinal epithelial cell is an example of:
   a. Diffusion
   b. Facilitated diffusion
   c. Primary active transport
   d. Secondary active transport
   e. Antiport
4. The Na$^+$/K$^+$-ATPase is:
   a. An electrogenic pump
   b. An example of primary active transport
   c. Contributing to the membrane potential

d. All of the above
e. None of the above
5. The decay of an action potential results from:
   a. The rapid closing of voltage-gated Na^+ channels
   b. The opening of voltage-gated Na^+ channels
   c. The opening of voltage-gated K^+ channels
   d. b and c are correct
   e. a and c are correct
6. Myelinated axons demonstrate increased:
   a. Electrical capacitance
   b. Amplitude of the action potential
   c. Velocity of the action potential
   d. a and c
   e. All of the above

7. Ionic selectivity of transmembrane channels is determined by:
   a. Polarity of amino acids residues lining the channel pore
   b. Waters of hydration
   c. The concentration gradient of an ion species across the plasma membrane
   d. a and b
   e. All of the above
8. The CFTR controls which type of ion channel in epithelial cells?
   a. Sodium
   b. Chloride
   c. Potassium
   d. Calcium
   e. Na/K-ATPase

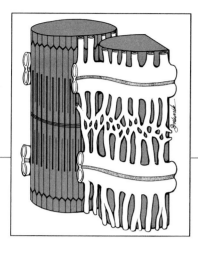

Chapter 3
Cytoskeleton

Clinical Case

A 3-year-old boy, Cary Debs, was evaluated by his pediatrician for clumsiness. Mom states he falls more than the other children his age, and he was evaluated in the emergency room after falling down the steps and losing consciousness. The boy can run only short distances without tiring. Over the last few months, he has been walking on his toes and developed a waddling gait. Past medical history and family history were negative. His physical examination was remarkable for decreased muscle tone and absent knee reflexes with hyperreflexia at the ankles. When the child was asked to stand from a supine position, he used his hands to climb up his legs (Gower's maneuver). A screening creatinine kinase was elevated at 4,000 IU; electromyography was abnormal. A muscle biopsy showed a deficiency of type 2B fibers and absent dystrophin.

An intriguing feature of eukaryotic cells is the ability of extracts containing cytosol, devoid of organelles, to roughly maintain the shape of the cell and even to move or contract, depending on how the extracts are prepared. This maintenance of structure by the cytosol arises from a complex network of protein filaments that traverse the cell cytoplasm, called the *cytoskeleton*. The cytoskeleton is not simply a passive feature of the cell

that provides structural integrity, it is a dynamic structure that is responsible for whole-cell movement, changes in cell shape, and contraction of muscle cells; it provides the machinery to move organelles from one place to another in the cytoplasm. In addition, recent studies have provided evidence that the cytoskeleton is the master organizer of the cell's cytoplasm, furnishing binding sites for the specific localization of ribonucleic acids (RNAs) and proteins that were once thought to diffuse freely through the cytoplasm.

Amazingly, the many activities of the cytoskeleton depend on just three principal types of protein assemblies: actin filaments, microtubules, and intermediate filaments. Each type of filament or microtubule is formed from specific association of protein monomers. The dynamic aspects of the cytoskeletal structures arise from accessory proteins that control the length of the assemblies, their position within the cell, and the specific-binding sites along the filaments and microtubules for association with protein complexes, organelles, and the cell membrane. Thus, although the protein filaments and microtubules define the cytoskeleton, the participation of accessory or regulatory proteins conveys its diverse activities. We will discuss the structures built from the interaction of proteins with the individual cytoskeletal assemblies, beginning with an examination of actin filaments. The initial focus will be on their well-defined role in muscle cell contraction, after which their participation in the membrane skeleton complex and structures formed in nonmuscle cells will be described.

The intermediate filaments and microtubule components of the cytoskeleton will then be discussed; finally, through a discussion of the neuronal cytoskeleton, we will consider how the different components of the cytoskeleton work together as an integrated network to organize the cytoplasm.

Microfilaments

Actin-Based Cytoskeletal Structures Were First Described in Muscle Tissue

Actin, first isolated from skeletal muscle, was originally thought to be a protein found exclusively in muscle tissue. However, actin is a component of all cells, representing 5–30% of the total protein in nonmuscle cells. Although present in all eukaryotic cells, actin isolated from nonmuscle cells (e.g., brain cells) is different from that found in skeletal muscle. Six different isoforms of actin have been described in human and animal cells: α-skeletal, in skeletal muscle; α-cardiac, in heart muscle; α-vascular, in smooth muscle of the vasculature; γ-enteric, in smooth muscle of the viscera; β-cytoplasmic and γ-cytoplasmic, preponderantly in nonmuscle cells. Actin is an extremely conserved protein, with greater than 80% identity of amino acid sequence between the different isoforms. The major difference in amino acid sequence occurs at the NH_2-terminal end of the actin isoform and seems to have little effect on the rate of actin monomers to polymerize into filaments (discussed in detail later).

Many of the other protein components that are common to actin-based cytoskeletal structures in all cells were also first isolated from muscle tissue. In muscle, these proteins demonstrate a rigorous organization, forming the specialized contractile machinery of the muscle cell. Therefore, we will examine the role of actin filaments and associated proteins in muscle cells to enhance our understanding of actin-based structures in nonmuscle cells.

Skeletal Muscle Is Formed from Bundles of Muscle Fibers

The organization of the skeletal muscle from the gross level to the molecular level is depicted in Fig. 3–1. Skeletal muscle is composed of long cylindrical, multinucleated cells that can be several centimeters in length. The individual muscle fibers are surrounded by a delicate loose connective tissue, called the *endomysium*, that carries the capillary network of blood supply for the muscle. Bundles of the individual muscle fibers are grouped together, forming the muscle fasciculi, which are bounded by a layer of connective tissue, the *perimysium*. The fasciculi are grouped to form the definitive muscle tissue that is covered by a thick, tough connec-

tive tissue layer, the *epimysium*. The three connective tissue layers of muscle tissue contain fibers of collagen and elastin and differ from each other primarily by their thickness. Skeletal muscle causes specific movements of the body by their attachment to tendons that are usually attached to the skeleton or bone.

The Functional Unit of Skeletal Muscle Is the Sarcomere

Each skeletal muscle cell, or myofiber, contains many bundles of regularly arranged filaments, called *myofibrils*. It is the highly structured arrangement of filaments within the myofibrils that give skeletal muscle its characteristic striped or striated appearance.

Skeletal muscle is the best biological example of the relation of structure, as viewed through the microscope, with function. Longitudinal sections of skeletal muscle, viewed under the light and electron microscopes, demonstrate an ordered banding pattern (Fig. 3–2). These are called the *A band, I band*, and the *Z disk* or *Z line*. The A band is the dark-staining region of the myofilaments and contains the thick filaments composed of the protein myosin II, as well as overlapping thin filaments. The light-staining I band contains the thin filaments of which the main protein component is actin. The Z disk appears as a dark line that bisects the I band. In electron micrographs of skeletal muscle, the dark-staining A band is observed to have distinct regions, termed the *H band* and *M line*. The H band is a zone of lighter staining within the central region of the A band, which is bisected by a dark-staining M line. This region of the A band is where the assembly of the myosin thick filaments occurs.

The segment of the myofibril between two Z disks, containing a complete A band and two halves of adjoining I band regions, is called the *sarcomere*. The sarcomere is the functional contractile unit of the myofibril. The myosin thick filaments mark the A band, which is equidistant from the two Z disks of the sarcomere. The thin filaments of the sarcomere are joined to the Z disk and extend through the light-staining I band region and partially into the A band, where they interdigitate with the myosin thick filaments. The Z disk functions to anchor the thin filaments of the sarcomere. Cross-sections through different portions of the sarcomere provide additional information about the organization of the thick and thin filaments (see Fig. 3–1). A cross-section through the I band shows only thin filaments, arranged in a hexagonal pattern. A section through the H zone of the A band demonstrates only thick filaments; whereas a section through the M band zone of the A band reveals a network of coiled filaments, representing the assembly of the bipolar myosin thick filaments. The segment of the A band at which the thin filaments interdigitate with the myosin thick filaments shows that each thick filament is surrounded by six thin

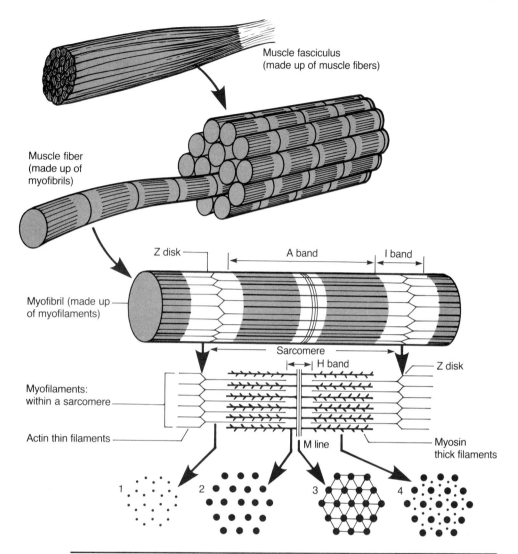

Figure 3–1. Organization of a skeletal muscle. A skeletal muscle consists of bundles of fibers called *fasciculi*. Each fasciculus consists of a bundle of long, multinucleated muscle fibers that are the cells of the muscle tissue. Within the muscle cells are the myofibrils, which are composed of highly organized arrangements of myosin II (thick) and actin (thin) filaments. The extreme structural organization of the myofilaments is the basis for the striated appearance of skeletal muscle. The myofilaments are organized into the functional units of skeletal muscle, the sarcomere, which extends from one Z disk to the next. Actin thin filaments extend from the Z disk (*light-staining I band*) toward the center of the sarcomere, where they interdigitate with the myosin thick filaments (*dark-staining A band*). Cross sections through the sarcomere near the Z disk (*1*) show the ~8-nm actin thin filaments, whereas sections in the regions of the A band (*4*) demonstrate that each ~15-nm thick filament is surrounded by a hexagonal array of six actin thin filaments. Sections through the sarcomere near the center in the segment of the A band referred to as the *H band* show the organization of the myosin thick filaments (*2*); whereas cross-sections through the center of the H band demonstrate a network of filaments that participate in the assembly of the thick filaments to form the *M line* (*3*). (Modified from Bloom W, Fawcett DW. *A Textbook of Histology*, 10th ed. Philadelphia: WB Saunders, 1975.)

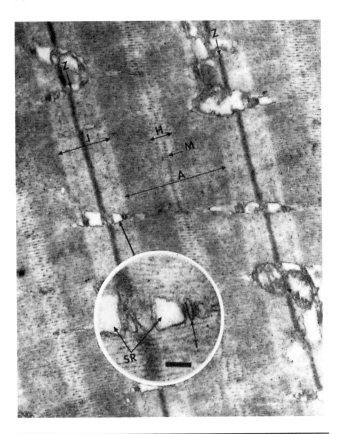

Figure 3–2. Electron microscopy of skeletal muscle. A longitudinal section through a skeletal muscle cell demonstrates the regular pattern of cross-striations derived from the myofibrils. As shown in this low-magnification electron micrograph, the skeletal muscle cell has many myofibrils aligned in parallel. In this repeating structure one can easily discern the Z disk. A, I, and H bands, and M line of the sarcomere (bar, 0.3 μm). The insert shows terminal cisternae of the sarcoplasmic reticulum (*SR*) and associated t-tubule (*T*) (bar, 0.1 μm). (Courtesy of Dr. Phillip Fields.)

filaments. This arrangement of thick and thin filaments is an essential structural feature of the sarcomere for the sliding of filaments during contraction.

Thin Filaments Are Built from the Proteins Actin, Tropomyosin, and Troponin

All eukaryotic cells appear to contain filaments 8 nm in diameter, called *microfilaments*, that are polymers of the protein actin. These filaments, referred to as *filamentous* or *F-actin*, are built from polymerization of a globular actin monomer, called *G-actin*, which has a relative molecular mass (M_r) of 43,000 (43 kDa). Each F-actin microfilament appears as two helically intertwined chains of G-actin monomers for which a complete turn of the helix occurs over a distance of 37 nm, or 14 G-actin monomers (Fig. 3–3).

Each G-actin monomer must have an adenosine triphosphate (ATP) molecule bound to polymerize onto

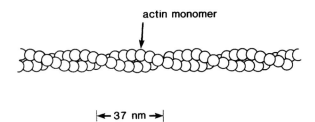

$\longmapsto$ 37 nm $\longrightarrow$

Figure 3–3. The structure of filamentous (F) actin. F-actin is a helical filament composed of polymerized G-actin monomers (*spheres*). The filament undergoes a complete turn of the helix every 14 G-actin monomers, or 37 nm.

an actin filament. If we added ATP and Mg^{2+} (or physiologic salt concentrations) to G-actin at high enough concentration, it would spontaneously polymerize to F-actin. The polymerization would have several stages (Fig. 3–4). First would be a lag phase when three G-actin monomers form an actin trimer, which can then serve as a seed or nucleation site for the polymerization of G-actin monomers onto the actin filament during the polymerization phase. Finally, a steady-state phase is reached, during which the rate of addition of G-actin monomers onto the filament equals the rate at which these monomers leave the filament. Actin microfilaments have a polarity, with a fast-growing plus (+) end and a slow-growing minus (–) end. At each end of the actin filament, there is a critical concentration of G-actin, at which the rate of addition to that end matches the rate of monomer removal from the same end. For the plus end, this concentration of G-actin is approximately 1 μm; it is 8 μm at the minus end. Therefore, at concentrations of G-actin between 1 and 8 μm in the presence of ATP and Mg^{2+}, a treadmill is formed by which actin is being added to the plus end and subtracted from the minus end. If no energy was supplied to this system, this would be a perpetual motion machine, which is thermodynamically impossible. However, shortly after the addition of each G-actin monomer to the actin filament, ATP is cleaved to adenosine diphosphate (ADP), with release of inorganic phosphate. This raises several interesting questions. First, the concentration of actin within the cytoplasm of muscle and nonmuscle cells is greater than 100 μm, suggesting that almost all of the actin within the cells of your body would be filamentous actin (as this is far above the critical concentration for the plus or minus end). However, most cells have a mechanism for maintaining a pool of G-actin monomers. Why is this the case? Second, does treadmilling occur within the living cell? The answer is yes, as we will discuss later.

Although F-actin is the preponderant protein of the skeletal muscle thin filament, these filaments also contain two other proteins, tropomyosin and troponin (Fig. 3–5). Tropomyosin is a long rod-shaped molecule (~41 nm in length), so-called because of its similarities with myosin—specifically, the rodlike tail domain of the

Figure 3–4. The polymerization of actin. The polymerization of actin occurs in three stages:. **1:** A lag phase in which an actin trimer nucleation site is formed. **2:** A polymerization phase, during which G-actin monomers are added preferentially at the plus end of the actin filament. **3:** A steady state, at which actin monomers are being added at the plus end at the same rate they are being removed at the minus end.

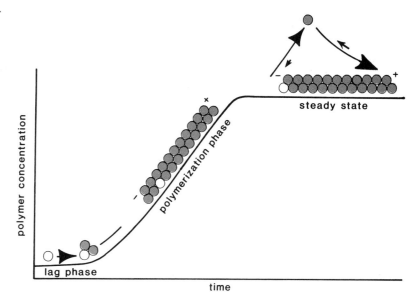

Figure 3–5. Formation of actin thin filaments and their arrangement in the sarcomere. Globular actin monomers polymerize through head-to-tail association to form the helical filamentous (F-actin) form of actin. Thin filaments are built from the specific association of the F-actin filaments with the rodlike tropomyosin molecule, which lines the grooves of the actin filament, and the troponin polypeptide complex. In the sarcomere, the thin filaments are anchored at the Z disk through their interactions with binding proteins, principally cap Z and α-actinin. The exact structure of the Z disk is unknown; the protein interactions shown in this diagram are based on the in vitro capabilities of the isolated cap Z and α-actinin proteins. As illustrated, the specific protein interactions of the Z-disk proteins immobilize the thin filaments at their plus (+) ends; thereby maintaining the polarity of the actin thin filaments in the sarcomere.

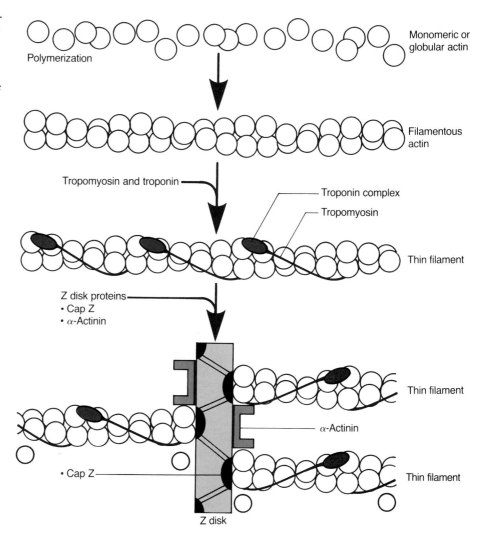

myosin molecule. Tropomyosin is formed from a dimer of two identical subunits. The individual subunit polypeptides are α-helical, and the two α-helical chains wind around each other in a coiled–coil to form the rigid rod-shaped molecule. Tropomyosin binds along the length of the actin filament, lining the grooves of the helical F-actin molecule, thereby stabilizing and stiffening the filament.

The other major accessory protein of the skeletal muscle thin filament is troponin. Troponin is a complex of three polypeptides: troponins T, I, and C. These polypeptides are named for their apparent functions within the troponin complex: troponin T for its tropomyosin binding, troponin I for its inhibitory role in calcium regulation of contraction (discussed later), and troponin C for its calcium-binding activity. The troponin complex is elongated, with the I and C subunits forming a globular head region and the T polypeptide forming a long tail domain. The tail domain formed from the T-subunit binds with tropomyosin, which is thought to position the complex on the actin thin filament. Because there is only one molecule of the troponin complex for every seven actin monomers in an actin filament, the positioning of the complex by the specific interactions of the T-subunit with the tropomyosin molecule is critical for its ability to regulate contraction.

Thick Filaments Are Composed of the Protein Myosin

Myosin was also first described in muscle cells but is now known to be a ubiquitous component of nonmuscle cells. The major form of myosin found in most cells, including skeletal muscle, is referred to as *myosin II*. Myosin II has an M_r of approximately 460 kDa, with two *identical* heavy chains of M_r 200 kDa, which form a coiled–coil helical tail and two globular heads (Fig. 3–6). For a coiled–coil helix to form, the myosin heavy chains must have a heptad amino acid repeat sequence—a, b, c, d, e, f, g, a, b, c, d, e, f, g—with hydrophobic amino acids in

positions a and d. Because an α-helix makes a complete turn every 3.5 amino acids, such a repeat would create a hydrophobic stripe that slowly rotates around the helix. To bury this hydrophobic stripe away from the aqueous environment, two such α-helices would wind around each other into a coiled–coil. The myosin molecule also contains two pairs of light chains with M_r of 20 and 18 kDa. These light chains are found associated with the myosin heads. If purified myosin is proteolytically cleaved with the enzyme papain, the globular heads (called *SF1 fragments*) can be separated from the myosin tails (see Fig. 3–6). The myosin tails brought to physiologic ionic strength and pH will spontaneously form thick filaments, similar to those found in skeletal muscle. The SF1 heads contain all of the myosin ATPase activity required for muscle contraction. If the purified heads are added to preformed F-actin and viewed by electron microscopy, the SF1 fragments look like arrowheads that all face in one direction. The pointed end of the arrowheads face the minus, or slow-growing, end of the filament, and the barbed end faces the plus, or fast-growing, end (Fig. 3–7). The polarity of actin filament interaction with myosin SF1 fragments (see Fig. 3–7) is important for muscle contraction.

Thick filament formation arises from association of the tail or rod segment of the myosin molecule, as demonstrated by the aggregation of isolated tail domains produced by proteolytic cleavage of myosin II. The association of the myosin II heavy chain dimers is due to hydrophobic interactions of the rodlike tail segments, and the formation of filaments depends on interactions between the coiled myosin II tail domains. In muscle, the rodlike fibrous tails of 300–400 myosin II dimers pack together to form the bipolar 15-nm-diameter-thick filaments. This association of the myosin II tail segments results in the formation of a filament that has a bare central zone composed of an antiparallel array of myosin II tails (Fig. 3–8). The globular myosin II head segments protrude from the filament at its terminal regions in a helical array, with a periodicity of 14 nm.

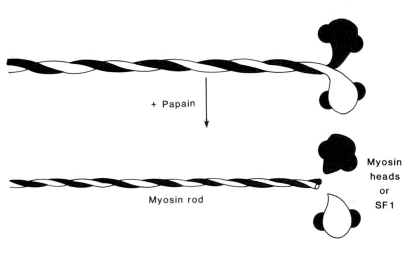

+ Papain

Myosin rod

Myosin heads or SF 1

Figure 3–6. The structure of myosin II and its cleavage by papain. Myosin II is a 150-nm long fibrous protein, with two globular heads. Treatment of myosin II with the proteolytic enzyme papain releases the two myosin heads, or SF1 fragments, from the myosin rod.

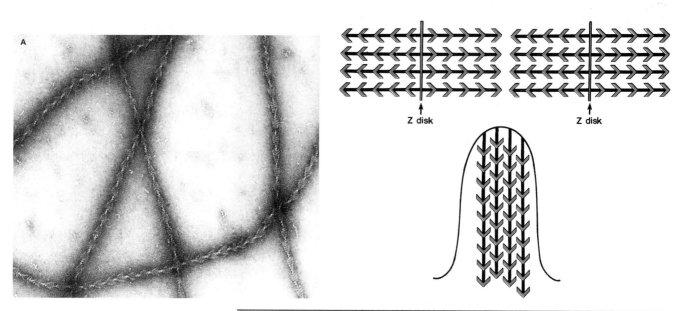

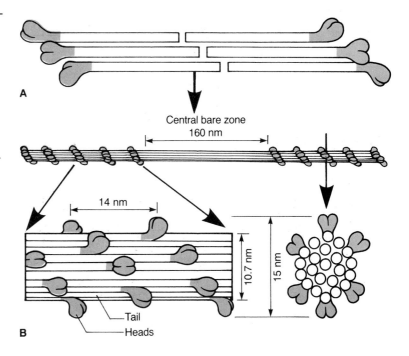

Figure 3–7. Actin filaments have a polarity. The polarity of actin filaments can be visualized by labeling with myosin SF1 fragments. **A:** This is an electron micrograph of *in vitro* actin filaments that have bound myosin SF1 fragments. The myosin fragments bind to the actin filaments demonstrating their polarity. The myosin heads look like arrowheads which all point to the minus (−) ends of the actin filament, the barbed ends facing the plus (+) ends of the filaments. (Photo courtesy of Dr. Roger Craig, University of Massachusetts.) **B:** In a sarcomere the barbed or plus (+) ends are attached to the Z disk. When actin filaments are bound to the cytoplasmic surface of the plasma membrane, it is the plus end that is associated with the membrane. The example shown here is the attachment of actin filaments to the tip of the microvillus.

Figure 3–8. Formation of myosin thick filaments. **A:** Thick filament formation is initiated by the end-to-end association of the rodlike tail domains of myosin II molecules. **B:** This results in the formation of the bipolar thick filament, with globular heads at either end separated by a 160-nm central bare zone consisting of myosin II tail domains. At the filament ends the myosin globular head domains protrude from a 10.7-nm diameter central core at intervals of 14 nm. The successive myosin heads rotate around the fiber, which forms a filament containing six rows of myosin head domains to contact the adjacent thin filaments of the sarcomere.

Thick filaments then display a high degree of structure—being symmetric about the bare central zone—and the polarity of the filament given by the arrangements of the globular head segments is reversed on either side of this zone.

Because the arrangement of the myosin II head domains in the intact thick filament is antiparallel (i.e., they are reversed in polarity on either side of the central zone of the filament), the directionality of their F-actin binding indicates that the thick and thin filaments in each half of the sarcomere have the same orientation because of the polarity of the thin filaments. This polar arrangement of thick and thin filaments forms the structural basis of filament movement within the sarcomere, which is its primary function in skeletal muscle.

Accessory Proteins Are Responsible for Maintenance of Myofibril Architecture

In vertebrate skeletal muscle, the structural orientation of the thick and thin filaments is crucial for contraction. Thus, the maintenance of this structure is very important for muscle function. Several proteins (but probably not all that are necessary) that interact with the thick and thin filaments and play a role in the maintenance of myofibril structure have now been identified.

The thin filaments terminate and are anchored to the Z disk structures of the sarcomere. This immobilizes the thin filaments with their plus ends at the Z disk, and their minus ends extending to the center region of the sarcomere. Therefore, a *sarcomere unit* (defined as the distance between two adjacent Z-disk structures) contains actin filaments that extend from each Z disk and exhibit polarity that is opposite on either side of the central region of the sarcomere. Cap Z, a two-subunit protein (M_r 32,000 and 36,000) that binds selectively to the plus ends of actin filaments, is one of the proteins that helps with the anchoring of thin filaments to the Z disk. Because it binds to the fast-growing or plus end of the actin filament, cap Z is thought to prevent growth and depolymerization of F-actin, causing the filaments of the myofibril to be very stable structures. Its localization at the Z disk suggests that cap Z may assist in the immobilization of the thin filaments (see Fig. 3–5), perhaps by interactions with other proteins of the Z disk. The major component of the Z disk is the protein α-actinin, a fibrous protein composed of two identical subunits (M_r 190,000). The NH$_2$-terminal domain of α-actinin bears a strong resemblance to NH$_2$-terminal domains of other cytoskeletal proteins (principally members of the spectrin and dystrophin gene family) that function to bind and cross-link actin filaments. It is

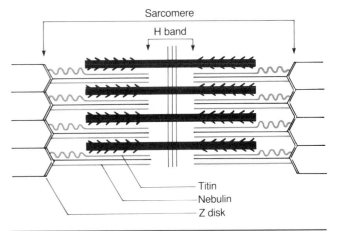

Figure 3–9. Titin and nebulin: accessory proteins of the skeletal muscle sarcomere. The location of the proteins titin and nebulin within the sarcomere is shown. Titin, a large protein that has elastic properties and links the myosin thick filaments to the Z disks, helps maintain their location in the sarcomere. Nebulin, a large filamentous protein anchored at the Z disk, is in close apposition to the actin thin filaments. Their close association with the thin filaments suggests that the nebulin fibers serve to organize the actin filaments of the sarcomere.

the NH$_2$-terminal domain of α-actinin that provides the ability of this protein to bind tightly to the sides of actin filaments, allowing the bundling together of adjacent thin filaments at the Z disk. Although the exact structure of the Z disk is unknown, evidence now indicates that it contains two sets of overlapping actin filaments of opposite polarity that originate in the two sarcomeres adjacent to the Z disk, and the thin filaments are anchored to the disk structure by interactions with proteins such as cap Z and α-actinin.

In skeletal muscle, there are mechanisms that maintain the relative position of the myofilaments and regulate the length of the polymerized filaments. Two proteins, titin and nebulin, appear to be important for these functions (Fig. 3–9). Titin, a very large fibrous protein, appears to connect the thick filaments to the Z disk. Thus, titin functions to keep the myosin thick filaments centered within the sarcomere structure. It may act as an elastic band to keep filaments in an appropriate orientation, also inhibiting the structural deterioration of the sarcomere during muscle contraction. Another very large fibrous protein, nebulin, forms a long inextensible filament that extends from the Z disk adjacent to the thin filaments. Because of their exacting length and their proximity to the actin filaments, the nebulin fibers may regulate the number of actin monomers that polymerize into thin filaments and aid in the formation of the regular geometric pattern of thin filaments during muscle formation.

Figure 3–10. Sliding filament model of muscle contraction. Muscle contraction occurs by the sliding of the myofilaments relative to each other in the sarcomere. **A:** In relaxed muscle, the thin filaments do not completely overlap the myosin thick filaments, and there is a prominent I band. **B:** With contraction, there is a movement of the thin filaments toward the center of the sarcomere and, because the thin filaments are anchored to the Z disks, their movement causes shortening of the sarcomere. The sliding of thin filaments is facilitated by contacts with the globular head domains of the bipolar myosin thick filaments.

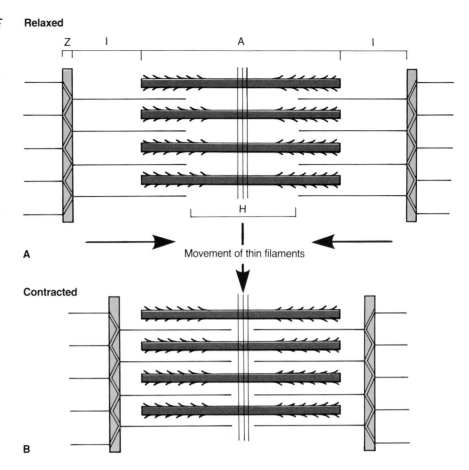

Muscle Contraction Involves the Sliding of the Thick and Thin Filaments Relative to Each Other in the Sarcomere

Measurements of sarcomere and A and I band lengths from electron micrographs of contracted and resting muscle firmly established the mechanism of muscle contraction—the sliding of actin thin and myosin thick filaments passed each other within the sarcomere unit. These measurements demonstrated that the lengths of the individual filaments do not change as a muscle contracts, yet the distance between two adjacent Z disks becomes shortened in contracted muscle relative to relaxed muscle. When the length of a sarcomere decreases in contracted muscle, the I band region shortens, whereas the length of the A band remains unchanged (Fig. 3–10). Because the lengths of the thick and thin filaments do not change, the change in length of the I band could occur only if the thin filaments were to slide past the thick filaments. Therefore, the reversed polarity of the thick and thin filaments relative to the center line of the sarcomere (defined by the M line) would cause a shortening of the sarcomere during contraction by the sliding of thin actin filaments, which are attached to the Z disk, past the thick myosin filaments toward the center of the sarcomere. This model of mus-

cle contraction, called the *sliding filament model*, was first proposed in 1954 and became the seminal observation leading to the dissection of molecular mechanisms of contraction.

Adenosine Triphosphate Hydrolysis Is Necessary for Cross-Bridge Interactions with Thin Filaments

Skeletal muscle contraction requires the interactions of myosin II head groups with the thin filaments. These interactions are governed by binding and hydrolysis of the high-energy molecule ATP by the ATPase activity resident in the globular myosin II head domain.

The ATP-driven interactions between myosin II and actin are illustrated in Fig. 3–11. When a myosin II head binds a molecule of ATP, it causes a weakening of the myosin–actin interaction. There is a dissociation of the myosin II headgroup binding to the thin filament (step 1). The cleavage of ATP to ADP and Pi creates an "activated" myosin II head that has undergone a change in structure, facilitated by the flexible hinge regions of the molecule, such that the myosin II head is perpendicular with an adjacent actin thin filament (step 2). The conversion between these two stages is reversible, because the ADP and Pi remain bound to the myosin II head,

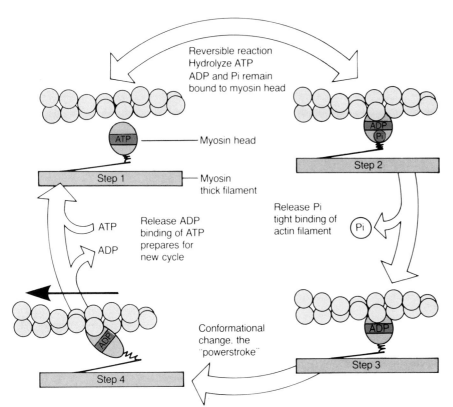

Figure 3–11. Illustration of the ATP-driven myosin–actin interactions during contraction. The binding of ATP to a myosin head group causes release from the actin filament (*step 1*). The hydrolysis of ATP to ADP + P: readies the myosin head to contact an actin filament (*step 2*). The initial contact of the myosin with an actin filament causes the release of P: and a tight binding of the actin filament (*step 3*). This tight binding induces a change in conformation of the myosin head, such that it pulls against the actin filament, the powerstroke (*step 4*). This change in conformation is accompanied with the release of ADP. The binding of an additional ATP causes a release of the actin filament and a return of the myosin head to a position ready for another cycle.

and the energy released from ATP hydrolysis is stored in the strained bonds resulting from the rotation of the myosin II head group. The activated myosin II molecule then comes into contact with a neighboring actin subunit, and this binding triggers the release of Pi which, in turn, strengthens the myosin–actin interaction (step 3). This strong binding causes a conformational change in the myosin II head, generating a "powerstroke," which pulls the actin filament relative to the fixed myosin II filament, resulting in contraction (step 4). The product of this step is the so-called rigor complex, in which the actin–myosin linkage is inflexible, and the thick and thin filaments cannot move past each other. If no ATP is available to the muscle (e.g., after death), the muscle will remain rigid, owing to the tight myosin–actin interactions. This condition is referred to as a *rigor mortis*. Under normal circumstances, a molecule of ATP will displace the bound ADP, causing release of the actin filament from the myosin head group, effectively relaxing the muscle and returning to step 1 of the cycle. The hydrolysis of the newly bound ATP then prepares the muscle for further rounds of myosin–actin interactions.

Each cycle of myosin–actin interaction would result in movement of an actin thin filament by about 10 nm. To achieve the rapid rates of contraction for intact muscle fibers, there must be a coordination of multiple myosin II head group interactions to provide a concerted movement of filaments, and a mechanism by which these interactions are regulated. Each thick filament is formed from the aggregation of multiple myosin

II rod domains, which results in a bipolar filament with each side of the filament containing approximately 300–400 head groups protruding in a spiral fashion. This arrangement provides multiple contacts of a thick filament (called *cross-bridge interactions*) with a thin filament. Along the length of the thin filament, there will be myosin II cross-bridges at various points in the myosin–actin cycle (see Fig. 3–11), such that the collective actin cross-bridged contacts ensure the smooth and rapid movement of the thin filament relative to the thick filament. To effectively coordinate the sliding of filaments from entire groups of myofibrils, leading to muscle contraction capable of producing mechanical work, the interactions of myosin II head group cross-bridges are regulated at the cellular level by a transient rise in calcium. The calcium-based regulation of muscle contraction occurs by overcoming a block of myosin–actin interactions by the troponin–tropomyosin complexes on the thin filament. The specific interactions involved in this regulation are discussed in the following section.

Calcium Regulation of Skeletal Muscle Contraction Is Mediated by Troponin and Tropomyosin

When myosin is mixed with filaments made from purified actin, the myosin ATPase activity is stimulated to its maximal activity, independent of calcium addition to the reaction. If thin filaments, which contain actin, tropomyosin, and troponin, are added to purified

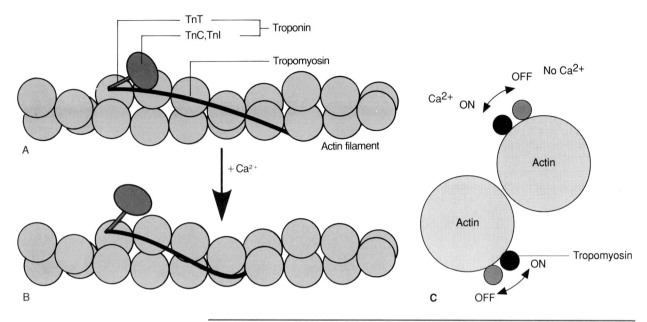

Figure 3–12. Diagram of Ca²⁺-mediated movements of troponin and tropomyosin filaments during muscle contraction. **A:** In the relaxed muscle, the tropomyosin filament is bound to the outer domains of seven actin monomers along the actin filament. The troponin complex is bound to the tropomyosin by the rod-shaped TnT polypeptide. **B:** In the presence of Ca²⁺, TnC binds the calcium, causing the globular domain of troponin (TnC and TnI) to move away from the tropomyosin filament. **C:** This movement permits the tropomyosin to shift to a position that is farther inside the groove of the helical actin filament, allowing the myosin heads to make contact with the released sites of the actin monomers.

myosin, the stimulation of the myosin ATPase activity is wholly dependent on the presence of calcium. The basis of calcium-dependent hydrolysis of ATP in this reaction is a reversal of the inhibition of the actin–myosin interaction caused by the position of tropomyosin and troponin on the thin filaments.

Each rodlike tropomyosin molecule contacts seven actin monomers and lines the grooves of the F-actin helix. Bound to a specific site of each tropomyosin molecule is the troponin complex that comprises three polypeptides; troponins (Tn) T, I, and C. The elongated troponin T molecule (M_r 37,000) binds the COOH-terminal region of tropomyosin and links both the TnI and the TnC to the tropomyosin. Troponin I (M_r 22,000) binds the TnT as well as actin and, in concert with tropomyosin, causes a change in the conformation of F-actin such that it interacts only weakly with myosin head groups. This weak interaction cannot activate the myosin ATPase activity. Along with TnI, the TnC ($M_r \approx 20,000$) subunit forms a globular domain of the troponin complex. Troponin C, the calcium-binding subunit, has a structure and function similar to that of the intracellular calcium receptor protein calmodulin. The binding of calcium ions at all four of the calcium-binding domains of TnC releases the TnI tropomyosin inhibition of actin activation of the myosin ATPase, thereby allowing con-

traction of the myofibril. The binding of calcium by TnC results in a shift or movement of the tropomyosin toward the center of the actin helix (Fig. 3–12), which exposes a region of the actin monomer, allowing the binding of myosin head groups in such a way that activation of the myosin ATPase activity occurs. The hydrolysis of ATP permits cycling of cross-bridge interactions and the sliding of filaments. The myosin-activating sites of F-actin are blocked by the troponin–tropomyosin complex in the resting, but not in the active, state of the myofibril. Thus, the contraction of skeletal muscle is regulated by the concentration of intracellular calcium ions.

Intracellular Calcium in Skeletal Muscle Is Regulated by a Specialized Membrane Compartment, the Sarcoplasmic Reticulum

In the resting or relaxed state, the concentration of calcium ions in skeletal muscle cells is very low. Thus, to have contraction and relaxation cycles of muscle, there must be a mechanism by which the internal calcium ion concentration is regulated. Moreover, the concerted contraction of a muscle to produce work depends on the simultaneous contraction of all of its constituent myofibers and their myofibrils. Therefore, the rapid changes in calcium ion concentration that are needed

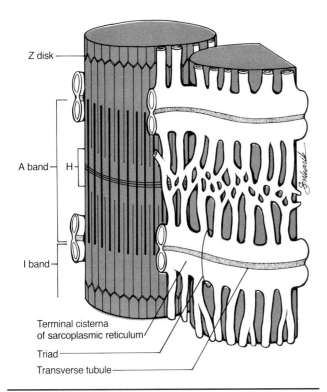

Z disk

A band — H

I band

Terminal cisterna
of sarcoplasmic reticulum

Triad

Transverse tubule

Figure 3–13. Diagram of part of a skeletal muscle fiber, illustrating the organization of the SR and t-tubule networks. The SR is a specialized smooth ER that in muscle serves as a store for Ca^{2+} ions. The SR forms a membranous tubule network that surrounds the myofibrils. At the A–I band junctions of the sarcomere, the SR forms a more regular channel, referred to as the *terminal cisternae*. Two terminal cisterna are separated by a second tubule system, the t-tubules, which are special invaginations of the sarcolemma. These three membrane-bound tubules form a structure known as the *triad*; a t-tubule flanked on either side by a terminal cisternae of the SR, at the region of the A–I junction of the sarcomere. (Modified from Cormack DH. *Ham's Histology*, 9th ed. Philadelphia: JB Lippincott, 1987.)

along the entire length of the myofibril for contraction must be maintained by mechanisms other than simple diffusion, which would be too slow for simultaneous contraction of myofibrils in skeletal muscle cells. To deliver calcium in a uniform fashion throughout the muscle cell, there is a special membrane-bound tubule system, derived from the endoplasmic reticulum (ER) in these cells.

Electron microscopy of skeletal muscle shows a network of smooth membranes, called the *sarcoplasmic reticulum* (SR), surrounding the myofibrils. The SR forms a network of membrane-limited tubules and cisternae that surround the outer regions of the A band of each myofibril (Fig. 3–13). In addition, the SR forms a more regular structure, called the *terminal sac* or *terminal cisternae*, which is a membrane-limited channel that surrounds the A-I junction of each individual myofibril. The

terminal cisternae are in close proximity to a specialized channel formed from delicate invaginations of the sarcolemma (plasma membrane of the muscle cell), called the *transverse tubules* (t-tubules). The t-tubules, in association with terminal cisternae from adjacent myofibrils, form a triad structure (see Fig. 3–13). These structures are important for the coupling of external stimuli (e.g., signals from motor neurons) to muscle contraction.

The sarcoplasmic reticulum forms a membranous compartment that occupies 1–5% of the total muscle volume and serves as a reservoir of calcium ions sequestered away from the myoplasm and myofibrils. For its role in maintenance of calcium ion concentration, the SR membrane contains numerous proteins for the transport of calcium, including a Ca^{2+}-ATPase protein that pumps calcium from the cytosol into the lumen of the SR. For each 1 mol of ATP hydrolyzed by the ATPase activity of the calcium pump, 2 mol of calcium are sequestered into the lumen of the SR. This active transport mechanism is responsible for the maintenance of the low calcium ion concentration in resting muscle. The stored calcium is released from the SR into the sarcoplasm as the action potential spreads along the sarcolemma. The action potential, stimulating Ca^{2+} release, travels through the t-tubule system. A voltage-sensitive protein sensor located in the t-tubule membrane, termed the *dihydropyridine-sensitive receptor* (DHSR), feels the action potential and translates its presence to the SR through a direct interaction with an SR calcium channel, the ryanodine receptor. These proteins, the DHSR and ryanodine receptor, are analogous to proteins found in other cells whose function is to release calcium from internal stores, the so-called IP$_3$ receptor pathway (described in Chapter 8). The large complex formed by these proteins in muscle when viewed in the electron microscope is often referred to as "feet" on the SR (Fig. 3–14). The net effect is the release of a pulse of calcium into the sarcoplasm by the transit of an action potential. The released calcium stimulates contraction through binding the troponin complex on the thin filaments. Following contraction, the calcium is actively transported into the lumen of the SR by Ca^{2+}-ATPase, returning the muscle to the relaxed state (Fig. 3–14).

Within the lumen of the SR are proteins that function to bind and store the internalized calcium ions (see Fig. 3–14). The best-characterized example is calsequestrin. Although the binding affinity of Ca^{2+} by calsequestrin is low, each molecule of the protein binds 40–43 Ca^{2+} ions. Thus, calsequestrin, along with other proteins that have similar properties, effectively lowers the SR luminal concentration of calcium from 20–30 mM (if all the Ca^{2+} ions were free in solution) to about 0.5 mM. The result of binding Ca^{2+} ions in the SR lumen is to greatly reduce the concentration gradient against which the membrane Ca^{2+} pump must act.

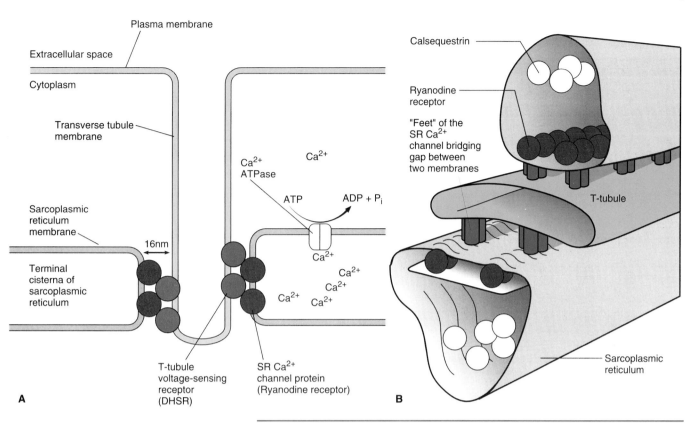

Figure 3–14. Model of Ca^{2+} ion regulation by the SR in muscle. **A:** An illustration of the association of the t-tubules and the terminal cisterna of the SR. The SR Ca^{2+} channel is shown to make direct contact with the voltage-sensing Ca^{2+} channel of the t-tubule. When depolarized, the t-tubule voltage-sensing protein (DHSR) undergoes a change in conformation and, because of its close association with the SR Ca^{2+} channel, causes the SR channel to open and release calcium to the cytoplasm. This Ca^{2+} release occurs with essentially no delay due to the direct interaction of the DHSR and the Ca^{2+} channel of the SR, the ryanodine receptor. The Ca^{2+} ions in the cytoplasm are returned to the lumen of the SR by the Ca^{2+}-ATPase pump in the SR membrane. **B:** A view of the t-tubule and SR terminal cisternae associations. The t-tubules and SR terminal cisternae are in close proximity; "feet" of the SR channel protein are shown bridging the gap between the t-tubule and SR membranes. Inside the lumen of the SR is the protein calsequestrin that weakly binds the internalized Ca^{2+} ions, lowering the effective internal concentration of free Ca^{2+} ions. (Modified from: **A,** Agnew, WS. *Nature* 1988;344:299–303; **B,** Eisenberg BR, Eisenberg RS. *J Gen Physiol* 1982;79:1–17.)

There Are Three Types of Muscle Tissue

In the preceding section, we focused on the contractile apparatus found in skeletal muscle. There are two other major types of muscle present in the vertebrates. Cardiac muscle forms the walls of the heart and is also found in walls of the major vessels adjacent to the heart. Smooth muscle is found in the hollow viscera of the body (e.g., the intestines) and in most blood vessels. All three types of muscle use actin–myosin structures for contraction by a sliding filament mechanism. However, there are some fundamental differences in the structural organization of the contractile apparatus and the regulation of contraction in the different muscle cells.

Myocardial Tissue: Striated Muscle Built from Individual Cells

Cardiac tissue consists of long fibers that, like skeletal muscle, exhibit cross-striations under the light microscope. The striated appearance of cardiac muscle derives from the highly organized arrangement of actin and myosin filaments of the contractile apparatus. Although cardiac muscle is similar in appearance to striated skeletal muscle, there are two main histologic criteria that distinguish these two muscle types.

First is the positioning of the nuclei within the cells. In skeletal muscle, nuclei are located at the periphery of the cell, just under the sarcolemma; whereas, in cardiac muscle, the nuclei are found at the central regions of the cell. Thus, cardiac cells have a bare or cleared zone surrounding the nucleus—the perinuclear space—which arises from the myofilaments arranging themselves such that they detour around the nuclear compartment.

The second major criterion that distinguishes cardiac from skeletal muscle is the appearance of dark-staining disk structures in cardiac muscle, the *intercalated disk*. These are specialized junctional complexes separating one cardiac muscle cell from another. Thus, cardiac muscle fibers are built from an arrangement of single cells, unlike skeletal muscle fibers that are built from the fusion of individual cells into a multinucleated fiber. Although, at the light microscope level, the intercalated disk appears as straight lines, demarcating one cell from another, the view of these structures in the electron microscope reveals that they take an irregular steplike path, such that part of this cell–cell junction is horizontal and part is longitudinal. Therefore, the individual cardiac muscle cells interdigitate with each other, forming the myocardial muscle fibers (Fig. 3–15). This arrangement of cell–cell contact allows myocardial muscle to contain straight fibers and fibers that branch to effectively construct a hollow organ capable of pumping blood.

Different regions of the intercalated disk contain specific junctional complexes (see Fig. 3–15). In the transverse (or vertical) sections of the intercalated disk, there are two junctional complexes. The first is the desmosomes, which are sometimes referred to as the *macula adherens*. These junctions function as the "spot welds" that hold the adjacent cardiac cells together. In this region of the plasma membrane is a second type of junctional complex, termed the *faciae adherens*, which functions to connect the thin filaments of adjacent cells and to hold them in register with the myosin thick fila-

ments (see later). In the longitudinal regions of the intercalated disk are junctional contacts called gap junctions (sometimes referred to as *nexus*). These contacts allow the cardiac cells to exchange small cytoplasmic solutes. The gap junction contacts permit electrical coupling of the cardiac muscle cells, such that there is a synchronization of contraction among these cells.

Some of the muscle cells within the heart, the Purkinje fibers, are specialized to carry electrical impulses. These cells are grouped into bundles that form two branches, one to each ventricle. Histologically, these cells are larger and more irregular in shape than are the surrounding cardiac muscle cells. The Purkinje cells contain large glycogen deposits and have smaller bundles of myofibrils at their periphery. These special conducting fibers are responsible for the final distribution of electrical stimulus to the myocardium.

The Contractile Apparatus of Cardiac Muscle Is Similar to That of Skeletal Muscle

Cardiac muscle owes its striated appearance to the arrangement of thick and thin filaments that make up the contractile apparatus. Electron micrographs of cardiac muscle reveal a banding pattern of myofibrils similar to that observed for skeletal muscle. Like skeletal muscle, these bands are referred to as the *A band*, *I band*, and *Z disk*. The dark-staining A band is the region of the myofilament containing the thick filaments composed of myosin and overlapping thin filaments. The I band contains the thin actin filaments and is bisected by the Z disk, to which the actin filaments are anchored. One notable difference in the structure of myofilaments in cardiac cells, compared with skeletal muscle cells, is the termination of some actin thin fila-

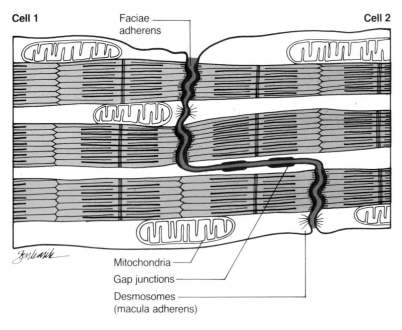

Cell 1   Faciae adherens   **Cell 2**

Mitochondria
Gap junctions
Desmosomes
(macula adherens)

Figure 3–15. Diagram of an intercalated disk between two cardiac muscle cells. The intercalated disk is a steplike structure that allows the interdigitation of cardiac muscle cells. In the transverse sections of this structure are the desmosomes, which hold the cells together, and the junctional complexes, the faciae adherens, which function as Z-disk structures to anchor actin thin filaments from adjacent cells. In the longitudinal sections of the intercalated disk are the gap junctions. These junctional complexes allow communication between the cells such that adjacent cardiac cells are electrically coupled.

Figure 3–16. Electron micrograph of cardiac muscle. This electron micrograph of cardiac muscle cells shows the regular arrays of the myofibrils into the sarcomeres. In this arrangement, one can easily see the Z disk; the A, I, and H bands; and the M line of the sarcomere structure. The insets show a higher magnification of the junctional compartments of the intercalated disk: (**1**) the macula adherens or desmosomes, (**2**) the fasciae adherens, and (**3**) gap junctions. Notice the thin filaments that terminate in the fascia adherens complex (bars: main figure, 0.2 μm; inserts, 0.05 μm). (Courtesy of Dr. Phillip Fields.)

ments at the region of the intercalated disk (Fig. 3–16). The faciae adherens complex in the transverse segment of the intercalated disk functions to anchor actin thin filaments at the cell periphery. Although the molecular details of how this junctional complex binds and arranges actin thin filaments are unknown, these complexes function as a Z disk, in that they maintain the exacting arrangement of six actin filaments surrounding each myosin thick filament.

Cardiac muscle thick filaments are made from the protein myosin II, which has a subunit structure similar to that found in skeletal muscle. Cardiac myosin II has two heavy chains of approximately 200,000 M_r that assemble by a rodlike tail domain and fold into a globular head domain at their NH_2-terminus. There are four light chains, two pairs of M_r 18,000–20,000, with one polypeptide from each set bound with each head seg-

ment of the molecule. Associated with the globular head domain of cardiac myosin II is an actin-activated ATPase activity that functions in cross-bridge formation and contraction. However, the isozymes of myosin expressed in cardiac muscle have a lower ATPase activity than those in skeletal muscle.

The thin filaments of cardiac muscle are built from actin, tropomyosin, and troponin. Although these proteins form the same complex as that found in skeletal muscle, they are different from the polypeptides found in their skeletal muscle counterpart: that is, they are cardiac-specific isoproteins. Cardiac muscle thin filaments exhibit the same stoichiometry and structure as those discussed for skeletal muscle. Thus, in cardiac muscle, there is an arrangement of six thin filaments surrounding each thick filament, and contraction or cross-bridge formation in cardiac muscle is regulated by Ca^{2+} by the thin filament-based troponin–tropomyosin complex.

The Smooth Muscle Cell Does Not Contain Sarcomeres

Smooth muscles are made of individual cells that can vary notably in size, from 20 μm in length in the walls of the small blood vessels to 200–300 μm in length in the intestine. The smooth-muscle cell is characterized by its fusiform shape. The cells are thickest at their midregion and taper at each end. Smooth muscles are built from sheets of cells that are linked together by various junctional contacts that serve as sites of cell–cell communication (e.g., gap junctions) and mechanical linkages. Cells of the smooth muscle are very active in the synthesis and deposition of connective tissue matrix, which serves to embed the cells and acts in limiting the distension of the hollow viscera.

Smooth muscle cells do not contain a highly ordered array of thick and thin filaments; thus, they do not appear striated. In electron micrographs of smooth muscle, numerous dense-staining regions, known as *dense bodies*, are found throughout the cytoplasm of the cell. The major protein component of the dense body is the actin-binding protein, α-actinin, which indi-

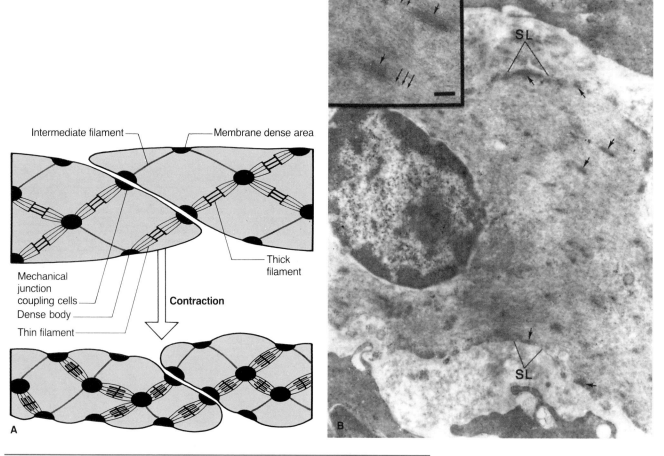

Figure 3–17. Organization of cytoskeletal and myofilament elements in smooth muscle. A: Smooth muscle cells contain small contractile elements that are not organized as in striated muscle. Numerous actin thin filaments are anchored into dense bodies within the smooth-muscle cytoplasm, which are the functional equivalent of the striated muscle Z disk. Intermediate filaments of desmin and vimentin form linkages between the dense bodies and the cytoskeleton of the cell. These links are important for contraction, which pulls the plasma membrane inward and changes the shape of the cell. B: In this electron micrograph of a smooth muscle cell, the dense bodies are seen throughout the cell and near the sarcolemma (SL). At higher magnification (*insert*) myofilaments (*small arrows*) are observed as they emanate from the dense bodies (*large arrows*) (bar, 0.29 μm). (**B,** Courtesy of Dr. Phillip Fields.)

cates that they serve as the functional equivalent of a skeletal muscle Z disk. Indeed, actin thin filaments are found anchored to the dense bodies. Two proteins, desmin and vimentin, belonging to the intermediate filament class of proteins (discussed later), are expressed at high levels in smooth muscle cells. The filaments formed from these proteins are prominent in these cells and appear to serve as links between the dense bodies and the cytoskeletal network of the cell. These links aid in contraction by maintenance of the dense body positioning (Fig. 3–17), allowing movement of the cell by an inward pulling of the plasma membrane.

The Contractile Apparatus of Smooth Muscle Contains Actin and Myosin

Actin and myosin can be isolated from smooth muscle cells, and *in vitro* these proteins demonstrate a sliding filament mechanism for contraction. However, the regulation of contraction in smooth muscle follows a path very different from that observed for striated muscle.

The thin filaments of smooth and striated muscle have very similar structures, except that the calcium regulatory protein troponin is not present in smooth muscle. The cellular content of actin and tropomyosin is higher in smooth muscle than in striated muscle (by about twofold). This, in combination with a reduced quantity of myosin in smooth muscle compared with striated muscle, produces a higher ratio of thin to thick filaments in smooth muscle (~12 thin per 1 thick) than that observed in the striated muscles (~6 thin per 1 thick).

Smooth muscle contains numerous thin filaments that approximately align along the long axis of the cell. These thin filaments are embedded into the cytoplasmic densities (dense bodies) and exhibit the same polarity relative to their attachment points. The thin filaments have their plus (+) ends at the dense body and their minus (−) ends extending into the cellular cytoplasm. Thus, although the filaments in smooth muscle are not as highly organized as those found in striated muscle, the polarity of actin thin filaments in smooth muscle is such that contraction by myosin cross-bridge cycling would

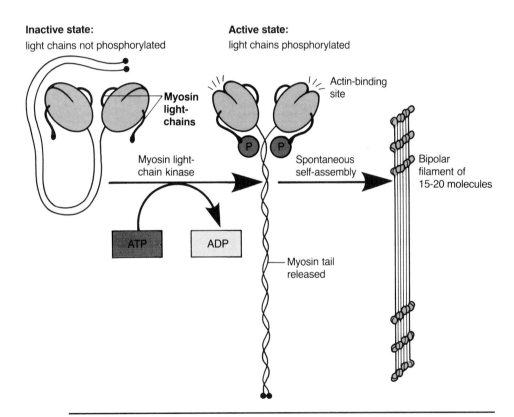

Figure 3–18. **Model for assembly of smooth-muscle myosin thick filaments.** Dephosphorylated myosin isolated from smooth muscle cells is in an inactive state and does not readily form thick filaments because of the conformation of the tail domain that binds with the globular head domain. The phosphorylation of the 18-kDa light chain of myosin has two effects: It causes a change in the conformation of the myosin head, exposing its actin-binding site; and it releases the myosin tail from its inactive conformation, allowing the myosin molecules to assemble into bipolar thick filaments. (Modified from Alberts B, Bray D, Lewis J, et al. *Molecular Biology of the Cell*, 2nd ed. New York: Garland Publishing, 1989.)

cause a pulling of dense bodies toward one another. This is critical for smooth-muscle contraction, in that it would cause an inward pulling of the plasma membrane, creating force generation by essentially reshaping the cell. A change in shape of several coupled cells would generate the force of smooth-muscle contraction.

Myosin isolated from smooth muscle has properties different from those of striated muscle. Similar to skeletal muscle, smooth-muscle myosin consists of two heavy chains and four light chains. Two polypeptides of myosin light chains are associated with each globular head domain of the smooth-muscle myosin. However, smooth-muscle myosin will form filaments under only certain conditions. When myosin isolated from smooth-muscle cells is dephosphorylated, it remains fully soluble. Analysis of soluble myosin by sedimentation assays and electron microscopy reveals that the dephosphorylated myosin folds up into a compact unit, with the tail domain reaching toward the globular head domain. In this configuration, the isolated myosin resists the formation of thick filaments, and its actin-activated ATPase activity is essentially blocked. On phosphorylation of the 18-kDa light chain of smooth-muscle myosin by the enzyme myosin light chain kinase, the tail segment is released from the head segment (Fig. 3–18). The resulting released myosin tails can form bipolar thick filaments. Moreover, the freeing of the tail domain to form bipolar filaments allows the activation of the head domain ATPase (permitting cross-bridge formation).

Smooth-Muscle Contraction Occurs Via Myosin-Based Calcium Ion Regulatory Mechanisms

Smooth muscle cells lack troponin, the Ca^{2+} regulatory protein found in the thin filaments of striated muscle, yet micromolar increases in intracellular Ca^{2+} concentrations are required for smooth-muscle contraction to occur. The Ca^{2+} regulation of smooth-muscle contraction occurs by changes in the phosphorylation state of the myosin molecule. The regulation of smooth-muscle contraction is said to be myosin-based. When stimulated, the Ca^{2+} concentration in the smooth-muscle cytoplasm increases, and the released Ca^{2+} first encounters the Ca^{2+}-binding protein, calmodulin. Calmodulin is present in all cells, and it is referred to as a *modulator protein*. The calmodulin molecule lacks enzymatic activity but exerts its effects by binding Ca^{2+}, and the Ca^{2+}–calmodulin complex is then able to bind with other proteins and modulate their activity. One such calmodulin-regulated protein is smooth-muscle myosin light chain kinase (SmMLCK). Without Ca^{2+}–calmodulin, the SmMLCK is in an inactive state. After the binding of Ca^{2+}–calmodulin, SmMLCK is active and phosphorylates the 18-kDa regulatory light chain of smooth-muscle myosin II (Fig. 3–19). This phos-

Figure 3–19. Mechanisms that regulate smooth-muscle contraction and relaxation. ▶ **A:** Regulation by Ca^{2+}–calmodulin: As intracellular Ca^{2+} rises, excess Ca^{2+} is bound by calmodulin, and the Ca^{2+}–calmodulin complex binds to and activates myosin light chain kinase. The activated kinase phosphorylates the regulatory light chain of myosin at site *X*, which leads to contraction. As the intracellular Ca^{2+} concentration falls below ~10^{-7} M, there is a dissociation of the Ca^{2+}–calmodulin complex from myosin light chain kinase, rendering it inactive. Under these conditions, the myosin light chain phosphatase, which is not dependent on Ca^{2+} for activity, dephosphorylates myosin, causing relaxation. **B:** Regulation by cAMP: Stimulation of β-adrenergic receptors by catecholamines, such as epinephrine, causes the stimulation of adenylate cyclase and an increase in intracellular cAMP concentrations. This stimulates the cAMP-dependent protein kinase that phosphorylates myosin light chain kinase at sites *A* and *B* near the calmodulin-binding domain of the molecule. This causes the myosin light chain kinase to have a lower affinity for calmodulin, rendering it inactive, such that it does not phosphorylate the regulatory light chain of myosin, causing relaxation. Dephosphorylation of the myosin light chain kinase restores its ability to bind Ca^{2+}–calmodulin for contraction. **C:** Diacylglycerol-mediated regulation: Diacylglycerol and Ca^{2+} stimulate the activity of protein kinase C, which phosphorylates myosin light chain kinase at sites different from those of the cAMP-dependent kinase (sites *C* and *D*). In addition, protein kinase C phosphorylates the regulatory light chain of myosin at a position different from the myosin light chain kinase. Both of these events render the proteins inactive and cause relaxation. **D:** Caldesmon regulation: At low concentrations of Ca^{2+} (<10^{-6} M), caldesmon binds to tropomyosin and actin, inhibiting the binding of myosin, thereby keeping the muscle in a relaxed state. When the intracellular Ca^{2+} concentration rises, the Ca^{2+} is bound by calmodulin and the Ca^{2+}–calmodulin complex binds with caldesmon, releasing it from the actin filament and allowing contraction. (Modified from Adelstein RS, Eisenberg, E. *Annu Rev Biochem* 1980;49:92–125; Rasmussen, H, Takuwa, Y, Park, S. *FASEB J* 1983;1:177–185.)

(a) Ca^{2+}-calmodulin regulation

(c) Diacylglycerol-mediated regulation

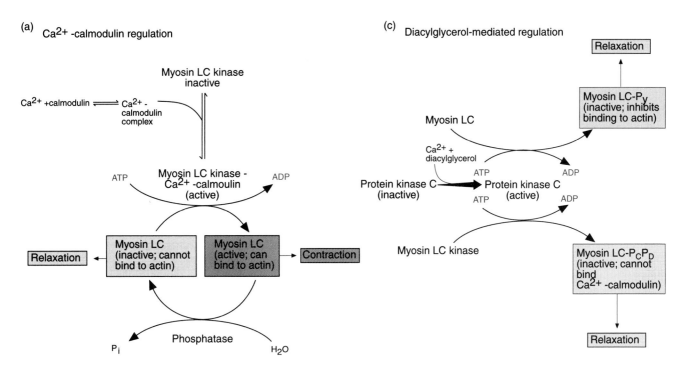

(b) cAMP-mediated regulation

(d) Caldesmon regulation

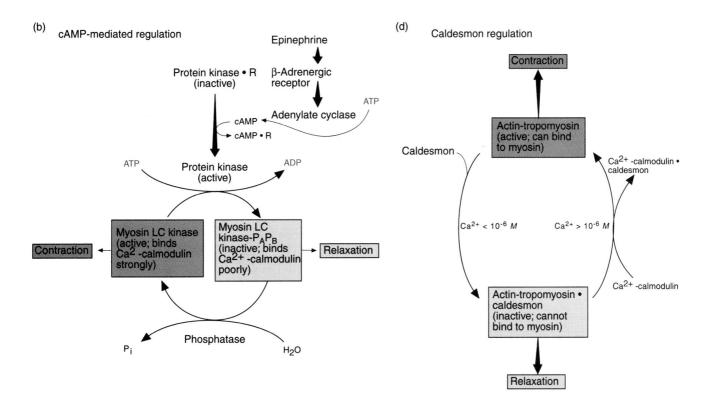

phorylation permits the myosin II to aggregate into thick filaments and allows cross-bridge formation of the thick filaments with the thin filaments of the smooth muscle. Thus, the phosphorylation of myosin light chains is an obligatory event for cross-bridge formation and cycling in smooth muscle. During relaxation, Ca^{2+} ion concentration decreases, and a net dephosphorylation occurs. The reduction in intracellular Ca^{2+} concentration causes an inactivation of the SmMLCK (by a reversal of Ca^{2+}–calmodulin binding). The regulatory activity of the requisite phosphatase enzyme(s) that dephosphorylates the myosin light chain is not well defined.

Smooth-Muscle Contraction Is Influenced at Multiple Levels

Because it can be stimulated by a variety of sources—neuronal as well as hormonal inputs—smooth-muscle contraction can be regulated by several mechanisms. These include regulation by cyclic adenosine monophosphate (cAMP), diacylglycerol, and the protein caldesmon (see Fig. 3–19). Each of these pathways effects a negative regulation; they serve to maintain a relaxed state of the smooth muscle. For example, activation of β-adrenergic receptors on smooth muscle cells causes a rise in intracellular cAMP levels which, in turn, activates cAMP-dependent protein kinase. One of the targets for cAMP-dependent protein kinase in smooth muscle is SmMLCK, and phosphorylation of the MLCK results in a lower affinity of the kinase for the Ca^{2+}–calmodulin complex. As a result, the SmMLCK does not phosphorylate myosin, and the myosin (and smooth muscle) remains in its relaxed state. Other hormones relax smooth muscle by activation of protein kinase C, which is mediated by Ca^{2+} and 1,2-diacylglycerol. The activation of protein kinase C allows it to phosphorylate SmMLCK, causing it to remain in an inactive state.

In addition to hormonal regulation of contraction, smooth muscle cells contain Ca^{2+}-binding proteins that interact with the actin thin filaments, thereby affecting contraction. Caldesmon is an elongated calmodulin-binding protein. In the absence of Ca^{2+}, caldesmon will bind to the actin filaments of smooth muscle, restricting the ability of actin and myosin to interact. In the presence of increased Ca^{2+} concentrations, the Ca^{2+}–calmodulin complex binds with caldesmon, causing a release of the protein from the thin filaments. Thus, the Ca^{2+}–calmodulin complex modulates contraction in smooth muscle by affecting myosin head group phosphorylation, in addition to releasing the caldesmon block on actin thin filaments. This dual control by Ca^{2+}–calmodulin allows the cell to regulate the duration and frequency of contractions.

Actin-Myosin Contractile Structures Are Found in Nonmuscle Cells

In nonmuscle cells the actin:myosin ratio is about 100:1. Thick filaments and microfilaments form within the cytoplasm, but they are in equilibrium with pools of nonpolymerized myosin and G-actin. Although the nonmuscle thick filaments are shorter than those of skeletal muscle, and the myosin and actin filaments do not form the highly structured array found in skeletal muscle, they are still responsible for contraction in nonmuscle cells. Figure 3–20 gives two examples.

During telophase, the last stage in mitosis, a contractile ring forms on the cytoplasmic membrane surface at the cleavage furrow. This contractile ring contracts (like a belt pulled tightly around the waist), forming a cleft between two cells that are separating. Just before telophase, actin filaments begin to form at the site that will become the cleavage furrow. In addition, free myosin begins to polymerize at the same site and, together with actin and the actin-binding protein α-actinin, forms the contractile ring. Because the actin filaments that are attached to the plasma membrane have mixed polarity, the short myosin filaments can use the energy of ATP hydrolysis to cause a contraction that pulls the dividing cell into a dumbbell shape. Before cell division, the actin and myosin filaments rapidly depolymerize.

A second example of nonmuscle contraction is pulling on the plasma membrane, created by stress fibers formed within fibroblasts. Fibroblasts are cells that synthesize and are in contact with extracellular matrix proteins throughout much of the connective tissue that surrounds the organs of your body. The plasma membrane of the fibroblast makes contact with extracellular matrix proteins, both within a tissue culture dish and within the body's connective tissue, at sites called *focal contacts* or *adhesion plaques*. At these contact sites, an integral transmembrane protein of the integrin family binds an extracellular matrix protein, such as fibronectin, at the outer cell surface; this interaction pulls outward on the plasma membrane. The fibroblast is not pulled apart because the same integrin binds actin bundles called *stress fibers* at the cytoplasmic membrane surface. The stress fibers contain interdigitated actin bundles of mixed polarity that are linked together in a parallel array. The stress fibers bind through actin-binding proteins, called *talin* and *vinculin*, as well as a plus end-capping protein to the integrin. Again, it is the plus end of the actin bundle that binds end-on to the plasma membrane at the focal contact, as well as to plus end-capping proteins on that side of the fibroblast that is not attached to the substrate. The stress fibers also contain short myosin filaments, and they exert a contractile force on the actin bundles, which results in an inward pulling on the

Figure 3–20. Nonmuscle actin and myosin have contractile functions. Two examples of a contractile function for nonmuscle actin and myosin are demonstrated. An assembly of actin and myosin creates a contractile ring (*top*) that draws in the center of a cell, leading to cell division. A simplified presentation of stress fibers (*below*) that interact with the plasma membrane at focal contacts, and because of the contractile activity of actomyosin, cause flattening of substrate-attached fibroblasts.

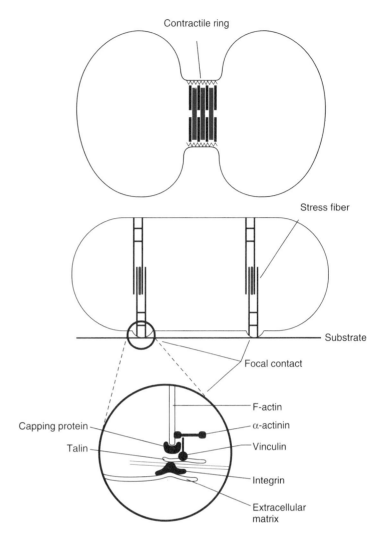

plasma membrane that counteracts the outward pull of the extracellular matrix and, in culture, leads to flattening of the fibroblast. The stress fibers rapidly assemble in response to fibroblast attachment to a substrate, and they rapidly depolymerize when the cells are detached. The depolymerization of actin bundles causes the cells to round up. A third example is the actin and myosin filaments that associate with the adhesion belt, characteristically located below the tight junction of an epithelial cell. Adjacent cells in an epithelial cell layer are held 15–20 nm apart by a Ca^{2+}-dependent transmembrane protein called *uvomorulin* or *E-cadherin*. Uvomorulin also binds, through the actin-binding proteins, α-actinin and vinculin, to the sides of actin filament bundles that form an adhesion belt around the cytoplasmic membrane surface. Myosin filaments and this circumferential F-actin contract, thereby mediating an important process in human development: the folding of epithelial cells into tubes.

In the neural plate, this contraction causes an apical narrowing, which leads to the plate rolling up to form the neural tube during human development.

Bundles of F-Actin Form a Structural Support for the Microvilli of Epithelial Cells

Epithelial tissue, which lines the surfaces of the body, the internal organs, body cavities, tubes, and ducts, contains absorptive cells with numerous fingerlike projections, called *microvilli*, on their apical surface. These microvilli increase the surface area of the epithelial cell's apical plasma membrane, thereby permitting a greater absorption of important nutrients. The microvilli, which are approximately 80 nm wide and 1 μm long, need a stable cytoskeletal scaffolding to maintain their shape and upright position. A very stable and highly structured core of 20–30 bundled actin filaments, which run parallel to the microvilli and attach to the cytoplasmic surface of the plasma membrane, serves as this scaffolding (Fig. 3–21). The actin filaments are bundled by two proteins, named *fimbrin* and *villin*. Actin-bundling proteins are characterized by having two binding sites for F-actin. As they bind to the sides of actin filaments in a helical staircase, they group the filaments into parallel bundles. Villin has an interesting second function: at Ca^{2+} concentrations greater than 10^{-6} M, villin becomes an actin-severing protein. (This class of proteins will be discussed later.) The actin bundles are attached at their plus end to the tip of the microvilli plasma membrane by undefined proteins. The lateral attachments of the actin bundles to the side wall of the microvilli's plasma membrane is through a complex containing calmodulin and myosin I (minimyosin). The core bundles of microvillar actin filaments end just below the apical plasma membrane's surface in a region of the epithelial cell, called the *terminal web* because it contains a meshwork of actin filaments, actin-binding proteins, and intermediate

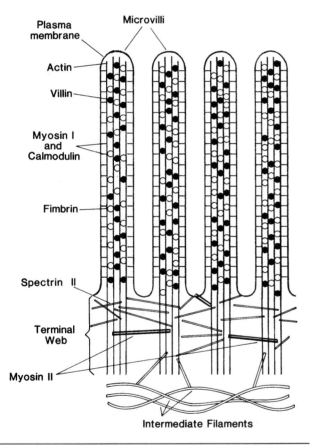

Figure 3–21. Bundled actin filaments have a structural function within microvilli. Actin bundles are attached at their plus ends to the tip of microvilli. The actin filaments are bundled by the proteins villin and fimbrin, and the bundles are attached to the side walls of the microvilli plasma membrane by association with myosin I and calmodulin. Within the terminal web, nonerythroid spectrin (spectrin II) and myosin II link adjacent actin bundles to each other and to intermediate filaments.

filaments. The actin-cross-linking protein, spectrin II (or nonerythroid spectrin), and short myosin filaments run perpendicular to and attach adjacent actin core bundles. These attachments of the core bundles to spectrin II and myosin are thought to hold the microvilli upright. Spectrin II also cross-links the actin core bundles to intermediate filaments.

The Gel–Sol State of the Cortical Cytoplasm Is Controlled by the Dynamic Status of Actin

The cytoplasm of human and animal cells has regions that have the characteristics of a pseudoplastic gel and other regions that liquefy into the sol state. Gel–sol transformations of the cytoplasm are essential for altering the shape of cells and controlling their movement. The gel–sol conversion within the cytoplasm is regulated by the dynamic state of actin and its interaction with actin-binding proteins (Fig. 3–22).

For example, in the cortical cytoplasm just below the plasma membrane, there is a thick three-dimensional matrix of actin filaments that excludes organelles from this region of the cell cytoplasm. Long actin filaments tend to self-associate, causing a highly viscous solution. In the cortical cytoplasm, however, these actin filaments are cross-linked into a three-dimensional meshwork by long fibrous actin cross-linking proteins. The two most prevalent actin cross-linking proteins are spectrin II (nonerythroid spectrin) and filamin, both of which are long fibrous proteins with two well-separated actin-binding sites at their ends. On occasion, it is essential that a region of the cortical cytoplasm becomes liquefied. For instance, when a macrophage contacts a bacterial cell, the cortical actin network must locally disassemble so that the cell surface can restructure to engulf the microorganism. This is carried out by a local increase in the cytoplasmic Ca^{2+} concentration to 10^{-6} M, which stimulates a Ca^{2+}-sensitive actin-severing protein, called *gelsolin*, to cut the actin filaments into short protofilaments. In the process, the gelsolin molecule binds to the plus end of the severed actin filaments and caps that end.

Although the concentration of actin within nonmuscle cells is far greater than the critical concentration for the plus and minus ends of F-actin, only 50% of the actin is in polymerized form in most cells. The actin within nonmuscle cells is in a dynamic state, undergoing polymerization and depolymerization as required. The reason for the pool of G-actin within nonmuscle cells is a group of small actin-binding proteins of the profilin family. Profilin is a 15-kDa protein that forms a one-to-one complex with a G-actin monomer. The complex, called *profilactin*, maintains the G-actin pool by preventing the nucleation step that initiates polymerization; also, it can add to only the plus end of existing fil-

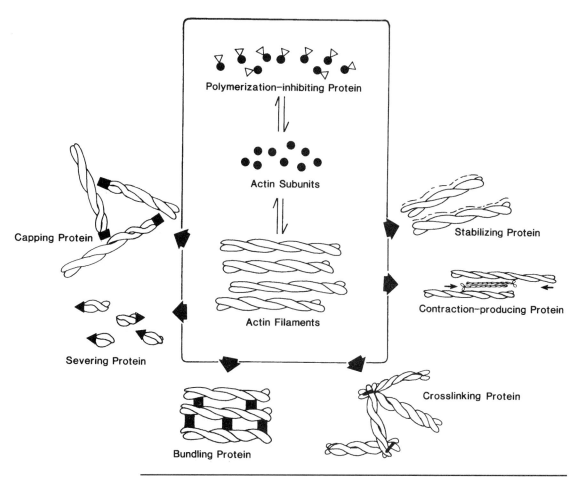

Figure 3–22. The various roles of actin-binding proteins. Summary of the ways in which various actin-binding proteins regulate the cellular organization of actin. (Modified from Widnell CC, Pfenninger KH. *Essential Cell Biology.* Baltimore: Williams & Wilkins, 1990.)

aments. Because the plus end is frequently capped *in vivo*, profilin blocks polymerization onto the capped plus-end actin filaments. When rapid polymerization is required, as in the formation of filopodia on the leading edge of motile cells or in activated platelets, the profilin becomes dissociated from the actin monomers, allowing polymerization to proceed. Cell motility and the role of the cytoskeleton will be discussed in Chapter 10; here, we will only briefly mention the lamellipodia and filopodia (microspikes) that form on the leading edge of a motile cell. As a cell moves, a sheetlike projection forms on the leading edge of the motile cell, called a *lamellipodium*, along with stiff protrusions, called *filopodia*, that can be 0.1 µm wide and 5–10 µm long. The lamellipodium makes contact with the substrate and pulls the cell along, whereas the filopodium serves as a feeler, helping direct the cell. The lamellipodium contain a sheet of highly organized and rapidly polymerized actin filaments, which are primarily perpendicular to the leading edge of the cell and make contact through their plus end with the leading-edge plasma membrane. The filopodia are formed by rapidly poly-

merizing bundles of actin filaments, which are oriented perpendicular to the plasma membrane and force the membrane outward by rapid addition of G-actin monomers to the plus end at the membrane interface. This rapid burst of actin polymerization in the filopodia of motile cells and activated platelets is caused by dissociation of the G-actin–profilin complex. Interestingly, at the leading edge of motile cells, actin treadmilling is occurring, with G-actin being added to the membrane-associated plus end, while it is being subtracted from the distal minus end.

Cytochalasins, a group of chemicals excreted by various molds, block cell movement. The cytochalasins bind to the plus end of microfilaments and block further polymerization, inhibit cell motility, phagocytosis, microfilament-based trafficking of organelles and vesicles, and the production of lamellipodia and microspikes. Phalloidin, an alkaloid isolated from the toadstool *Amanita phalloides*, stabilizes microfilaments and does not allow depolymerization. These chemicals also block cell movement, indicating that both actin filament assembly and disassembly are required for cell motility.

Table 3–1
Actin-Binding Proteins

Protein	Functions
Tropomyosin	Stabilizes filaments
Fimbrin	Bundles filaments
α-Actinin	Bundles filaments
Filamin	Cross-links filaments
Spectrin I/II	Cross-links filaments in membrane skeleton
Gelsolin	Fragments filaments
Myosin II	Slides filaments in muscle
Myosin I	Moves vesicles on filaments
Cap Z	Caps plus ends of filaments
Profilin	Binds actin monomers

Some actin-binding proteins stabilize actin filaments. Capping proteins bind to the plus or minus ends of actin filaments and block polymerization at that end. We have already mentioned that cap Z (a heterodimer with subunits of 36 and 32 kDa) is a plus end-capping protein found in the Z line of the skeletal muscle sarcomere. Other proteins line the actin filaments and prevent the breakage and the erosion of actin monomers. Tropomyosin, for example, binds along the grooves of the actin filament, associating with and covering seven actin monomers in the chain, thereby stabilizing the actin filament.

Table 3–1 summarizes the functions of the various proteins that interact with actin, either G-actin monomers or F-actin filaments. We have discussed intracellular actin-based assemblies (e.g., within the cytoplasm of the cell) and now will consider the participation of actin filaments in structures found at the cytoplasmic surface of the cell membrane, referred to as the *spectrin membrane skeleton*.

The Spectrin Membrane Skeleton

The spectrin membrane skeleton, first described in erythrocytes, but now known to be a ubiquitous component of nonerythroid cells, is essential for maintaining cellular shape and membrane stability, and for controlling the lateral mobility and position of transmembrane proteins within biological membranes.

The Erythrocyte Spectrin Membrane Skeleton's Structure and Function Is Understood in Exquisite Detail

The spectrin membrane skeleton was first described and is best understood in the mammalian erythrocyte. The human erythrocyte's spectrin membrane skeleton maintains the biconcave shape of the erythrocyte, gives it its properties of elasticity and flexibility, stabilizes the plasma membrane, and controls the lateral mobility of integral membrane proteins. These are important properties for an 8-μm diameter biconcave disk that must continuously deform as it passes through capillaries as small as 2 μm in diameter.

The major proteins of the skeleton are spectrin I, actin, and protein 4.1 [a nomenclature based on migration on sodium dodecylsulfate–polyacrylamide gel electrophoresis (SDS–PAGE)]. Spectrin I is composed of two large subunits of approximately 280 (α) and 246 kDa (β). The simplest form of spectrin is an antiparallel αβ heterodimer; however, on the cytoplasmic surface of the erythrocyte membrane, it is an (αβ)2 tetramer, formed by head-to-head interaction of two heterodimers. Each end of the spectrin tetramer contains an actin-binding site, and spectrin I cross-links the actin filaments into a two-dimensional meshwork that covers the cytoplasmic surface of the plasma membrane (Fig. 3–23). The actin filaments are very short, approximately 14 actin monomers long (~33 nm); therefore, they are called *actin protofilaments*. The actin protofilaments are stabilized by tropomyosin, and each protofilament binds six spectrin tetramers, forming a hexagonal array. The spectrin–F-actin complex is stabilized by protein 4.1, which also binds to the ends of the spectrin tetramer. The spectrin skeleton is attached to the bilayer by two types of linkages. A peripheral protein, called *ankyrin*, binds to the spectrin β-subunit toward the junctional end of the heterodimers and links spectrin to the cytoplasmic NH_2-terminal domain of band 3. Protein 4.1, in addition to stabilizing the spectrin–actin interaction, binds to a member of the glycophorin family, thereby serving as a link to the bilayer.

Because the spectrin membrane skeleton is responsible for an erythrocyte's normal biconcave shape, genetic defects in these proteins cause abnormal red cell shapes and stability. Hereditary spherocytosis (HS) is a common hemolytic anemia in white populations, in whom the erythrocytes are spherical and fragile (Fig. 3–24). All patients with the common dominant form of HS show a small reduction in spectrin content, but more importantly, they also contain dysfunctional proteins, such as a spectrin molecule that cannot bind protein 4.1 and, therefore, cannot form a stable spectrin 4.1–actin complex. In hereditary elliptocytosis, in which the red cells are elliptical and fragile, the most prevalent defect is a spectrin dimer that cannot form tetramers. The primary genetic defect in sickle cell anemia is in the hemoglobin molecule, but a subset of sickle cells lock into an irreversibly sickled cell, which is a major factor leading to the sickle cell crisis. The event leading to the irreversibly sickled cell is a post-translational modification in the actin molecules of the spectrin skeleton.

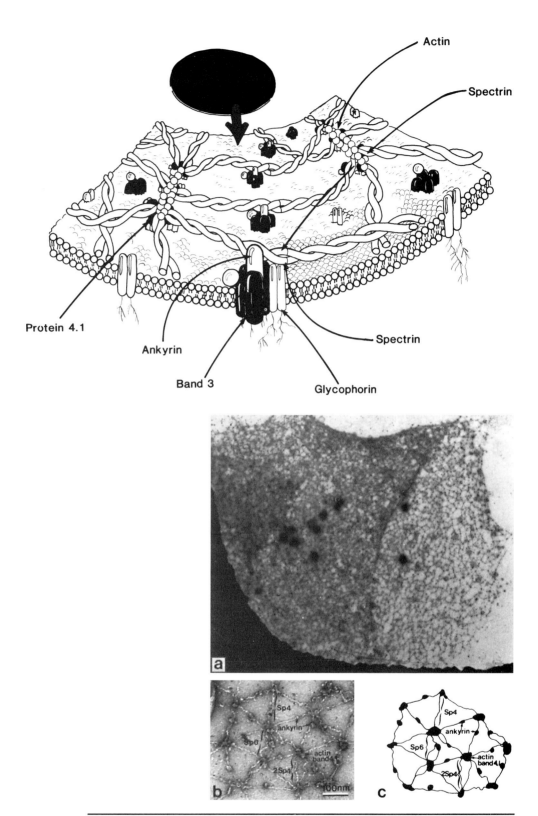

Figure 3–23. Protein interactions in the erythrocyte's spectrin membrane skeleton.
Top: Schematic drawing of the protein interactions within the membrane skeleton.
Bottom: Spread membrane skeleton examined by negative-staining electron
microscopy (**a**) and (**b**). At higher magnification, we see a hexagonal lattice of junctional complexes containing actin protofilaments and protein 4.1, cross-linked by
spectrin tetramers (Sp4), three-armed spectrin molecules (Sp6), and double spectrin
filaments (2Sp4). Ankyrin is attached to spectrin filaments 80 nm from their distal
end. (**c**) Schematic diagram of the hexagonal lattice shown in **b.** (**Top,** modified
from Goodman SR, Krebs KE, Whitfield CF, et al. *CRC Crit Rev Biochem*
1988;23:196; **bottom,** Liu SC, Derick LH, Palek J. *J Cell Biol* 1987;104:527–536,
with permission.)

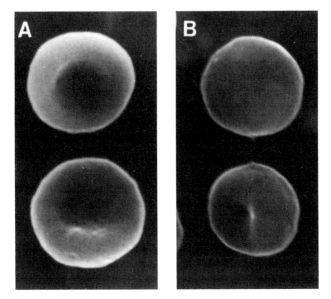

Figure 3–24. Scanning electron microscopy of normal and HS erythrocytes. A: Biconcave erythrocytes from a normal subject. B: A spherocyte and stomatospherocyte from an HS subject. (Goodman SR, Shiffer K. *Am J Physiol* 1983;244: C134–141.)

We Now Know That Spectrin Is an Ubiquitous Component of Nonerythroid Cells

Until 1981, spectrin and the membrane skeleton were thought to be components found only in erythrocytes. That year, Goodman and co-workers demonstrated that spectrin-related molecules were ubiquitous components of nonerythroid cells. This led to the important question of the function of these nonerythroid spectrin molecules. Not only spectrin, but also ankyrin, protein 4.1, and band 3 analogues were found lining the membranes of nonerythroid cells. Furthermore, two isoforms of nonerythroid spectrin have been extensively characterized. One isoform contains α-spectrin, linked to an alternately spliced form of erythroid β-spectrin (spectrin IΣ2). This isoform is found in brain, skeletal muscle, and cardiac muscle. The sec-

ond isoform contains a nonerythroid α- and β-spectrin that share approximately 60% sequence identity with erythroid spectrin. This isoform, called *spectrin II*, is the product of a distinct set of genes and is the most universal form of spectrin.

The spectrins I and II that are found in nonerythroid cells line the cytoplasmic surface of the plasma membrane and organelle membranes, and probably control the membrane contour and stability. However, nonerythroid spectrins are also multifunctional cross-linkers within the cytoplasm of nonerythroid cells and tissues. Spectrin II also cross-links actin rootlets within the terminal web region of the epithelial cell cytoplasm. Within neurons, spectrins I and II link actin filaments to microtubules, neurofilaments, organelles, and synaptic vesicles within the cytoplasm. The synaptic vesicle–spectrin interaction within the presynaptic terminal is central to regulated exocytosis (discussed in Chapter 8).

Spectrins I and II, α-Actinin, and Dystrophin Form the Spectrin Supergene Family of Actin-Binding Proteins

The complete sequences for spectrins I and II have been determined. Both α-and β-subunits contain triple-helical repeat units of approximately 106 amino acids, separated by flexible nonhelical regions throughout most of their sequence. These repeats share approximately 20–40% sequence identity. Interestingly, the NH$_2$- and COOH-terminal ends of α- and β-spectrin I and II do not contain the typical repeat structure. Furthermore, a 140-amino acid stretch at the NH$_2$-terminus of the β-subunit has been demonstrated to represent the actin-binding domain of spectrin (Fig. 3–25). Whereas spectrins I and II share only 60% sequence identity throughout the α- and β-sequence, they are more than 90% identical in the actin-binding domain.

α-Actinin (an actin-bundling protein) and dystrophin [the protein missing in subjects with Duchenne's muscular dystrophy (DMD)] have sequences that are highly related to spectrin I and II. α-Actinin, a 190-kDa

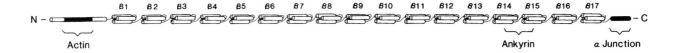

β Spectrin II

Figure 3–25. Structure of β-spectrin II. β-Spectrin II is presented as an example of the structure of the members of the spectrin supergene family. β-Spectrin II has a nonhelical actin-binding domain at its NH$_2$-terminus. There are 17 triple-helical spectrin repeats separated by flexible hinge regions. The COOH-terminus contains a nonhelical region involved in association of the α-spectrin subunit. (Ma, et al. *Mol Brain Res* 1993;18:87–99.)

dimer, is composed of two identical antiparallel subunits. Dystrophin is an 800-kDa homodimer, with two antiparallel 400-kDa subunits. Both α-actinin and dystrophin contain the spectrin triple-helical repeat units, with about 10–20% identity with the spectrin repeats. Both proteins contain a nonhelical region at their NH_2-terminus, with 60–80% identity with the actin-binding domain of β-spectrin. This finding was important in determining the function of dystrophin and the etiology of Duchenne's muscular dystrophy. Because of its sequence identity with the actin-binding domain of spectrin, dystrophin has been proposed to function in anchoring actin filaments to the plasma membrane in skeletal muscle. The common structure found for spectrins I and II, α-actinin, and dystrophin, has led to the concept that they are descendants of a common ancestral gene and to their being called the *spectrin supergene family*.

Intermediate Filaments

Intermediate filaments are 10 nm in diameter and, therefore, intermediate in thickness between microfilaments and myosin thick filaments, or microfilaments and microtubules. Although much work is required to determine the functions of this ropelike filament, its role appears to be primarily structural. That is to say, the major function of intermediate filaments is to provide resistance to mechanical stress placed upon the cell. Intermediate filaments within muscle cells link together the Z disks of adjacent myofibrils. Neurofilaments within the axon serve as a structural support to resist breakage of these long slender processes. Intermediate filaments of epithelial cells interconnect spot desmosomes, thereby stabilizing epithelial sheets.

A Heterogeneous Group of Proteins Form Intermediate Filaments in Various Cells

The protein monomers that constitute intermediate filaments (IFs) differ from the components of microfilaments and microtubules (discussed later) in several important ways. The intermediate filaments in various human and animal cells are composed of a heterogeneous group of proteins, but microfilaments are always composed of actin, and microtubules are always composed of tubulin. The intermediate filament subunits are fibrous proteins, although both G-actin and tubulin are globular. Almost all of the intermediate filament subunits are incorporated into stable intermediate filaments within various cells, whereas the same cells contain a substantial pool of unpolymerized G-actin and tubulin. No energy in the form of ATP or guanosine triphosphate (GTP) hydrolysis is required for intermediate filament polymerization. Intermediate filaments have no polarity, whereas microfilaments and microtubules have plus and minus ends.

Intermediate filaments are composed of a heterogeneous class of subunits (Table 3–2). The keratin filaments found in epithelial cells always contain an equal number of subunits of acidic (type I) and neutral basic (type II) cytokeratins. In humans, there is a genetic disease, epidermolysis bullosa simplex, that arises from mutations in the keratin genes expressed in the basal cell layer of the epidermis. This disrupts the normal network of keratin filaments in these cells, and people afflicted with these keratin gene mutations are keenly sensitive to

Table 3–2
Intermediate Filaments of Human Cells

Intermediate Filament	Subunits (M_r)	Cell Type
Keratin filaments	Type I acidic keratins Type II neutral/basic keratins (40–65 kDa)	Epithelial cells
Neurofilaments	NF_L (70 kDa) NF_M (140 kDa) NF_H (210 kDa)	Neurons
Vimentin-containing filaments	Vimentin (55 kDa) Vimentin + glial fibrillary acidic protein (50 kDa) Vimentin + desmin (51 kDa)	Fibroblasts Glial cells Muscle cells
Nuclear lamina	Lamins A, B, and C (65–75 kDa)	All nucleated cells

NF, Neurofilament.

mechanical injury; even a gentle squeeze can cause disruption of this cell layer and blistering of the skin. Vimentin, desmin, glial fibrillary acid protein (GFAP), and neurofilament light chain (NF$_L$) are capable of forming homopolymeric intermediate filaments, but when together in a single cell (e.g., muscle cells, glial cells), they may copolymerize. However, in an epithelial cell where cytokeratins can be coexpressed with vimentin, they do not copolymerize but, instead, form separate intermediate filaments. Within axons and dendrites, NF$_L$, NF$_M$, and NF$_H$ (subscripts L, M, and H stand for light, medium, and heavy chains) copolymerize to form the neurofilaments. The cell type specificity of intermediate filament proteins has been useful to pathologists, who use fluorescent intermediate filament type-specific monoclonal antibodies to identify the tissue of origin of metastatic cancer cells. The nuclear lamina is composed of the IF-related proteins lamin A, lamin B, and lamin C.

These proteins and the square lattice that they form on the inner nuclear envelope will be discussed in the section of Chapter 4 dealing with the nucleus.

How Can Such a Heterogeneous Group of Proteins All Form Intermediate Filaments?

It is truly remarkable that proteins of the IF class, which range in M_r from 40 to 210 kDa, are all capable of forming intermediate filaments. The molecular basis for this common morphology is shown in Fig. 3–26. All IF proteins contain a subunit-specific NH$_2$-terminus of variable size, a homologous central α-helical region of approximately 310 amino acids (with three nonhelical gaps), and a subunit-specific COOH-terminus of variable size. Only the homologous 310-amino acid α-helical region is a portion of the 10-nm IF core. The variable regions extend from the core and are responsible

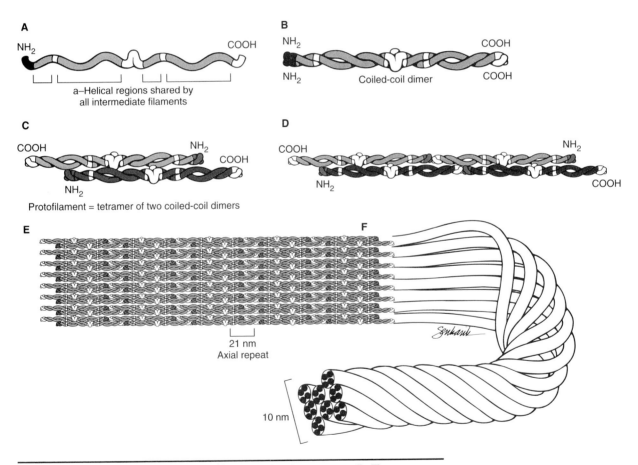

Figure 3–26. Assembly of intermediate filaments. **A:** IF monomers. **B:** Two monomers form a parallel coiled–coil dimer. **C:** Two dimers form an antiparallel tetramer by side-to-side interaction. **D:** The two dimers forming a tetramer are staggered, which allows the formation of higher order structure. **E:** The tetramers continue to associate in a helical array up to eight tetramers (protofilaments) wide. **F:** The intermediate filaments became longer and wind into a ropelike structure. (Modified from Alberts B, Bray D, Lewis J, et al. *Molecular Biology of the Cell*, 3rd ed. New York: Garland Publishing, 1994.)

for cross-linking intermediate filaments to other cyto-skeletal structures. In the formation of the intermediate filaments, the first step is that the 310-amino acid α-helical region of two monomers wind around each other into a parallel coiled coil. The IF proteins contain the heptad repeat within the 310-amino acid α-helical region required for coiled-coil formation. Next two dimers link side by side in an antiparallel conformation to form a tetramer. Because the tetramers have an antiparallel conformation, the intermediate filaments have no polarity. The IF tetramers attach laterally to each other in a staggered array until there are eight tetramers (32 monomers) making up the wall of the intermediate filament. The eight tetramers are wound to form the ropelike structure of the intermediate filament.

Although microfilaments have actin-binding proteins to allow their association with other cytoskeletal structures, and microtubule-associated proteins (MAPs) play a similar function for microtubules, the variable NH_2- and COOH-terminal subunit-specific regions of IF proteins enable their attachment to other cytoskeletal structures. In anaphase cells, the intermediate filaments normally form a tight weave around the nucleus and then spread in wavelike fashion toward the plasma membrane. If the microtubules of an anaphase cell are depolymerized with colchicine (ColBENEMID) or de-mecolchicine (Colcemid), the intermediate filaments collapse around the nucleus: obviously, the intermediate filaments are highly integrated with microtubules. When antibodies against spectrin II were microinjected into fibroblasts, the intermediate filament network again collapsed around the nucleus, even though there was no obvious effect on microtubules or microfilament stress fibers. This suggests that spectrin II may also play an important role in linking intermediate filaments to other cytoskeletal structures. Indeed, immunoelectron-microscopy has demonstrated such a role for spectrin II within the terminal web of epithelial cells and within the axons and dendrites of mammalian neurons. All of the studies described above suggest an association of intermediate filaments with the nuclear envelope. In addition, intermediate filaments attach to the plasma membrane by interactions with ankyrin and with spectrin II, perhaps by an intermediate filament-associated protein called *plectin* (M_r 300,000).

Microtubules

Microtubules Are Polymers Composed of Tubulin

Microtubules are the third type of cytoskeletal structure, and they have been implicated in a variety of cellular phenomena, including ciliary and flagellar motility, mitotic and meiotic chromosomal movements, intracellular vesicle transport, secretion, and several other cellular processes. Their principal component is the protein tubulin, a heterodimer composed of non-identical α and β subunits, with each subunit having an M_r of nearly 50 kDa. In addition, several other proteins are associated with microtubules, and it is these accessory proteins that are responsible for many of the characteristics of microtubule-based motility. The MAPs will be described later in this section. Through the electron microscope, microtubules are hollow cylinders with an outer diameter of 24 nm. When viewed in cross-section, the wall of each microtubule is seen to be composed of 13 tubulin dimers (Fig. 3–27), which represent 13 protofilaments composed of tubulin subunits. As microtubules assemble, the tubulin molecules are added to the growing microtubule to form the 13 protofilaments. The individual protofilaments are organized such that α and β subunits alternate along the length of the protofilament, which provides a microtubule with an inherent polarity.

Tubulin is one of the most highly conserved proteins known. The significance of this is as yet unclear, but it is presumed that this results from the many essential functional subdomains within the tubulin molecule. Not only are there regions that are necessary for subunit interactions during microtubule assembly, but tubulin also contains regions for GTP binding, for interacting with microtubule-associated proteins, and sites for binding to several different drugs. Pharmacologic agents that are bound by tubulin, such as colchicine, vinblastine sulfate (Velban), nocodazole, and paclitaxel (Taxol), disrupt the normal dynamic behavior of microtubules. Because proper microtubule functioning is essential for spindle formation and cell division, microtubule inhibitors are commonly used for cancer chemotherapy.

Microtubules Undergo Rapid Assembly and Disassembly

Cytoplasmic microtubules are labile structures that have the capacity to undergo rapid assembly and disassembly. This characteristic is important for many microtubule functions. For example, cytoplasmic microtubules must be broken down rapidly as the cell enters mitosis. Likewise, the disassembled microtubules must be reformed to assemble the mitotic spindle. The ability of the spindle microtubules to be broken down by the cell during mitosis also appears to be essential for chromosomal separation. If dividing cells are cultured in the presence of taxol, a drug that blocks microtubule disassembly, chromosomal segregation is blocked. The ability of the microtubule cytoskeleton to reorganize rapidly may be important for many other cellular events, such as cell migration and the establishment of cellular polarity.

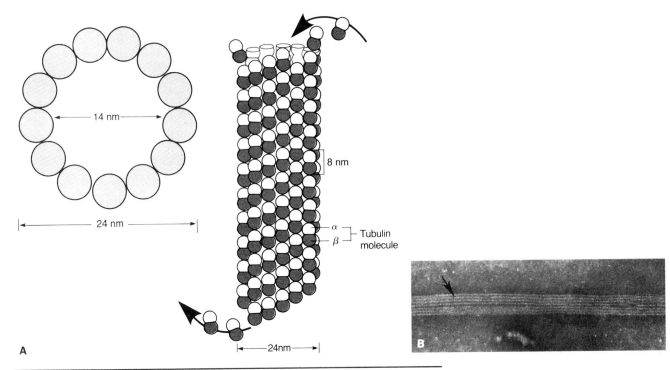

Figure 3–27. The morphology of cytoplasmic microtubules. **A:** Schematic representations of microtubules in cross and longitudinal sections. **B:** Negative-stained whole-mount EM of a microtubule showing tubulin protofilaments (*arrow*). (**B,** courtesy of Dr. P. Fields.)

Similar to actin filaments, growing microtubules have an inherent structural polarity. This polarity occurs because of the orientation of the tubulin subunits in the microtubule polymer. When growing microtubules are analyzed *in vitro*, subunits add to one end of the elongating polymer faster (the plus end) than to the other end (the minus end) (see Fig. 3–27). Inside cells, however, the minus end of the microtubule is capped, owing to its association with the centrosome complex; therefore, only the events that occur at the plus end will be considered. Current ideas concerning microtubule dynamics focus on the binding of GTP by tubulin subunits during microtubule assembly and the subsequent hydrolysis of the bound GTP to guanosine diphosphate (GDP) (Fig. 3–28). For the tubulin monomer to add to an elongating microtubule polymer, the tubulin subunit must bind to GTP. The GTP-tubulin can then add to the growing end of a microtubule and, some time after adding to the microtubule, the bound GTP is hydrolyzed to GDP. In effect, this results in the presence of either a small GTP-tubulin cap or a GDP-tubulin cap on the end of a microtubule, with the remainder of the microtubule polymer being composed of GDP-tubulin. As long as a microtubule continues to grow rapidly, tubulin subunits will be added to the tubule faster than the nucleotide can be hydrolyzed, and the GTP cap will remain intact. This is important

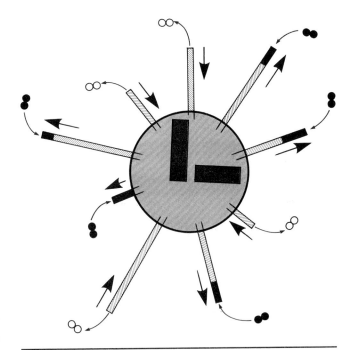

Figure 3–28. Dynamic instability of microtubules. A schematic demonstrating dynamic instability. The GDP-tubulin regions of microtubules are shown as *hatched* regions, and GTP-tubulin microtubule caps are shown as *solid* regions. Microtubules that are assembling from the centrosome contain GTP caps, whereas those that are catastrophically disassembling contain only GDP-tubulin.

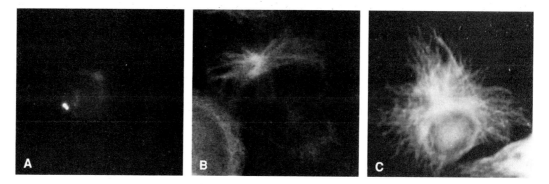

Figure 3–29. The centrosome nucleates cellular microtubules. Antitubulin immuno-fluorescent staining of cultured mammalian cells demonstrating that the centrosome is the microtubule-organizing center in mammalian cells. **A:** Mammalian cells were experimentally treated so that all of the microtubules were disassembled. Only the microtubule-containing centrosomes could be identified. **B:** When the experimental treatment was reversed, a starlike array of microtubules began to form off the centrosome. **C:** With time, the microtubules elongated until they eventually filled the cytoplasm. (Courtesy of R. Balczon.)

because tubulin subunits add to GTP-capped micro-tubules much more efficiently than they bind to GDP-tubules. If the rate of microtubule assembly slows, GTP hydrolysis can catch up, the GTP cap will be lost, and the entire length of the microtubule polymer will comprise GDP-tubulin. The GDP-capped microtubules are unstable and tend to lose GDP-tubulin subunits from the end of the microtubules, which results in microtubule shortening. Moreover, this rate of loss is very rapid. The addition of GTP-tubulin to GDP-tubules is not completely efficient; once a microtubule begins to break down, it usually disassembles completely. This formation and catastrophic breakdown of microtubules is called *dynamic instability*. Dynamic instability provides an explanation for how a cell is able to reorganize its microtubule cytoskeleton so rapidly.

By Capping the Minus Ends of Microtubules, the Centrosome Acts as a Microtubule-Organizing Center

Unlike other cytoskeletal filaments, which appear to be nucleated and oriented haphazardly throughout the cytoplasm, cytoplasmic microtubules are all nucleated by the centrosome complex. If cultured cells are fixed and processed for antitubulin immunofluorescence microscopy, a starlike array of microtubules is observed that originates near the nucleus and radiates throughout the cytoplasm (Fig. 3–29). At the focal point of the astral microtubule array is the centrosome. Ultrastructurally, the centrosome is composed of a centriole pair and an osmiophilic cloud of amorphous material, called *pericentriolar material*, which surrounds the centrioles (Fig. 3–30). The individual centrioles of the centriole pair are oriented at right angles to each other, and each

centriole comprises nine triplets of short microtubules (0.4–0.5 µm in length).

Experimental analysis shows that the microtubule nucleating capacity of the centrosome complex is contained within the pericentriolar material and not in the centriole pair. The nature of the microtubule-nucleating centers within the centrosome is unknown, but the pericentriolar material must contain protein components that bind to tubulin subunits in such a way that the plus ends of microtubules are oriented toward the

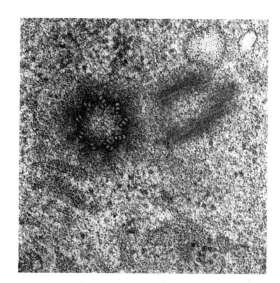

Figure 3–30. The morphology of the centrosome complex. An electron micrograph of a centrosome complex. The centrioles of a centriole pair are oriented at right angles to each other. The centrioles are composed of nine short triplet microtubules and are surrounded by weakly staining pericentriolar material. (Courtesy of R. Balczon.)

periphery of the cell. The most likely candidate as the centrosome component that nucleates microtubules is a protein called *gamma tubulin*. Gamma tubulin is a specialized member of the tubulin superfamily of proteins that is localized specifically to the pericentriolar material. It has been proposed that gamma tubulin forms a nucleation complex within the centrosome that allows microtubule formation. Moreover, by binding to the minus ends of microtubules, the gamma tubulin within the centrosome caps these microtubule ends. This means that all assembly and disassembly phenomena must occur at the plus ends of microtubules. In addition, the capping of the minus ends of the microtubules provides an explanation of how microtubule formation could occur inside a cell. Like actin microfilaments, free microtubules *in vitro* have an assembly, or plus end, and a disassembly, or minus end. Microtubules will assemble *in vitro* only if the concentration of tubulin is so high that tubulin subunits are adding to the assembly end more quickly than tubulin monomers are dissociating from the disassembly end. Apparently, the concentration of tubulin inside a cell is below this critical concentration necessary for spontaneous assembly. Therefore, the rate of tubulin loss from the minus end of a microtubule that is free in the cytoplasm would exceed the rate of addition of tubulin subunits to the plus end. As a result, microtubules cannot form freely in the cytoplasm and must be nucleated by the centrosome. By capping the minus end of the microtubule, the presence of a centrosome permits a cell to maintain its cytoplasmic tubulin concentration at levels that are too low to support spontaneous microtubule assembly. Because tubulin levels are so low, the only microtubules that can form in a cell under physiologic conditions are those capped at their minus ends by their associations with a centrosome.

The Behavior of Cytoplasmic Microtubules Can Be Regulated

The picture that forms when considering dynamic instability is one of a cytoplasm that is constantly changing because of the rapid turnover of microtubules. At any one instance, many microtubules would be rapidly growing, whereas others would be quickly and catastrophically disassembling. Although true to some degree, in reality, the average life span of cytoplasmic microtubules is about 10 minutes. Microtubules can exist for periods that are longer than might be expected because the cell has several mechanisms for stabilizing cytoplasmic microtubules.

One adaptation that cells can use for stabilizing microtubules is to cap the plus end. Because cytoplasmic microtubules are capped on their minus ends by the centrosome, if they were capped on their plus ends

they would not have a free end available for disassembly. This mechanism is, in fact, used by cells during mitotic spindle assembly. As a cell enters into mitosis, numerous microtubules begin to form off the centrosomes. Most of these microtubules rapidly disassemble, owing to dynamic instability. However, some of the growing microtubules make contact with the kinetochore regions of the mitotic chromosomes and are capped on their plus ends by the proteins of the kinetochore. This microtubule capping selectively stabilizes these microtubules and is an important event in spindle morphogenesis.

Other molecular and biochemical modifications can result in the increased stability of cytoplasmic microtubules. These changes in microtubule behavior can be caused either by posttranslational modifications of tubulin or by the interaction of microtubules with any one of several MAPs. The principal posttranslational modification of tubulin in cellular microtubules is the removal of the COOH-terminal tyrosine of the α-tubulin subunit by a detyrosinating enzyme that is present in cells. This detyrosination occurs after tubulin has been incorporated into a microtubule and results in the stabilization or maturation of cytoplasmic microtubules. However, detyrosination may have no direct effect on microtubule kinetics; microtubules that are formed *in vitro*, using either tyrosinated or detyrosinated tubulin, show no differences in their inherent stability. Therefore, it is conceivable that detyrosination may act as a signal that induces the binding of a second protein to microtubules, which results in the increased stability that is observed in detyrosinated cytoplasmic microtubules. When a detyrosinated microtubule is disassembled, a cellular cytoplasmic enzyme is responsible for adding a tyrosine back onto the COOH-terminus of the α-tubulin polypeptide.

The interaction of tubulin with MAPs also results in considerable modifications in the behavior of microtubules. A protein is characterized as being a MAP if it binds to and copurifies with microtubules during their isolation from cellular homogenates. The individual types of MAPs appear to vary among cell types. However, considerable information on the functions of MAPs has been obtained by studying those that have been isolated from neural tissue. Two major classes of MAPs have been identified in neurons: the high M_r MAPs, a small family of proteins of 200–300 kDa; and the tau proteins, a group of polypeptides of 40–60 kDa. Experimental analysis has demonstrated that these proteins bind to tubulin monomers and assist with microtubule nucleation. In addition, they appear to be involved in the tight bundling of microtubules that is characteristic of the microtubular configurations seen in nerve axons and at other selected cellular sites.

Microtubules Are Involved in Intracellular Vesicle and Organelle Transport

One of the important functions of microtubules is the intracellular transport of organelles and vesicles. For the microtubule cytoskeleton to fulfill this role, microtubules must have a means of generating a force that allows such motile behavior to occur. Experimental analysis has allowed the identification and purification of the ATPases that appear to be involved in force generation.

Two specific microtubule-dependent ATPases have been identified that appear to be important in cytoplasmic transport. One of these enzymes, kinesin, is a large multisubunit protein that is involved in translocating vesicles along microtubules from the minus, or centrosome, end toward the distal plus ends. This allows the transport of vesicles from deep within the cytoplasm, where they are produced by budding from the Golgi apparatus, to the cell cortex where secretion can occur. The other enzyme that appears to be involved in cellular motile events is the cytoplasmic form of the ciliary enzyme dynein. Cytoplasmic dynein (or MAP 1C) is a very high M_r multisubunit protein complex that translocates structures along microtubules from their plus to their minus ends. Whether these two ATPases can explain all forms of intracellular microtubule-based motility has not been determined. Regardless, according to current thinking, microtubules appear to play a relatively passive role in most types of intracellular movement, with the active roles being performed by the microtubule-dependent ATPases (Fig. 3–31). A good analogy for visualizing these events would be to consider a railroad: the microtubules would serve as the tracks and the locomotory forces responsible for transporting the vesicular cargo would be generated by ATPases, such as kinesin and dynein.

Cilia and Flagella Are Specialized Organelles Composed of Microtubules

Cilia and flagella are specialized cellular appendages that extend from the surfaces of several different cell types. Cilia are prominent in the respiratory tract and on the apical surface of the epithelial cells that line the oviduct. In the respiratory tract, cilia are involved in clearing mucus from the respiratory and nasal passages, whereas those that line the oviduct are involved in transporting ova toward the uterus. The major type of flagellated cell in humans is the spermatozoon. For the mature sperm cell, the beating flagellum provides the force that allows the sperm to swim.

Cilia and flagella are very similar ultrastructurally. At the core of one of these organelles is the axoneme, a complex structure composed of microtubules and various other proteins that allows ciliary and flagellar bending to occur. When viewed in cross-section, the axonemal microtubules are arranged in a distinctive nine-plus-two array (Fig. 3–32). The term *nine-plus-two* refers to the orientation of the microtubules that make up the axoneme. In axonemes, there are two complete central microtubules (the central pair) that are surrounded by a circumferential ring of nine doublet microtubules. The outer doublets are arranged so that each doublet pair is composed of one complete microtubule (the A tubule), which consists of 13 protofilaments, and an incomplete microtubule (the B tubule), which is composed of only 11 protofilaments. The B tubule shares a portion of the A tubule wall. The nine-plus-two array of microtubules traverses the axoneme, extending from the specialized centrioles, called *basal bodies*, which are located at the base of the cilium or flagellum, all of the way out to near the very tip of the cilium or flagellum.

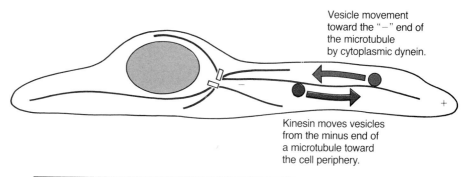

Vesicle movement toward the "−" end of the microtubule by cytoplasmic dynein.

Kinesin moves vesicles from the minus end of a microtubule toward the cell periphery.

Figure 3–31. **Vesicle transport along microtubule tracks.** A schematic demonstrating how vesicles might be transported along cytoplasmic microtubules by microtubule-dependent ATPases. Microtubules are polar structures with defined plus and minus ends. Vesicles are thought to be transported in the anterograde direction (from the minus to the plus end of a microtubule) by kinesin. Vesicles and organelles are thought to be translocated in the retrograde direction (from the plus to the minus end of the microtubule) by cytoplasmic dynein.

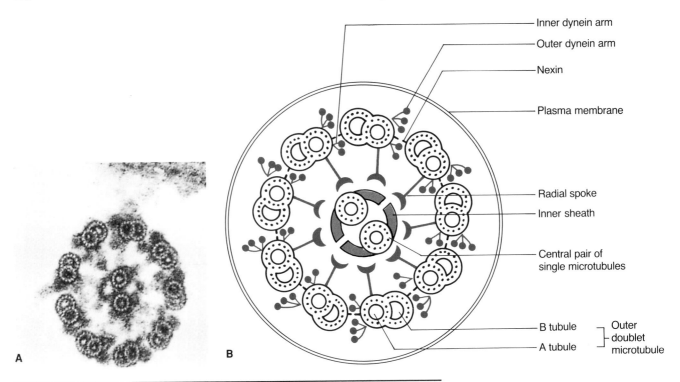

Figure 3–32. Axonemal nine-plus-two array of microtubules and other axoneme proteins. **A:** An electron micrograph showing the nine-plus-two organization of microtubules in an axoneme. (**A,** courtesy of Dr. W.L. Dentler.) **B:** A schematic showing the organization of the proteins in an axoneme. (Modified from Alberts B, Bray D, Lewis J, et al. *Molecular Biology of the Cell,* 2nd ed. New York: Garland Publishing, 1989.)

In addition to the microtubules, several other important proteins can be found in axonemes. These accessory proteins are absolutely essential for normal ciliary function. Extending from the A tubule of each doublet toward the B tubule of the neighboring doublet are two proteinaceous arms (see Fig. 3–32). These arms are actually the enzyme dynein, the ATPase that is responsible for ciliary and flagellar motility. Also extending between the A tubule and the B tubule of the neighboring doublet is a protein called *nexin.* This attaches neighboring doublets to each other. Finally, a radial spoke extends off the A tubule of each doublet and makes contact with an electron-dense sheath surrounding the central pair of microtubules, thereby connecting the doublet microtubules to the central pair. The dynein arms, nexin links, and spoke proteins exhibit a periodicity along the entire length of the axoneme. In addition to these prominent proteins, numerous other minor proteins are present in the axoneme.

Axonemal Microtubules Are Stable

Most cellular microtubules are labile structures that can be assembled and disassembled rapidly. Axonemal microtubules, however, are stable structures that resist breakdown. One of the modifications of axonemal microtubules that may contribute to this increased stability is the enzymatic acetylation of a lysine residue on α-tubulin subunits. Cytoplasmic microtubules, which are turning over rapidly, are nonacetylated. Like detyrosinated tubulin, acetylated tubulin shows *in vitro* kinetic behavior similar to unmodified tubulin, suggesting that the acetylation of axonemal α-tubulin subunits may serve as a signal for other proteins that bind to the microtubule wall to stabilize the axonemal tubules.

Microtubule Sliding Results in Axonemal Motility

Axonemes can be isolated from several sources quite easily. This is accomplished by shearing cilia or flagella from cells, selectively removing the residual plasma membrane that surrounds the axonemes, and extracting the axonemes in a buffer that contains a very mild detergent. When isolated sperm flagellar axonemes are incubated in a buffer that contains ATP, the axonemes will continue to beat in a relatively normal fashion; therefore, all of the information that is required for ciliary and flagellar motility is contained within the structure of the axoneme alone. Biochemical and genetic

Figure 3–33. Dynein cross-bridge cycles lead to the bending of cilia and flagella. A schematic showing how the binding of ATP causes a conformational change in (A) the structure of the ATPase dynein. B: The ATP binding causes dynein to release from the wall of the adjacent B tubule of a microtubule pair and then reattach to that same microtubule pair farther down the length of that microtubule doublet. C: Cleavage of ATP to ADP results in a force that causes the two microtubule pairs to slide past one another. *In vivo*, the sliding of the microtubule doublets relative to one another is converted to bending by the nexin cross-links and spoke proteins.

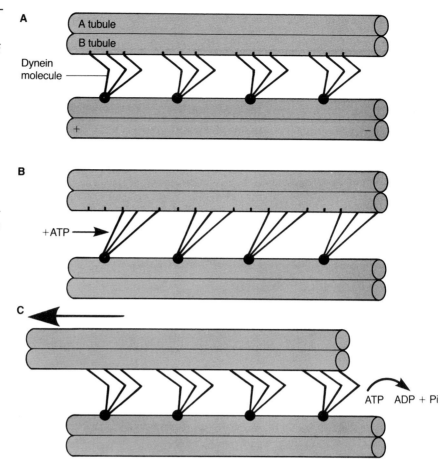

studies have been used to dissect the mechanism of axonemal motility.

When isolated axonemes are treated with a proteolytic enzyme, the nexin cross-links and radial spokes are selectively digested, whereas the microtubules and dynein arms remain intact. If these protease-treated axonemes are incubated with ATP, the axoneme elongates up to nine times its original length. Microscopic analysis has shown that this is because the nine outer doublet microtubule pairs actively slide past one another. That these treated axonemes lack cross-links and spokes, suggests that the dynein arms are the ATPase that drives axonemal motility. Moreover, this result suggests that nexin and the spoke protein are able to convert the activity of dynein into the bending that results in ciliary and flagellar motility. In fact, this is true. When isolated axonemes are low-salt extracted, the dynein arms are released, whereas the microtubules, nexin links, and spoke proteins remain intact. Such extracted axonemes will not beat when ATP is added. However, if ATP is added to the salt extract that contains the dynein arms, the ATP is actively hydrolyzed. Therefore, the following mechanism can be visualized for ciliary and flagellar activity (Fig. 3–33). In the presence of ATP, dynein undergoes a conformational change. The net effect of this change is that the dynein

arm releases from the B tubule of the adjacent microtubule doublet pair and then reattaches to that same doublet pair farther down the length of the doublet. This cycle is repeated when another ATP binds to the dynein arm. However, this "walking" of the dynein arms down the length of the microtubule wall is resisted by the nexin cross-links and radial spokes, and these two structures convert the sliding of the adjacent microtubule pairs into a bending motion.

Such a complex pathway must be tightly regulated because, at any one instance, dynein arms on one region of an axoneme must be active, while in other areas, the dynein arms must be relaxed for ciliary and flagellar beating to occur. Although the detailed mechanisms of regulation are unknown, several mutants in ciliary and flagellar motility have been identified in lower eukaryotes. These mutants should provide important information about the regulation of axonemal beating. In addition, several human diseases are the result of mutations in one of the genes encoding axonemal proteins. As would be expected, males with such conditions are sterile because their sperm are immotile. In addition, patients afflicted with one of these conditions show chronic respiratory tract problems because the respiratory cilia are unable to clear mucus from the bronchiole and nasal passages.

The Neuronal Cytoskeleton

For probably no other mammalian cell type are the demands placed on its cytoskeleton more compelling than in the support of neuronal growth and metabolism. Neurons exhibit unparalleled variation in both the dimensions of their cell bodies and their process-bearing architecture that determine specific patterns of synaptic connections. Despite such polymorphism, nearly all neurons elaborate two fundamentally distinct cytoplasmic processes that are responsible for the directional flow of information within the dendrite and axon (Fig. 3–34).

Dendrites convey input toward the cell body, from where they typically emanate in a stellate array. Alternatively, dendrites can arborize from a single trunk stemming eccentrically from the cell body, as is characteristic of the Purkinje cell in the cerebellum. In both configurations, dendrites extend a relatively short distance but branch extensively, thereby enabling neurons to receive input from a sizable tissue field near the cell body. Neuronal output is usually conveyed along a single axon that extends a much greater distance than dendrites do before branching distally to supply multiple target cells of innervation. The axon of a ventral motor neuron can project as far as 1 meter before synapsing at the neuromuscular junction. Such an example serves to emphasize a general property of all neurons; namely, that the volume of cytoplasm and area of plasma membrane invested in their processes far exceed those of the cell body by as much as several orders of magnitude. The various components of the neuronal cytoskeleton are confronted with the task for forming, maintaining, and modifying these specialized structures.

The Shape of Neurons Is Determined by Their Cytoskeleton

Electron-microscopic techniques reveal that the neuronal cytoplasm is filled along its entirety with a highly cross-linked gel packed with cytoskeletal structures (Fig. 3–35). Two conspicuous filamentous structures are aligned along the axial plane of cytoplasmic processes: microtubules, common to all cell types; and neurofila-

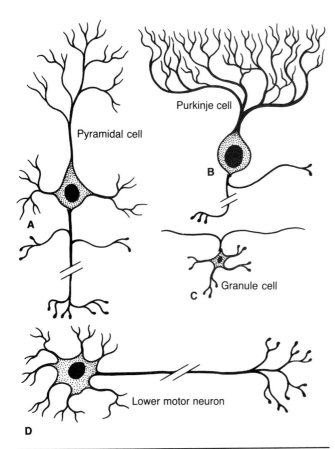

Figure 3–34. Branching pattern of different neurons. Examples are given for neurons whose cell bodies reside in the (**A**) cerebral cortex and (**B,C**) cerebellar cortex. D: Lower motor neurons in the brain stem or spinal cord send their axon through cranial or spinal nerves, respectively, to supply musculature. Dendrites and cell bodies are shown (*black*); most axons (*red*) have been shortened. (Modified from Barr ML, Kiernan JA: *The Human Nervous System*, 6th ed. Philadelphia: JB Lippincott, 1993.)

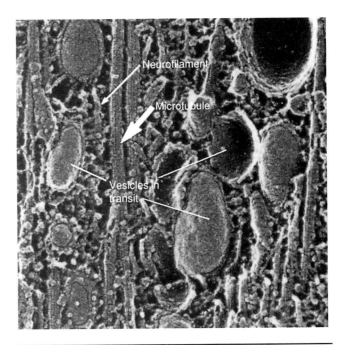

Figure 3–35. Rapid freeze deep etch electron-microscopy of the axonal cytoskeleton. Numerous vesicles of different shapes and sizes are interspersed among microtubules (*thick arrow*) and neurofilaments (*thin arrow*). The small particulate structures filling the remainder of the axoplasm are linker proteins attached to the longitudinal fibers. (Hirokawa N. *J Cell Biol* 1982; 94:129.)

ments, a neuron-specific subclass of intermediate filaments with a smaller diameter, approximately 10 nm. Both are extensively cross-linked to one another in a heterologous fashion, as well as to filaments of the same type. Actin microfilaments, a third filament type, predominate in the cortical cytoplasm as a meshwork located immediately adjacent to the plasma membrane and concentrated, in particular, at the tip, or growth cone, of formative neurites. When neurons are stripped of all noncytoskeletal proteins by detergent treatment, left behind are only the undisturbed cytoskeletal elements, which bear a striking resemblance to the original shape of the complete cell and its process outgrowth. This observation provides tangible evidence for the important role that cytoskeletal structures play in constructing the unique neuronal morphology.

Neuronal Metabolism Requires Stable and Dynamically Unstable Filamentous Structures

The neuronal cytoskeleton imparts mechanical strength to functionally mature cells. However, neurons elongate and retract their processes during development and can change their patterns of connectivity throughout life. This is made possible by the existence of actin and tubulin in both a monomeric and polymeric state within neurons. There exists a shifting equilibrium between microfilament and microtubule disassembly and assembly occurring in response to various physiologic stimuli, such as environmental signals affecting cell growth, or long-term changes in electrical activity. Regardless of the external stimulus, the depolymerization and polymerization of actin and tubulin, to regulate microfilament and microtubule assembly, proceeds by the same biochemical mechanisms governing these events in other cell types.

In the neuronal cell body, microtubules nucleate from the centrosome or microtubule-organizing center, where they attach at their minus ends. Those extending into the axon grow by polymerization at their distal, positive ends, at which the process of dynamic assembly and disassembly occurs. Almost all axonal microtubules orient with their positive ends directed toward the terminal. Because they average approximately 200 µm in length, microtubules in longer axons are arranged in tandem to span the greater distance. By contrast, microtubules in dendrites organize with mixed polarity; either the plus end or minus end can face the terminal. This fundamental difference in organization may underlie separate mechanisms by which axons and dendrites traffic substances away from the cell body toward synaptic junctions in the periphery.

Although less is understood about the precise function of neurofilaments, their frequency in mammalian axons is hypothesized to determine the axonal diameter among different neurons or along various segments of a single axon. Three neurofilament subunits of different M_r (210, 140, and 70 kDa) coassemble in the cell body and are transported down the axon during which, unlike microtubules and actin microfilaments, they undergo relatively little turnover.

Accessory Proteins of the Neuronal Cytoskeleton Are Functionally Diverse

Neurons require many homologous, if not identical, nonfilamentous cytoskeletal proteins used by other cell types, and some that are unique. One family of these accessory proteins, those that bind actin, is remarkably varied in function. Some members work by bundling and cross-linking microfilaments of actin into the meshwork that underlies the neuronal plasma membrane. Others regulate the length of actin polymers by capping the ends of microfilaments, which curtails further polymerization, or by cleaving the strands into smaller lengths. Another actin-binding protein, 4.1 (amelin), links microfilaments to other membrane skeletal proteins, such as brain spectrin, and intrinsic membrane proteins, functioning as it does in the erythrocyte. The neuron-specific protein, synapsin I, binds reserve synaptic vesicles to actin microfilaments in the presynaptic terminal until they are used for synaptic transmission (see Chapter 8).

A different class of accessory proteins, MAPs, function in pivotal roles of microtubule assembly, microtubule cross-linking, and transport functions. At least ten different MAPs are presently recognized and divided into high M_r (~200 kDa) and low M_r (the tau MAPs, 40–60 kDa) categories. Some MAPs and actin-binding proteins are referred to as *motor proteins* because they hydrolyze ATP to provide the energy necessary to move themselves along polymerized filaments. For example, myosin I (minimyosin) functions as a motor by associating with actin microfilaments to influence the process of growth cone motility and neurite extension (covered later in Chapter 6). Motor MAPs drive the process of axonal transport and are discussed in the following section.

Nerve Terminals Are Nurtured by Axoplasmic Flow

To provide for the constant turnover of macromolecules at nerve endings distant from the cell's biosynthetic machinery, a cytoskeleton-mediated transit system shuttles membrane and secretory materials back and forth between the cell body and axon terminal. Axoplasmic flow, the means of this replenishment, was discovered in 1948 by Weiss, who observed that ligation of the sciatic nerve produced a gradual distension of axoplasm on the proximal side of the constricture, i.e., nearest to the cell body. He concluded that axoplasm flows from the cell body into the axon at a constant rate. This seminal

Table 3–3

Major Rate Components of Axonal Transport

Rate Component	Velocity (mm/day)	Properties and Composition
Fast axonal transport		Membrane-bound organelles
Fast anterograde	200–400	Tubulovesicular structures, synaptic vesicles, membrane-associated proteins, neuropeptides, neurotransmitters, and associated enzymes
Fast mitochondria	50–100	Mitochondria, associated enzymes and lipids
Fast retrograde	100–200	Prelysosomal vesicles, multivesicular bodies, multilamellar bodies, growth factors, recycled proteins from fast anterograde transport
Slow axonal transport		Cytoskeletal and cytoplasmic elements
Slow component b	2–6	Actin, clathrin and associated proteins, spectrin, glycolytic enzymes, and calmodulin
Slow component a	0.1–1.0	Neurofilaments, microtubules, and associated proteins

experiment has ultimately led to the recognition that axoplasmic flow consists of multiple kinetic components, defined by velocity, direction, and the constituents of transport (Table 3–3).

Fast Axonal Transport Is a Major Form of Motility for Secretory and Membranous Components

Macromolecules obligated for fast axonal transport move from the cell body toward the nerve terminal at an average rate of 250–400 mm/day and are exclusively membrane-associated. Constituents of the plasmalemma (proteins, cholesterol, phospholipids, gangliosides), synaptic vesicles (and their component neurotransmitters, neuropeptides, and enzymes), and mitochondria belong in this category, although the latter travel at a slower rate of 50 mm/day. A requirement for entry to the fast-transport system is the preassembly of individual molecules into membranous organelles or vesicles beforehand. Packaging occurs in the perinuclear cytoplasm as newly synthesized neuronal proteins destined for secretion and membrane incorporation translocate from their synthetic sites, on bound polyribosomes, to the Golgi apparatus by way of the ER for modification and packaging. Vesicular structures, ranging in diameter from 50 to 80 nm, bud from the trans Golgi network and are directly routed to the axon. Thus, neurons commit such proteins for their journey to the axon by trafficking them through essentially the same biosynthetic course common to nonneuronal cells.

(For a detailed discussion of these events, the reader is referred to Chapter 4.)

Videomicroscopy of the living axon has revealed that organelles, viewed as particles, travel by fast transport with a jerky, or saltatory, motion rather than by continuous flow. Furthermore, fast transport is bidirectional; particle movement occurs both anterogradely (i.e., from the cell body toward the axon) and retrogradely (return flow) (Fig. 3–36). Retrograde transport removes aged and excess material from synaptic terminals for degradation in lysosomes in the cell body. It is also the avenue by which environmental agents sampled by endocytosis at nerve endings gain access to neuronal cell bodies and the central nervous system. Target cell-secreted peptides, such as nerve growth factor, are able to reach their subcellar sites of actions by this retrieval mechanism. Moreover, clinically relevant agents, such as tetanus toxin, polio, rabies, and herpes simplex virus, also enter neurons by this process to exert their deleterious affects.

Microtubule and Microfilament Motor Proteins Propel Membrane Vesicles Along Axons

In both directions, moving particles track along a relatively immobile substrate of fibrils consisting of discontinuous microtubules traversing the length of the axon. Structures that resemble cross-bridges between microtubules and particles under the electron microscope are thought to provide the energy of propulsion (Fig. 3–37). These linkages are individual MAPs that have ATPase

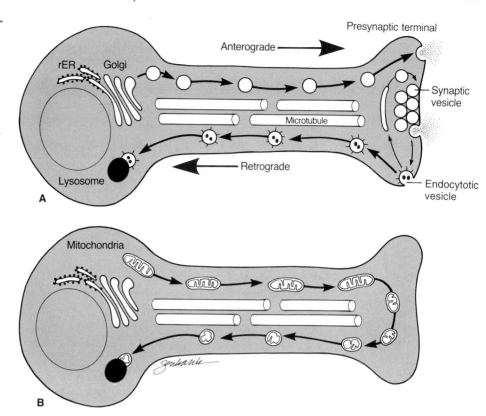

Figure 3–36. Pathways of fast axon transport. In this schematic diagram, (**A**) vesicular organelles and (**B**) mitochondria travel in a continuous cycle between their site of synthesis and assembly in the cell body and their destination in the axon terminal. Movement in either direction requires microtubules as guides. Anterogradely transported organelles include prepackaged synaptic vesicles and their neurotransmitters. Retrogradely transported structures include aged organelles (e.g., mitochondria) and endocytotic vesicles containing agents (viruses, toxins, growth factors) sampled from the extracellular fluid. Endocytotic vesicles are also recycled locally in the terminal into synaptic vesicles. (Modified from Lasek R, Katz M. *Prog Brain Res* 1987;71:49.)

activity. The two principal MAPs in axons are kinesin and cytoplasmic dynein. Kinesin specifically accommodates fast transport in the anterograde direction by moving vesicles selectively toward the positive end of microtubules. Kinesin is a flexible rodlike structure that contains a pair of globular head regions, incorporating both the ATP- and microtubule-binding sites, and a splayed tail region for organelle membrane attachment. Although the mechanochemical basis of how ATP hydrolysis drives motility remains poorly understood, video images of particle translocation *in vitro* predict that the cross-bridge and bound particles glide as a single complex along the microtubular track.

At velocities only slightly slower than that of fast anterograde transport, vesicular structures viewed within the same axon are also simultaneously returning from the terminus to the cell body. Retrogradely transported vesicles are larger and more variable in size (100–500 nm) than are anterogradely moving structures and require microtubules and a different ATP-generating system. Cytoplasmic dynein, related to the dynein motor proteins operating in cilia and flagella, functions similarly to kinesin but propels vesicles retrogradely, as it operates only toward the minus end of microtubules (Fig. 3–38).

Membrane vesicles approaching the axon terminal where microtubules are absent continue to move by following microfilaments. Vesicles in the terminal are no longer shuttled by kinesin or dynein but propelled instead by the microfilament motor protein, myosin I (Fig. 3–38).

Movement of the Cytosol Occurs by Slow Axonal Transport

Whereas substances destined for secretion and membrane incorporation travel axonally by fast transport, most of the cytosol moves by slow axonal flow. Movement of these components appears to be unidirectional and always away from the cell body. Furthermore, two velocities of slow transport can be distinguished by their respective protein constituents. Moving at the slower rate of about 1 mm/day are proteins consisting primarily of fibrillar elements in the cytoskeleton because these, too, must be slowly replenished and assembled at growing nerve endings. Included are α- and β-tubulin, the polypeptide subunits of microtubules, and the three subunit proteins of neurofilaments. These elements are transported as short polymers assembled in the cell body, then depolymerized for addition at the growing end of the axon. Collectively referred to as the *slow component a* (SCa), transported neurofilaments, microtubules, and associated proteins are thought to travel together as a preassembled network of cross-linked fibrils. Whereas only five major polypeptides constitute 70–80% of the protein traveling in SCa, as many as 200 minor species are conveyed by the faster component of slow transport, *slow component b* (SCb), at 2–4 mm/day. Included here are other cytoskeletal proteins, such as actin, spectrin II (spectrin is unusual, in that some also moves by fast transport in association with the intact mitochondrial membrane), the actin-binding

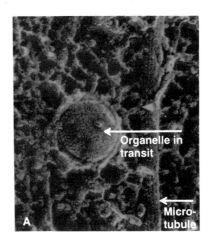

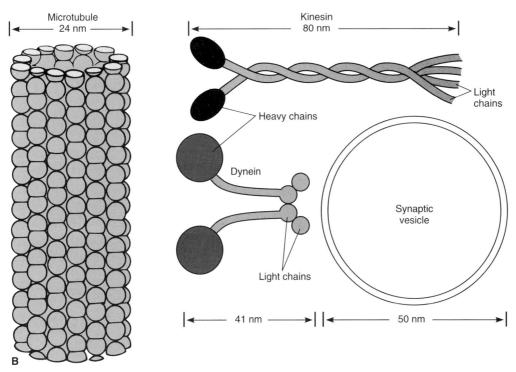

Figure 3–37. Microtubule motor proteins structures cross-link vesicles to micro-tubules in the axoplasm. **A:** Higher magnification (183,000x) of tissue examined as in Fig. 3–35 reveals rodshaped structures (*arrows*) linking rounded organelles to microtubules. **B:** Both proteins consist of two heavy and two light chains and are modeled in proportion to the diameter of a microtubule and synaptic vesicle. (**A**, Hirokawa N. *Cell* 1989,108:111; **B**, modified from Brady ST. *Neuron* 1991, 7:521.)

proteins, as well as clathrin and metabolic enzymes. These proteins are synthesized on free ribosomes and are collectively referred to as the *axoplasmic matrix*.

Dendrites Have a Distinct Transport System

Mechanisms that sort some proteins into axons and others into dendrites remain unclear. Differences in microtubule orientation and the discovery that differ-

ent MAPs compartmentalize in axons and dendrites implies that the cytoskeleton plays an important role in sorting. For example, MAP 2 is restricted to the cell body and dendrites, whereas another MAP, the tau protein, predominates in axons. Overaccumulation of a protein derivative of tau results in the formation of paired helical filaments, a neuropathological hallmark of many neurons affected by Alzheimer's disease. Neurons also possess an unknown mechanism for trans-

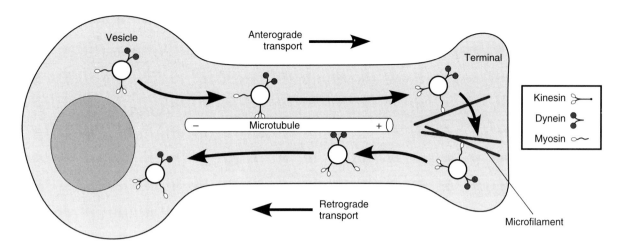

Figure 3–38. **Model for bidirectional axon transport mediated by motor proteins.** Membrane-bound vesicles packaged in the cell body attach to motor proteins kinesin, dynein, and myosin. Nearly all microtubules in the axon are polarized with their plus ends oriented toward the terminal. Kinesin, a plus end-directed microtubule motor, transports organelles anterogradely toward the terminal. Vesicles are able to traverse the microtubule-free axon terminal along actin microfilaments to which they are linked by myosin I, a microfilament-specific motor. Vesicles recycled from the axon terminal are transported retrogradely because the reverse polarity of the microtubule track activates dynein, a minus end-directed microtubule motor. Inactive dynein is transported on vesicles to the terminal, where it binds to aged organelles and endocytotic vesicles, priming them for transport to the cell body.

porting specific mRNAs (e.g., that for MAP 2) into dendrites, which, unlike axons, house ribosomes. Thus, for at least some proteins, it appears that the problems of dendritic transport are overcome by local synthesis in the periphery. The local translation of selective mRNAs in dendrites represents another way in which proteins become differentially distributed in neurons.

Neurofilaments Determine Axon Caliber

As explained in Chapter 2, the velocity of nerve impulses along axons is determined by the axonal diameter (caliber). Following the growth phase of neurite (axon and dendrite) elongation, a phase of radial growth ensues to increase the diameter of these processes as much as tenfold. Neurofilaments (NFs), representing only a minor component during elongation, become the predominant cytoskeletal elemental during radial growth and are believed to play a major role in regulating axonal caliber. The most abundant of the neuronal intermediate filament proteins, the three NF subunits, NF_L, NF_M and NF_H, heteropolymerize to form neurofilaments. Neurofilament proteins are distinct from the other classes of intermediate filament proteins because the COOH-terminal regions of two of them, specifically, NF_M and NF_H, form characteristic

sidearms that are heavily phosphorylated. Although a linear relationship exists between axon diameter and NF number, the mechanism leading to radial axonal growth is unclear but probably involves the regulation of NF phosphorylation and assembly.

Abnormal accumulation and assembly of NFs are a hallmark of the pathogenesis for specific types of neurodegenerative human motor neuron diseases, including amyotrophic later sclerosis (ALS). Studies of transgenic mice that overaccumulate NF subunits suggest that axonal swelling and degeneration are the pathologic consequences of abberrant NF aggregation and dysfunction in ALS.

Clinical Case Discussion

As discussed in this chapter, dystrophin, the defective gene product in Duchenne's muscular dystrophy, is a member of the spectrin supergene family. Duchenne's muscular dystrophy is an X-linked recessive condition that involves an abnormality in the dystrophin gene located on chromosome Xp21. Thirty to fifty percent of children have no family history of disease. There is also a milder type, Becker's muscular dystrophy, which is allelic. About 60–65% of families have a deletion in the dystrophin gene. Female carriers with the deletion can be identified. Dystrophin is a protein, localized to the cytoplasmic surface of the sarcolemma, that functions to anchor actin filaments to the plasma membrane in skeletal muscles. Males with low or undetectable levels of dystrophin have Duchenne's muscular dystrophy, whereas those with nearly normal levels but dystrophin of an abnormal size have Becker's muscular dystrophy.

The usual age of onset is 2–6 years old. Symptoms include clumsiness, easy fatigability, walking on toes, waddling gait, and lordosis. Patients usually have a positive Gower's maneuver—climbing up on legs when rising from a supine position. Typical findings on physical examination include pseudohypertrophy of the gastrocnemius muscle (90%), progressive weakness of the axial and proximal muscle groups that gradually involves the distal muscle groups. Reflexes are absent at the knees but present in the ankles. Muscle biopsy shows a characteristic deficiency of type 2B fibers and absent dystrophin in Duchenne's type. In Becker's type, there is type 2B fibers present and reduced dystrophin. The most common cause of death is pneumonia by age 20. Treatment of Duchenne's muscular dystrophy consist of prednisone (Deltasone, Prednicen, Sterapred) administration, which seems to delay the progression of the disease.

Suggested Readings

Microfilaments

Alberts B, Bray D, Lewis J, et al. *Molecular Biology of the Cell*, 2nd ed. New York: Garland Publishing, 1989.

Cooke R. The mechanism of muscle contraction. *CRC Crit Rev Biochem* 1986;21:53.

Pollard TD, Cooper JA. Actin and actin-binding proteins: a critical evaluation of mechanisms and functions. *Annu Rev Biochem* 1986;55:987.

The Spectrin Membrane Skeleton

Goodman SR, Krebs K, Whitfield C, et al. Spectrin and related molecules. *CRC Crit Rev Biochem* 1988;23:171.

Intermediate Filaments

Ip W, Hartzer MK, Pang SY, et al. Assembly of vimentin in vitro and its implications concerning the structure of intermediate filaments. *J Mol Biol* 1985;183:365.

Steinert PM, Roop DR. Molecular and cellular biology of intermediate filaments. *Annu Rev Biochem* 1988;57:593.

Microtubules

Johnson KA. Pathway of the microtubule dynein ATPase and structure of the dynein: a comparison with actomyosin. *Annu Rev Biophys Biophys Chem* 1985;14:161.

Olmsted JB. Microtubule-associated proteins. *Annu Rev Cell Biol* 1986;2:421.

Sullivan KF. Structure and utilization of tubulin isotypes. *Annu Rev Cell Biol* 1988;4:687.

The Neuronal Cytoskeleton

Brady ST. Molecular motors in the nervous system. *Neuron* 1991;7:521.

Lee MK, Cleveland DW. Neuronal intermediate filaments. *Annu Rev Neurosci* 1996;19:187–217.

Vallee RB, Bloom GS. Mechanisms of fast and slow transport. *Annu Rev Neurosci* 1992;14:59.

Review Questions

1. Regarding the neuronal cytoskeleton
   a. Dendrites lack microtubules
   b. Fast axon transport can be retrograde
   c. Axon diameter is determined primarily by neurofilaments
   d. Membrane vesicles moving toward the negative pole of axonal microtubules are transported anterogradely
   e. b and c
2. The characteristic finding which allows a definitive diagnosis of Duchenne's versus Becker's muscular dystrophy is:
   a. abnormal electromyocardiogram.
   b. X-linked recessive inheritance.
   c. absent dystrophin.
   d. elevated creatinine kinase levels.
   e. absent Gower's maneuver.
3. Increased Ca^{2+} levels in skeletal muscle regulates contraction via Ca^{2+} binding with
   a. Tubulin
   b. Tropomyosin
   c. α-skeletal actin
   d. Troponin C
   e. Myosin heavy chain
4. Intermediate filaments differ from microtubules and microfilaments by
   a. Their need of Mn^{2+} as a cofactor of polymerization
   b. Their ability to be made as heteropolymers
   c. Their need of calmodulin accessory protein for polymerization
   d. Their need of high cellular ATP concentrations for polymerization
   e. Their formation from monomers of normally globular subunits
5. During mitosis, some of the microtubules formed are stabilized by

a. Capping of the minus (−) ends by the cell membrane
b. Capping of the minus (−) ends by the kinetochore
c. Capping of the plus (+) ends by the cell membrane
d. Capping of the plus (+) ends by the kinetochore
e. The removal of the COOH-terminal tyrosine from the tubulin solvent

6. Vasodilation caused by epinephrine is most likely due to:
a. Decreasing the cAMP concentration within the vascular smooth muscle cell
b. Increasing the cAMP concentration within the vascular smooth muscle cell
c. Decreasing the Ca^{2+} within the vascular smooth muscle cell
d. Increasing the Ca^{2+} within the vascular smooth muscle cell
e. Increasing the activity of protein kinase C through diacylglycerol

7. Movement of organelles throughout a cell, based upon microtubules occurs via the actions of:
a. Actin and α-actinin
b. Kinesin an dynein
c. Myosin I and calmodulin
d. Uvomorulin and spectrin
e. Keratin and CAP Z

8. The red blood cell protein that cross-links actin filaments at the membrane is:
a. Spectrin I
b. Spectrin II
c. Fimbrin
d. Villin
e. Dystrophin

9. In nonmuscle cells, actin-myosin interactions are important for:
a. Separation of daughter cells in telophase
b. Flattening of fibroblast cells
c. Folding of epithelial cells into tubes
d. All of the above
e. None of the above

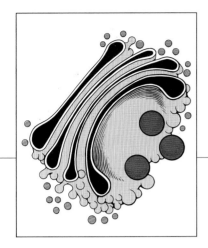

Chapter 4

Organelle Structure and Function

Clinical Case

A 35-year-old man, David Brown, decides to see his doctor for a routine annual physical. He has been in good health, gets plenty of exercise, and eats a well-balanced diet. Family history reveals that an uncle on his father's side died from a heart attack at 40 years of age. His father has an elevated cholesterol level, which is being controlled by dietary modifications. Physical examination was remarkable for a blood pressure of 150/100 mm Hg, a bruit over the femoral artery, and barely palpable popliteal and dorsal pedal pulses. The ankle:arm systolic pressure ratio was markedly decreased, suggesting arterial insufficiency in the lower limbs. Doppler studies confirmed decreased blood flow in the femoral artery. A complete chemistry profile demonstrated elevated total cholesterol (350 mg/dl) and low-density lipoprotein (LDL) cholesterol (200 mg/dl) levels, and low high-density lipoprotein (HDL) cholesterol (40 mg/dl). The remainder of his chemistries were normal. The patient was placed on a diet low in cholesterol and high in polyunsaturated fats.

The Cell Nucleus

The Nucleus Is Bounded by a Specialized Membrane Complex, the Nuclear Envelope

The sequestering of nearly all of the cellular DNA in the nucleus marks the major difference between eukaryotic and prokaryotic cells. Nuclei are generally spherical and are bounded by the nuclear envelope, which defines the nuclear compartment. The nuclear envelope is formed from two distinct lipid bilayers (Fig. 4–1). The inner nuclear membrane is in close contact with a meshwork of intermediate filaments, the nuclear lamina, which provides support for this lipid bilayer. Additionally, this membrane contains proteins that provide contact sites for chromosomes and nuclear ribonucleic acids (RNAs), either directly or through proteins of the nuclear matrix. The outer nuclear membrane is contiguous with the membrane of the endoplasmic reticulum (ER). Ribosomes that are actively synthesizing transmembrane proteins are often observed associated with the outer nuclear membrane. The outer nuclear membrane can be regarded as a specialized region of the ER. Proteins synthesized on ribosomes associated with the outer nuclear membrane are either destined for the inner or outer membranes or translocated across the membrane into the region between the inner and outer nuclear membranes, termed the *perinuclear space*.

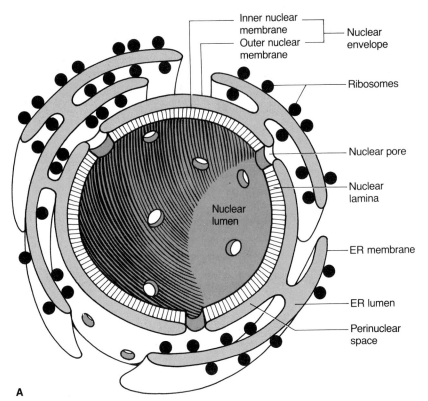

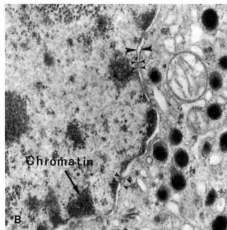

Figure 4–1. Illustration of the relation of the nuclear envelope with cellular structures. **A:** This diagram shows the double-membrane envelope that surrounds the nuclear compartment. The inner nuclear membrane is lined by the fibrous protein meshwork of the nuclear lamina. The outer nuclear membrane is contiguous with the membrane of the endoplasmic reticulum (ER). As illustrated in this diagram, the outer nuclear membrane often has ribosomes associated with it that are actively synthesizing proteins that first enter the region between the inner and outer nuclear membranes, the perinuclear space, which is contiguous with the lumen of the ER. The double membrane of the nuclear envelope is perforated with holes or channels of the nuclear pores. **B:** Electron micrograph of a nucleus from a luteal cell. *Large arrowheads* denote the inner and outer nuclear membranes of the nuclear envelope, which contains the nuclear pores (*small arrows*). (Courtesy of Dr. P. Fields.)

Within the confines of the nuclear envelope are the components of the genetic apparatus, which include deoxyribonucleic acid (DNA), RNA, and nuclear proteins. Figure 4–2 demonstrates the organization of components within a typical interphase nucleus. The varied proteins in the nucleus aid in the performance of nuclear functions. These proteins, which include structural proteins of the matrix and lamina, DNA and RNA polymerases, RNA-processing proteins, histones, and gene regulatory proteins, are synthesized in the cytoplasm on free ribosomes and are then brought into the nuclear compartment, where they perform their functions. This means that when proteins destined for the nuclear lumen are synthesized, they must pass the double-membrane barrier of the nuclear envelope to perform their task. This transport of materials to and from the nucleus is facilitated by structures that create holes

in the nuclear envelope, called the *nuclear pores*. However, nuclear transport is highly selective in that nearly all of the proteins made in the cytoplasm are excluded from the nuclear compartment, and only certain molecules are passed from the nucleus to the cytosol.

Nuclear Pores Allow Communication Between the Nucleus and Cytosol

The interior of the nucleus and the cytoplasm of the cell maintain contact or communication through the nuclear pores. In electron micrographs, the pores appear as highly organized disklike structures surrounding a central hole or cavity (Fig. 4–3). Nuclear pore structure is conferred by a set of protein granule subunits, which are arranged in an octagonal pattern and form the boundaries of the pore complex. These

Figure 4–2. Schematic model of a typical eukaryotic interphase nucleus. The organization of the internal nuclear compartment is illustrated. The inner nuclear membrane is in direct contact with the protein network of the nuclear lamina, which is often associated with a highly condensed DNA protein complex, referred to as the perinucleolar heterochromatin. Most of the nuclear compartment is filled with noncondensed DNA–protein complexes; the euchromatin. The most obvious structure within an interphase nucleus is the dark-staining nucleolus, which can be further divided into fibrillar and granular components.

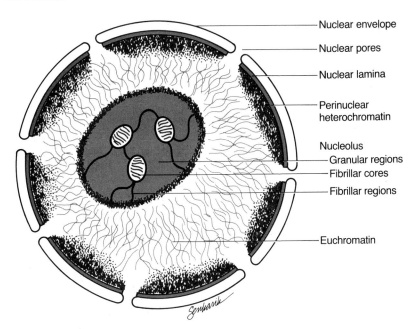

Nuclear envelope
Nuclear pores
Nuclear lamina
Perinuclear heterochromatin
Nucleolus
Granular regions
Fibrillar cores
Fibrillar regions
Euchromatin

eight protein subunits consist of radial arm segments joined together by a set of proteins referred to as *spokes*, which traverse the pore membrane. Sophisticated electron microscopy [field emission in-lens scanning election microscopy (FEISEM)] coupled with computer-enhanced three-dimensional (3-D) reconstruction of isolated nuclear pore complexes has indicated the presence of a transporter subunit(s) that line the central channel of the pore and, as detailed below, form the barrier between the nucleus and cytoplasm. On the cytoplasmic and nucleoplasmic faces of the outer and inner nuclear membranes are ring structures formed from eight bipartite subunits that are in apparent contact with the spoke proteins and the membrane phospholipids. These are important to the nuclear pore complex, not only for overall structure, but because they anchor filaments that extend into the cytoplasm and nuclear compartments. These nuclear pore ring attached filaments allow for direct coupling of the nuclear compartment with the cell cytoplasm through interaction with cytoskeletal and nucleoskeletal filaments. In addition, the nuclear pore ring filaments appear to participate in the recognition of molecules that need to be transported through the nuclear pore. Thus, the nuclear pore contains an intricate composition of proteins that form this important structure (Fig. 4–3). The pore complex penetrates the double membrane of the nuclear envelope, bringing together the lipid bilayers of the inner and outer nuclear membranes at the boundaries of each pore. Although this would seem to allow exchange of components (i.e., proteins, phospholipids, etc.) between these two membranes, evidence now indicates that these two membranes remain chemically distinct. Therefore, the protein components

of the nuclear pore complex must provide a barrier preventing bulk exchange between these two membranes.

Measurements of nuclear pore complexes have demonstrated that they are highly organized structures with an outside diameter of approximately 100 nm and an internal channel of 9–10 nm in diameter. The properties of transport through the nuclear pores have been addressed by injection of radiolabeled compounds into the cytosol and examination of the rate of their appearance into the nucleus. Such experiments have demonstrated that the nuclear pores are freely permeable to ions and small molecules, including proteins with a diameter smaller than 9 nm [≈60 kDa or less relative molecular mass (M_r)]. Nonnuclear proteins larger than 9 nm in diameter (greater than 60 kDa) are excluded from nuclear transit. However, nuclear resident proteins that are synthesized in the cytoplasm and are larger than 60 kDa are readily transported into the nucleus, indicating that there must be mechanisms for the selective transport of molecules across the nuclear envelope.

Recent evidence has demonstrated that selective transport of large molecules and complexes across the nuclear envelope occurs through the nuclear pore by a receptor-mediated process. The key experiments illustrating this concept have made use of the protein, nucleoplasmin, a 165-kDa M_r pentameric protein found in high concentrations in frog oocyte nuclei (Fig. 4–4). When purified nucleoplasmin is injected into the cytosol, it accumulates into the nucleus at a rate greater than can be explained by simple diffusion, suggesting that the protein was concentrated into the nucleus by a selective uptake mechanism. To determine how the nucleoplasmin was transported into the nucleus, small gold particles were coated with nucleoplasmin and,

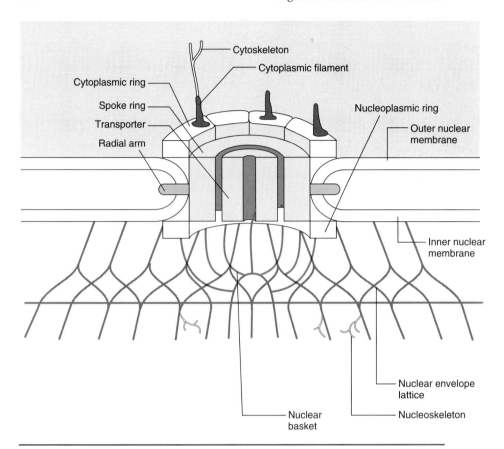

Figure 4–3. Diagram of nuclear pore complex that allows communication of the cell's cytoplasm with the internal nuclear compartment. A diagram showing the proteins which make up the ~100-nm, octagon disk-shaped nuclear pore complex is illustrated. The pore complex is anchored into the nuclear envelope (ONM, outer nuclear membrane; INM, inner nuclear membrane) by the radial arms (RA) and spoke rings (SR). Subunits that make up the transporter (T) are just inside the spoke ring and form the aqueous channel of the nuclear pore complex. Adjacent to these structures are the cytoplasmic (CR) and nucleoplasmic (NR) rings that anchor the connections of cytoplasmic filaments (CF) and nucleoplasmic filaments (Basket). It is through these filament structures that potential connections to the cytoskeleton (*outside*) and nuclear matrix (*inside*) are made, adding a physical connection between the nuclear compartment and the rest of the cell. (Adapted from Goldberg, M and Allen, T. *Current Opinion in Cell Biology* 1995;7:301–307.)

after microinjection of this complex into the cytosol, its location in the cell was monitored by electron microscopy. These experiments showed that shortly after injection, the nucleoplasmin gold complex was found at and within the nuclear pores and, at later times, the complex was accumulated in the nucleus. Thus, the nuclear pore complex participates in the selective transport of material across the nuclear envelope. The specificity of this uptake was demonstrated using nucleoplasmin that had been altered by proteolytic cleavage. Removal of the COOH-terminal regions of the protein yielded a pentameric core that was unable to enter the nucleus. Additionally, the unattached COOH-terminal tail segments of the protein accumulated into nuclei when injected into the cytosol. The same results were obtained when nuclei isolated from cytoplasmic

material were incubated in a solution containing intact nucleoplasmin or its COOH-tail fragments, but only if adenosine triphosphate (ATP) was also present in the incubation mixture (see Fig. 4–4). In the absence of ATP in the incubation medium, the COOH-tail fragments or intact nucleoplasmin would bind to the nuclear pore complex but would not enter the nucleus. When ATP was subsequently added, the bound proteins were transported into the nuclear lumen. These studies conclusively demonstrate that the COOH-tail segment of nucleoplasmin contains a signal sequence for binding a receptor in the nuclear pore complex and that transport of the protein into the nuclear compartment through the pore requires the energy released from ATP hydrolysis. The mechanism of nuclear localization signal mediated protein import appears to encompass a variety of pro-

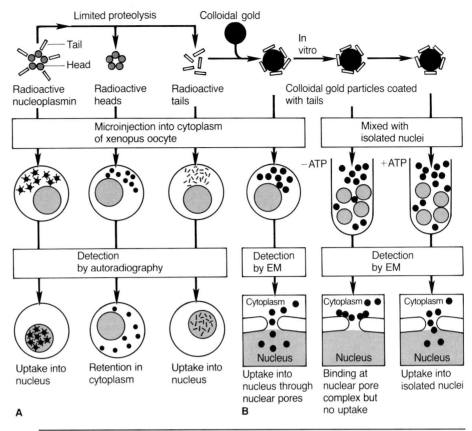

Figure 4–4. Scheme of experiments demonstrating the selective import of proteins into the nucleus. **A:** Microinjection of the pentameric nucleoplasmic protein into xenopus oocytes results in rapid concentration of the protein into the nucleus. However, separation of the tails from the core or head domains shows that only the tail fragments will concentrate into the nucleus, as the head domains remain dispersed in the xenopus oocyte cytoplasm. The participation of the nuclear pores in the selective uptake of proteins is demonstrated by electron microscopy of oocytes into which gold particles, by themselves too large to traverse the pores, that had been coated with the nucleoplasmin tail fragments, had been microinjected. **B:** When gold particles, coated with nucleoplasmin tail fragments (or peptides containing SV40 t-antigen nuclear import signal sequences), are mixed with isolated nuclei, the gold–protein complex is located inside the nuclei only when ATP is present in the incubation media. Without ATP, the gold–protein complex is found associated with the nuclear envelope; suggesting an energy-dependent, receptor-mediated mechanism for nuclear import. (**A,** Feldeherr C, Kallenbach E, Schultz N. *J Cell Biol* 1984;99: 2216–2222; **B,** Newmeyer DD, Forbes DJ. *Cell* 1988;52:641–653.)

teins, both soluble in the cytoplasm and located on the nuclear pore complex, that work in a multistep pathway (Fig. 4–5). First, a protein that has a nuclear transport signal, or nuclear localization sequence (NLS), binds a receptor complex. This is a multisubunit receptor, soluble within the cell cytoplasm, which functions to dock the protein to be transported with filaments extending from the cytoplasmic ring of the nuclear pore complex. Following the docking, more proteins associate with the complex; most importantly are GTPases and their activating proteins, which provide the energy of the protein translocation by the hydrolysis of guanosine triphosphate (GTP). Not all of the proteins that form the

translocation complex go through the pore to the nucleoplasm, and those factors that do are apparently recycled, by some mechanism, to the cell cytoplasm for further rounds of nuclear transport. A model for nuclear import, based on current data, is shown in Figure 4–5. Although many factors that participate in this process have been examined, there may be more that have not yet been identified.

Nuclear transport signal sequences have been identified from a variety of nuclear targeted proteins. Listed in Table 4–1 are nuclear resident proteins for which an import signal sequence has been identified. Each sequence yet identified is a small region of the total pro-

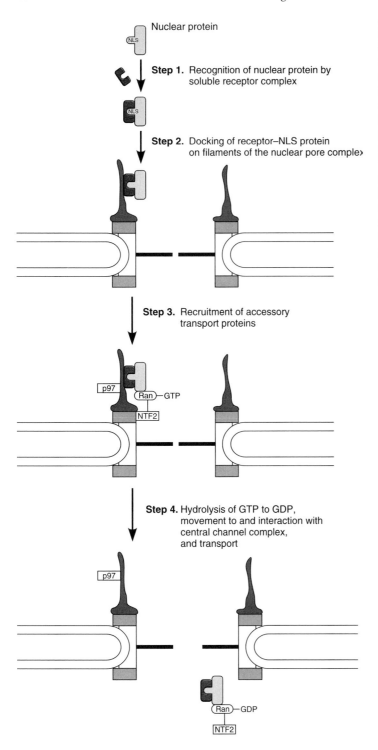

Nuclear protein

Step 1. Recognition of nuclear protein by soluble receptor complex

Step 2. Docking of receptor–NLS protein on filaments of the nuclear pore complex

Step 3. Recruitment of accessory transport proteins

p97
Ran—GTP
NTF2

Step 4. Hydrolysis of GTP to GDP, movement to and interaction with central channel complex, and transport

p97

Ran—GDP
NTF2

Figure 4–5. **Diagram of the steps necessary for protein import through the nuclear pore complex.** The protein to be transported to the nucleus is demonstrating its nuclear localization sequence (NLS), which is recognized by the nuclear transport receptor, a soluble protein complex within the cytosol (*Step 1*). This binding/recognition leads to docking of the complex upon the cytoplasmic filaments emanating from the cytoplasmin ring of the nuclear pore complex (*Step 2*). This docking signals the need for accessory proteins, specifically, p97, an apparent structural component, NTF2 to promote interactions between the transported complex and subunits of the central transporter channel, and Ran, a GTPase that catalyzes the hydrolysis of GTP to provide the energy needed for transport (*Step 3*). The hydrolysis of GTP to GDP allows for movement of the complex through the control channel to the nucleoplasm (*Step 4*).

tein (~4–8 amino acids in length), and most are basic; however, there is no apparent consensus of sequences, either in primary structure or in any location within the different proteins. Therefore, it is not known whether a single receptor is able to bind different amino acid signal sequences, or whether there are multiple receptor sites within the nuclear pore complex that bind with the individual signals. The same nuclear pore complexes are responsible for the transport of RNAs out of the

nucleus to the cytosol. Transport out of the nucleus is selective and appears to require a receptor in the nuclear pore complex that binds with the protein-RNA complex. Thus, it would seem likely that the nuclear pore complex contains multiple receptor sites that recognize a variety of signal sequences on protein complexes to be transported across the nuclear envelope. Whether this is a property of a single or multiple subunit protein within the pore complex remains to be elucidated.

Table 4–1
Nuclear Import Signal Sequences Derived from Various Nuclear Resident Proteins

Protein	Location of Signal Sequence	Amino Acids of Signal Sequence
SV40-large T antigen	Internal: residues 126–132	Pro-Lys-Lys-Lys-Arg-Lys-Val
Influenza virus nucleoprotein	COOH-terminus: residues 336–345	Ala-Ala-Phe-Glu-Asp-Leu-Arg-Val-Leu-Ser
Yeast mat α 2	NH$_2$-terminus: residues 3–7	Lys-Ile-Pro-Ile-Lys
Yeast ribosomal protein L3	NH$_2$-terminus: residues 18–24	Pro-Arg-Lys-Arg

Summarized from Dingwall C, Laskey R. *Annu Rev Cell Biol* 1986;2:367–390; Lyons RH, Ferguson B, Rosenberg M. *Mol Cell Biol* 1987;7:2451–2456; Newmeyer DD, Forbes DJ. *Cell* 1988;52:641–653.

In summary, the nuclear pore is an important channel of communication between the interior of the nuclear compartment and the cytoplasm of the cell. It displays properties of a molecular filter, in that ions and small molecules are freely permeable through the aqueous channel, whereas larger protein complexes are selectively transported through the pore. This selective transport is highly specific, owing to the presence of amino acid signal sequences within the protein complexes that cause them to bind with receptor-like proteins on the pore complexes. Then an energy-dependent process is responsible for the translocation across the nuclear envelope. The exact mechanism of transport through the pore channel and the components of the pore complex responsible for the translocation are unknown.

The Structure of the Nucleus Is Determined by Proteins of the Nuclear Lamina and the Nuclear Matrix

Lining the inner surface of the nuclear envelope in interphase cells is a protein meshwork, the nuclear lamina. This meshwork forms an electron-dense layer 30–100 nm thick that provides connections between the inner nuclear membrane and perinuclear chromatin. When examined by electron microscopy, the lamina appears as a square latticework built from filaments that are about 10 nm in diameter. These filaments are classified as intermediate filaments and are composed of three extrinsic membrane proteins, called *lamins A, B,* and *C,* which have M_r of 60–70 kDa. The mRNAs for lamins A and C are formed from alternately spliced transcripts of the same gene and encode identical proteins, except that lamin A contains a COOH-terminal extension of 133 amino acids. Lamin B is encoded from an mRNA that is synthesized by a gene distinct from the lamin A/C gene.

All three of the nuclear lamins are structurally similar to the intermediate filament (IF) class of proteins. Isolated lamins have a rodlike structure of approximately 52 nm in length and a globular head domain. Similar to other IF proteins, the formation of the long filaments is mediated by the globular head domain, although the exact interaction is unknown. Lamin B is different from lamins A and C, in that it contains membrane-binding sequences. The lamin B protein is post-translationally modified by the addition of an isoprenyl group, which allows membrane lipid attachment. The inner nuclear membrane contains a receptor molecule of about 58,000 M_r that binds specifically to lamin B. Lamins A and C then bind with the membrane-linked lamin B to mediate interactions with the lamina and chromatin. Therefore, in interphase cells, all three lamin proteins are found adjacent to the inner nuclear membrane, forming the nuclear lamina complex.

As the chromatin condenses in the prophase stage of mitosis, there is an apparent disappearance of the nuclear membrane and the nuclear lamina. Figure 4–6 summarizes a model to explain the participation of the nuclear lamins in the breakdown and reformation of the nuclear envelope during the cell cycle. Examination of cells by electron microscopy shows that, during prophase, the nuclear membrane fragments into smaller vesicles that remain associated with the ER. Lamin B is found tightly coupled with these vesicles, whereas lamins A and C are depolymerized and are dispersed throughout the cell. This depolymerization and subsequent breakdown of nuclear membrane is thought to be mediated by phosphorylation of the lamins by a lamin kinase, the cdk1 kinase discussed in Chapter 9. When the lamins are phosphorylated, they depolymerize, and the nuclear membrane breaks down into small vesicles, whereas the chromosomes condense. During telophase of the cell cycle, the nuclear membrane and associated structures reassemble around the separated daughter

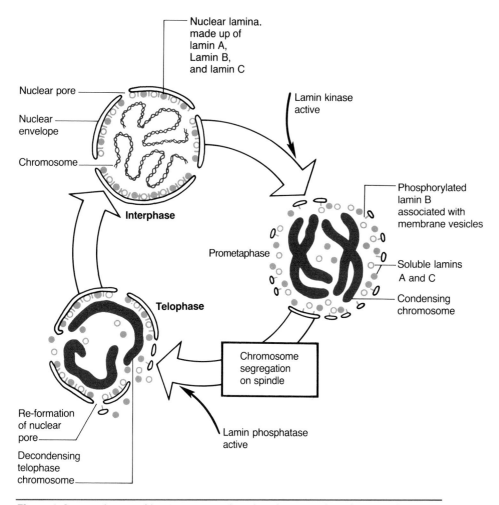

Figure 4–6. Correlation of lamin protein phosphorylation and nuclear envelope structure during mitosis. As cells proceed from interphase to prophase of the cell cycle, there is a condensation of the chromosomes and a breakdown of the nuclear envelope. Nuclear envelope breakdown is concomitant with the activation of lamin kinase, which phosphorylates the lamin proteins (**A, B,** and **C**) and causes the depolymerization of the nuclear lamina matrix. Lamin B remains associated with remnant membrane vesicles when phosphorylated, whereas lamins A and C are dispersed within the prometaphase cell. Coincident with the decondensation of chromosomes in daughter cells (*telophase*), the phosphates are removed from the lamin proteins by an activated lamin phosphatase. This allows the polymerization of the nuclear lamina, using the membrane-bound lamin B as a nucleation site and the formation of a nuclear envelope.

chromosomes. This reassembly of the nuclear envelope appears to be mediated by the lamins and is coincident with the removal of phosphates from these proteins. The reformation of the nuclear membrane closely follows the decondensation of the daughter chromosomes. As the chromatin becomes dispersed, it apparently induces the dephosphorylation of the lamins, allowing them to polymerize, which, in turn, causes the small vesicles associated with lamin B to fuse and form a normal interphase nuclear membrane. Although the interactions that lead to reassembly of the nuclear membrane are not yet clearly defined, it is thought that the phosphatase responsible for removal of phosphates from the

lamins is tightly associated with the chromatin, possibly a component of the internal nuclear structure called the *nuclear matrix.*

When isolated nuclei are subjected to extraction in high-ionic strength, neutral-detergent-containing buffers to remove most of the internal components, a fibrous network of proteins, the nuclear matrix, remains intact and roughly maintains the outward appearance of the nucleus. The *nuclear matrix* is defined biochemically as a structure containing 10% of the total nuclear protein, 30% of the nuclear RNA, 1–3% of the total DNA, and 3% of the nuclear phospholipid. Electron microscopy of nuclear matrix preparations show that they comprise mainly fibrillar elements that remain associated with identified nuclear structures, such as the nuclear pore complex and lamina.

Although defined structurally and biochemically, the function of the nuclear matrix is unclear. Perhaps its most obvious role would be to provide organization and structure to the internal nuclear compartment. Newly replicated DNA and the enzymatic components necessary for DNA synthesis are associated with the matrix, suggesting a role in the organization of the DNA replication machinery. Recent evidence indicates that genes that are being transcribed, and the products of their transcription [e.g., heterogeneous nuclear RNA (hnRNA)] are enriched in nuclear matrix preparations. Localization of RNA transcripts in the nucleus using fluorescent-labeled nucleic acid probes has shown that the RNAs follow tracks in the nuclear compartment, with a more intense fluorescent signal seen near the nuclear borders. Thus, following transcription, RNAs do not diffuse within the nucleoplasm, but are possibly bound to nuclear matrix fibers as they are spliced to form mature mRNAs bound for the cytoplasm. These experiments support the concept that the nuclear matrix plays an important role in the organization of the nuclear compartment. However, the nature of the individual nuclear matrix components and their interactions remain to be elucidated.

The Genome of the Cell Is Sequestered in the Nuclear Compartment

The nucleus contains almost all of the genetic information of the cell in the form of DNA. DNA is composed of four nucleotides: two are purines that have a double-ring structure (adenine and guanine), and two are pyrimidines that have a single-ring structure (thymine and cytosine). The basic structure of DNA, derived in 1953 by Watson and Crick, is that of two polynucleotide chains that are held together by hydrogen bonds between adenine and thymine (A-T base pairing) and guanosine and cytosine (G-C base pairing). The two chains are antiparallel or complementary and are coiled into a double helix of approximately 2 nm in diameter. The nucleotides are arranged in a nonrandom fashion, such that the genetic information is contained in the specific linear arrangement of bases. The information is stored in "words" consisting of three nucleotides, termed the *codons*, of the genetic code. A model of DNA as genetic information and its relation to chromosomes stored in a eukaryotic nucleus is presented in Fig. 4–7.

Figure 4–7. Relation of molecular details of the genetic code stored in DNA and chromosomes within the nucleus of the cell. Shown at the right of this diagram is a model of DNA in which two antiparallel strands (one is 5′ to 3′ top to bottom and the other is opposite) are held together by pairing of nucleotide bases. It is the strict arrangements of the nucleotides, in codons of three bases, that are the stores of genetic information. As indicated in this drawing, the 2-nm helix of DNA is associated with protein, forming the chromatin fibers of individual chromosomes that are housed in the nucleus of eukaryotic cells.

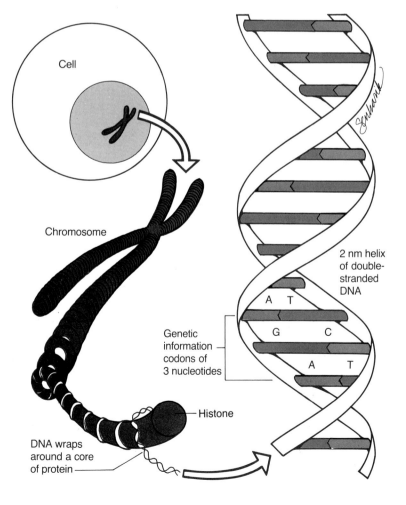

In eukaryotic cells, each DNA molecule is packaged into linearly arranged units, termed the *chromosome*, and the total genetic information stored within the chromosomes is referred to as the *genome* of the organism. The human genome contains about 3×10^9 nucleotide pairs that are packaged into 24 separate chromosomes (22 autosomes and 2 different sex-determinant chromosomes). In diploid somatic cells, there are two copies of each chromosome present; one inherited from the mother and one from the father, except for the sex chromosomes in males, in which the Y chromosome is from the father and an X chromosome from the mother. Thus, a diploid human cell contains 46 chromosomes and approximately 6×10^9 nucleotide pairs of DNA. For these chromosomes to remain as discrete functional units, they must possess the ability to replicate, separate into daughter cells at mitosis, and maintain their integrity between cell generations. Experiments examining chromosomal architecture and function have defined three domains or elements necessary for the maintenance and propagation of individual chromosomal units (Fig. 4–8).

To replicate, the DNA contains specific regions that function as the focal point for the initiation of DNA synthesis, termed the *DNA replication origin*. Each chromosome contains many replication origins dispersed throughout its length, which become activated in an asynchronous fashion during the S phase of the cell cycle. These are specific nucleotide sequences at which DNA synthesis begins, although not all of the origins are active at the same time. This suggests that there must be heterogeneity among these sequences, allowing the ordered replication of the chromosome.

A second sequence element that is responsible for attachment of the chromosome to the mitotic spindle during M phase of the cell cycle is termed the *centromere*. Each chromosome contains one centromere region in which the DNA interacts with a complex set of proteins, forming a structure called the *kinetochore*. This structure is responsible for the segregation of the chromosome into daughter cells on cell division.

The third sequence element that is required for maintenance of chromosomal structure is located at termini of the linear chromosome and is called the *telomere*. This is a specialized sequence that defines the ends of the linear chromosome and is built from sequence repeats enriched in guanosine and cytosine bases. These sequences are replicated by telomerase, which folds the guanosine-rich DNA strand to form a special structure that protects the ends of the chromosome.

The DNA within the cell nucleus is associated with a variety of nuclear proteins, and the DNA protein complex is referred to as *chromatin*. Proteins associated with the DNA can be divided into two general categories: the histone and the nonhistone chromosomal proteins. Nonhistone proteins are a heterogeneous class

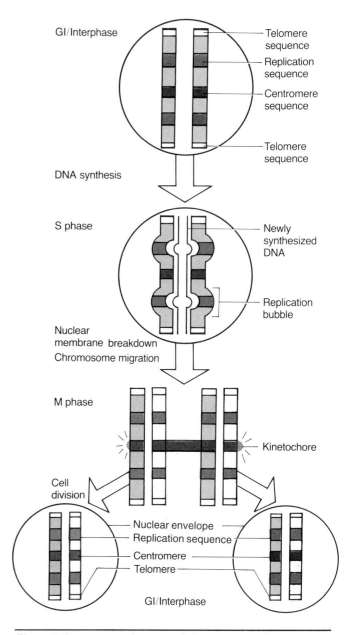

Figure 4–8. Sequence elements of chromosomes that are necessary for maintenance of structure and propagation. There are three sequence elements that are needed to maintain chromosomes as individual units in nuclei of eukaryotic cells. First is the telomere sequence, which caps the ends of the chromosomes, keeping degradative enzymes from attacking the units during interphase or G phase of the cell cycle. For duplication, chromosomes contain numerous origins of replication that serve as points of initiation for DNA synthesis during the synthetic, or S, phase of the cell cycle. Following S phase, the nuclear envelope breaks down and the chromosomes are segregated into the daughter cells by use of the kinetochore, which forms at the region of chromosomal constriction called the centromere DNA element.

of polypeptides that includes structural proteins [the high-mobility group of proteins (HMG)]; regulatory proteins (those that appear to have a direct role in gene regulation, e.g., *Fos, Myc*); and enzymes needed for nuclear function (RNA polymerases, DNA polymerases). The histones are found only in eukaryotic cells and are by far the most abundant proteins present in the nucleus. Histones are relatively small proteins that are rich in positively charged amino acids (arginine and lysine), which gives them an overall strong positive charge (basic) that enables them to bind tightly with the negatively charged (acidic) DNA molecules.

There are five types of histones, designated *H1, H2A, H2B, H3,* and *H4.* Four of the histones, H2A, H2B, H3, and H4, are termed the *nucleosomal histones,* as they are responsible for formation of the inner core of a DNA-protein complex called the *nucleosome.* The nucleosome is the basic unit of chromatin fiber and gives chromatin the beads-on-a-string appearance in electron micrographs. Examination of the structure of histone-DNA chromatin complexes has relied on digestion of the chromatin with nonspecific nucleases (Fig.

4–9). These studies have shown that the basic structure of chromatin can be resolved into a repeating unit, called the *nucleosomal bead.* Each nucleosome bead is formed from an octamer of proteins containing two copies each of the H2A, H2B, H3, and H4 histones, around which is wrapped about 150 nucleotide pairs of DNA. This is the amount of DNA that will make two complete turns around the octamer core of nucleosomal histones, forming a chromatin fiber that is approximately 11 nm in diameter.

Because it contains the simplest arrangement of DNA and protein, the 11-nm chromatin fiber can be considered the basic unit of chromatin packaging in the nucleus. However, only a small portion of the DNA is found packaged as an 11-nm fiber in an interphase cell and is probably limited to those regions of DNA that are actively transcribing gene sequences. When nuclei are treated very gently and examined by electron microscopy, most of the chromatin is found in a fiber that measures 30 nm in diameter. This 30-nm chromatin fiber is thought to represent the packaging of the nucleosomes by the remaining histone, H1. One model that

Figure 4–9. Outline of experiment examining chromatin repeating structure. The digestion of chromatin with nonspecific nucleases, such as DNase I, results in the release of repetitive units termed the *nucleosome.* This represents the "beads" of the chromatin fiber. Analysis of the components of the nucleosome demonstrates that the bead is made from DNA, a repeating size of 146 base pairs, wrapped around a core of protein. The protein core is made from two molecules each of the core or nucleosomal histones: H2A, H2B, H3, and H4.

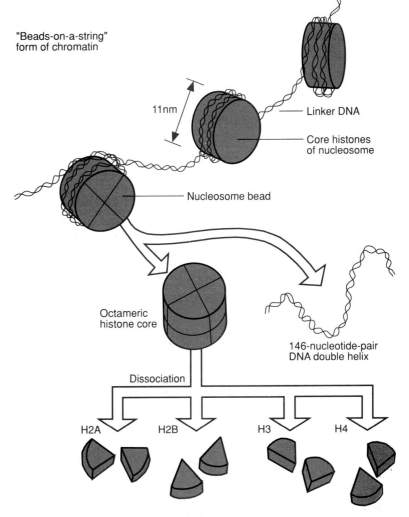

"Beads-on-a-string" form of chromatin

11nm

Linker DNA

Core histones of nucleosome

Nucleosome bead

Octameric histone core

146-nucleotide-pair DNA double helix

Dissociation

H2A H2B H3 H4

accounts for the formation of the 30-nm chromatin fiber is the cooperative binding of H1 molecules to nucleosomal DNA. Each histone H1 molecule binds through the central region of the molecule to a unique site on the nucleosome and extends to contact sites on adjacent nucleosomes (Fig. 4–10). This cooperative binding would compact the nucleosomes such that they are pulled together into regular-repeating arrays, forming the 30-nm chromatin fiber.

Chromatin inside a eukaryotic cell nucleus in interphase has been divided into two classes, based on its state of condensation. Chromatin that is highly condensed and considered to be transcriptionally inactive is referred to as *heterochromatin*. In electron micrographs of interphase nuclei, the heterochromatin is generally concentrated in a band around the periphery of the nucleus and around the nucleolus. The amount of heterochromatin present in the nucleus is correlated with the transcriptional activity of the cell. That is, little heterochromatin is present in transcriptionally active cells, whereas nuclei of mature spermatozoa, a transcriptionally inactive cell, contains predominantly highly condensed chromatin. In a typical eukaryotic cell, about 90% of the chromatin is thought to be transcriptionally inactive. This amount of inactive chromatin is much more than can be accounted for as the highly condensed heterochromatin. Therefore, heterochromatin is thought to be a special class of inert chromatin that may have

specialized functions. For example, the DNA near the centromere region is composed of repetitive DNA, and these sequences appear to constitute a major portion of the heterochromatin DNA.

The remaining 10% of chromatin that is transcriptionally active is found in a more extended dispersed conformation and is called *euchromatin*. Euchromatin is responsible for providing the RNA molecules that exit the nucleus and encode the proteins of the particular cell type.

Chromosomes are visible as distinct units in the light microscope when the chromatin is extensively condensed at mitosis. As a 30-nm fiber, the chromatin could not account for the degree of DNA condensation in metaphase chromosomes. Consequently, higher-order packaging units are required to achieve this state. From studies examining the appearance of specialized chromosomes, such as the lampbrush chromosome found in frog oocytes, it is thought that regions of the chromosome are present as extended loops of the 30-nm chromatin fiber held together at base of the loop by a specific protein-DNA complex. The model of chromatin condensation presented in Figure 4–11 shows that to account for the size of the typical human chromosome (~1.4 μm) in its most condensed state, the extended loop structures of the chromatin must be condensed again, possibly by drawing in the loop domains to form a tightly wound helical formation. Thus, to

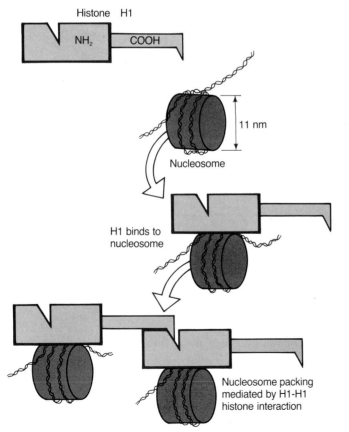

Figure 4–10. Model for histone H1 packaging of chromatin into a 30-nm fiber. A histone H1 molecule contains two distinct domains: a globular NH_2-domain and a COOH-terminal "arm" segment. In the presence of nucleosomes, the H1 molecule binds to a specific region of the nucleosome through its globular domain with the COOH-arm segment able to reach out to subsequent H1-containing nucleosomes. The COOH-terminal domain of histone H1 is then able to interact with specific sites on the adjoining H1 nucleosome by cooperative H1–H1 protein interactions.

Figure 4–11. Model of chromatin condensation needed to achieve the packaging observed of metaphase chromosomes. The first-order packaging involves the formation of the 11-nm chromatin fiber by association of the DNA helix with the core nucleosome proteins. To form the 30-nm fiber, there is the cooperative binding of histone H1, molecules pulling the nucleosomes into close apposition. The 30-nm fiber is representative of a looped section of the chromosome, and this folding results in a tenfold packing unit of ≈300 nm. The looped domains are thought to be arranged in a secondary loop, folding the chromatin into a 700-nm structure; however, the interactions that result in this packaging are not well defined.

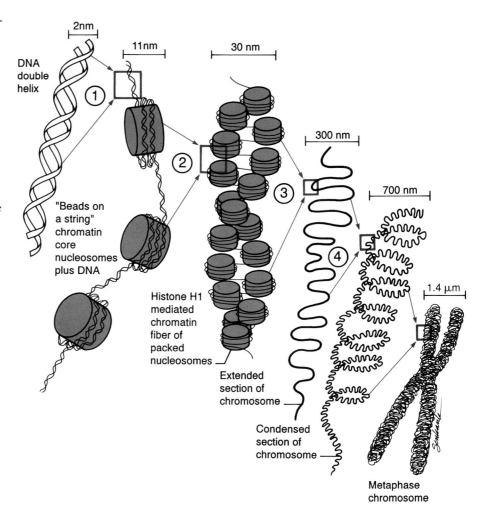

achieve the compaction necessary to fit the ~10^7 base pairs of DNA in the individual chromosomes into the ~1.4-µm chromosome seen at metaphase, there must be at least four orders of packaging of the DNA above the 2-nm double-helical chain of the DNA molecule (see Fig. 4–11).

The human genome contains 46 chromosomes in each diploid cell; one pair of the sex-determining chromosomes and 22 pairs of autosomal chromosomes. Cytologic methods, involving staining of fully condensed metaphase chromosomes with various stains or dyes, have been useful for the identification of individual chromosomes. For example, staining metaphase chromosomes with the Giemsa reagent results in a very characteristic pattern of bands on each chromosomal unit, termed *G banding*. Once the chromosomes have been stained, they can be examined under the microscope. The display of the chromosomes prepared in such a manner is referred to as the *karyotype* of the organism. Figure 4–12 shows an example of karyotype analysis, which is a Giemsa-staining pattern of metaphase chromosomes of a normal human female (46,XX karyotype). With the aid of these methods, it has been possible to correlate a variety of human syndromes with abnormalities in chromosome number. An example of Down's syndrome, which is characterized by the presence of an additional chromosome 21, is shown in Fig. 4-12. Other abnormalities that result in the loss or movement of a particular region of an individual chromosome in the genome (e.g., *cri du chat* syndrome, which results from the loss of a portion of the small arm of chromosome 5) can also be identified by these technologies. Thus, karyotype examination provides a powerful tool for the recognition of chromosomal abnormalities associated with particular genetic diseases and is a particularly useful technique for prenatal diagnosis of such disorders.

A Dense Nuclear Organelle That Specializes in the Formation of Ribosomal RNA Is the Nucleolus

When interphase cells are examined under the microscope, the most prominent feature observed in the nucleus is a dense structure, termed the *nucleolus*. The primary function of the nucleolus is the production and assembly of ribosomal subunits. In electron micrographs, four distinct regions of the nucleolus can be

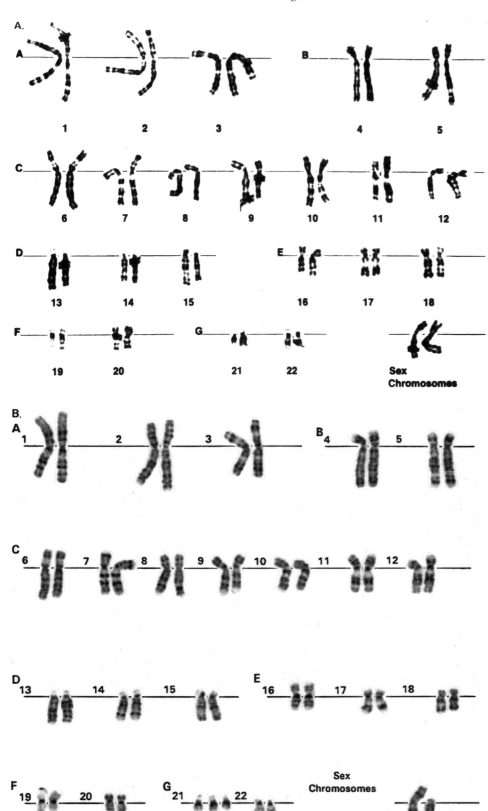

Figure 4–12. Giemsa-staining pattern of human metaphase chromosomes. **A:** An example of G-binding karyotype analysis. Shown are the aligned metaphase chromosomes of a normal human female, as illustrated by the 46,XX karyotype. **B:** An example of a karyotype from an individual with Down syndrome is shown with an extra chromosome 21. (Courtesy of Dr. Cathy Tuck-Muller, Department of Medical Genetics, University of South Alabama, College of Medicine.)

identified: a pale-staining fibrillar center, which contains DNA that is not being transcribed; a dense fibrillar core or center, which contains many RNAs in the process of transcription; a granular region where the maturing ribosomal particles are assembled; and a nucleolar matrix, a fibrous network that may participate in the organization of nucleolus (see Fig. 4–2).

The size and shape of the nucleolus is dependent on its activity. In cells that are actively synthesizing large amount of proteins, the nucleolus may occupy up to 25% of the total nuclear volume; whereas in dormant cells, it may be hardly visible. Examination of cells from different physiologic states have shown that the observed differences in nucleolar size are primarily due to differences in the granular component. That is, cells that are very active in protein synthesis contain more of the maturing ribosomal precursor particles in their nucleus. This increased size of the granular region probably reflects the time necessary to assemble the ribosomal RNAs (rRNAs) with proteins of the ribosomal subunits, because electron micrographs of cells containing large nucleoli demonstrate an increase in the number of active ribosomal genes and an increase in the apparent rate that each gene is transcribed.

In general, the nucleolus is visible only in interphase cells. Concomitant with the condensation of chromosomes as the cell approaches mitosis, the nucleolus is observed to decrease in size, then disappears as RNA synthesis stops. In humans, the rRNA genes represent clusters of DNA segments located near the tip of five different chromosomes (chromosomes 13, 14, 15, 21, and 22); thus, there are 10 different ribosomal gene loci in diploid somatic cells. Following mitosis, rRNA synthesis is restarted initially on small nucleoli located at the 10 ribosomal gene loci, which are often referred to as *nucleolar-organizing regions* (NORs). These small nuclear-organizing regions usually are not observed separately because of the rapid fusion of the NORs that forms the larger characteristic interphase nucleolus.

Ribosomal gene transcription occurs in the fibrillar region of the nucleolus, yielding a 45S precursor RNA molecule. Following transcription, this precursor molecule is almost immediately cleaved or processed into the 18S, 5.8S, and 28S RNAs found in the mature ribosome. The processed rRNAs are then assembled into mature ribosomal subunits in the granular region of the nucleolus. The 5.8S and 28S rRNAs are complexed with a third RNA species, called 5S rRNA, that is transcribed from nonnucleolus genes and a specific set of protein to form the large or 60S ribosomal subunit. The small, or 40S, ribosomal subunit is formed from the association of ribosomal proteins with the 18S rRNA. The completed subunits are transported to the cytoplasm through the nuclear pores, where they provide the machinery for translation.

Organelles Involved in Macro-molecular Synthesis and Turnover

Free and Membrane-Associated Ribosomes Catalyze the Synthesis of Differing Classes of Proteins

Ribosomes serve as catalysts of protein synthesis within the cytoplasm of eukaryotic cells. The ribosomal small subunit (40S) and large subunit (60S) form an 80S ribosome (Fig. 4–13), which is found either associated with the cytoplasmic surface of the ER or free within the cytosol. Ribosomes that are not ER membrane-attached are called *free*, despite the fact that many are associated with the cytoskeleton. There is no structural difference between membrane-bound and free ribosomes; both types come from the same pool of ribosomes and return to this same pool when translation is complete. Whether a ribosome becomes bound or free does not depend on the ribosome, per se, but rather upon the sequence of the protein being synthesized, as will be explained later. Proteins synthesized on membrane-associated ribosomes include integral transmembrane plasma membrane proteins, ER, Golgi complex, endosomal, and lysosomal proteins, and proteins to be secreted from the

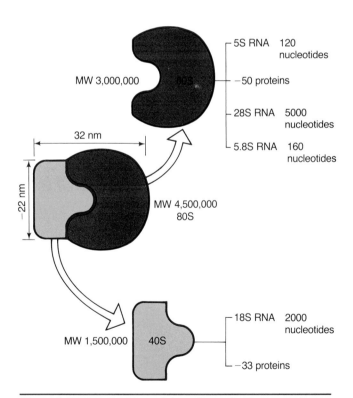

Figure 4–13. Composition of the eukaryotic ribosome. A summary of the protein and RNA composition of the 40S and 60S ribosomal subunits that associate to form the mature 80S ribosome. (Modified from Widnell CC, Pfenninger KH. *Essential Cell Biology*. Baltimore: Williams & Wilkins, 1990.)

cell. Free ribosomes or, more appropriately, non-membrane-associated ribosomes, are responsible for synthesis of cytosolic proteins, peripheral membrane proteins, or some proteins destined for the nucleus, mitochondria, and peroxisomes.

Study of mechanisms by which proteins synthesized within the ER become sorted and then targeted to their final destination is one of the most exciting fields in modern cell biology. To fully understand this subject, it may be helpful to begin with a few generalizations:

1. Most transmembrane proteins and membrane lipids are synthesized within the ER.
2. The targeting of membrane proteins to specific sites such as the ER, Golgi, lysosomes, or plasma membrane requires specific signal sequences within the newly synthesized protein.
3. In the absence of a signal sequence, the newly synthesized protein follows a default pathway. For proteins synthesized on ER-associated ribosomes, the default pathway is: ER→Golgi→secretory vesicles→cell surface. Proteins synthesized on free ribosomes have the following default pathway: ribosomes→cytosol (Fig. 4–14).
4. New plasma membrane is added by fusion of vesicles coming from the Golgi complex. To avoid a constantly increasing plasma membrane surface area, membrane is recycled by endocytosis.

The pathway used by secreted proteins was clearly demonstrated in a classic pulse chase experiment by Jamieson and Palade (Fig. 4–15). Pancreatic acinar cells secrete approximately 90% of newly synthesized proteins. To trace the position of newly synthesized protein over time, Jamieson and Palade incubated the radioactive amino acid [³H]leucine with slices of pancreas in culture for a brief time (pulse), followed by washing away the radiolabel and adding unlabeled amino acids for the duration of the experiment (chase). The time course of secretory protein passage was demonstrated by the technique of autoradiography and electron microscopy. In this technique, the radiolabeled tissue slice is overlaid with a photographic emulsion. Where radioactivity is incorporated within a cell, the position is marked by a silver grain that can be visualized by electron microscopy. By employing this approach, Jamieson and Palade demonstrated that the radioactive protein is initially localized within the ER (see Fig. 4–15). After 3 minutes, the radioactive protein was found in the ER; at 20 minutes, the protein was found in the Golgi complex; and after a 90-minute chase, most of the protein was in mature zymogen granules.

In the acinar cell, the mature secretory vesicles (zymogen granules) are found in the apical region (facing the lumen of the duct system). The secreted proteins are released to the outside by exocytosis, which results from fusion of the zymogen granules with the apical

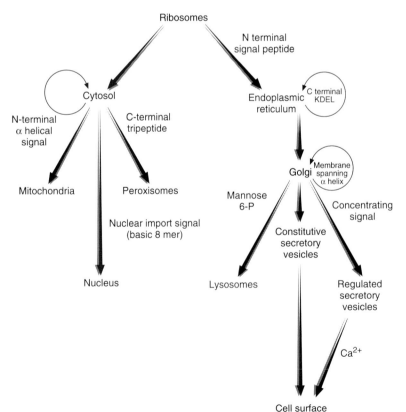

Figure 4–14. Concept map of protein trafficking. This concept map should help you understand the targeting signals that drive proteins to specific locations within the cell. Default pathways that require no signal (*black*) and signal-requiring pathways (*red*) are shown.

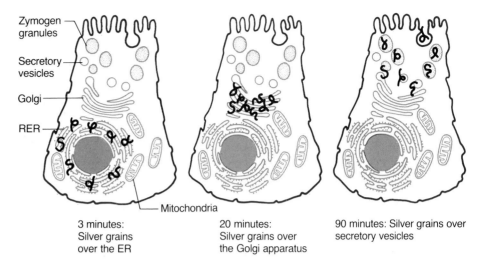

Zymogen granules
Secretory vesicles
Golgi
RER
Mitochondria

3 minutes: Silver grains over the ER

20 minutes: Silver grains over the Golgi apparatus

90 minutes: Silver grains over secretory vesicles

Figure 4–15. Pathway followed by newly synthesized proteins that are to be secreted. As described in the text, electron microscopy and autoradiography were used by Jamieson and Palade to track the pathway of newly synthesized proteins over time. Silver grains, produced when a photographic emulsion was placed over thin sections of pancreatic acinar cells, allowed the detection of ^{3}H-labeled proteins. In pancreatic acinar cells, 90% of ^{3}H-labeled proteins are secretory proteins. (Modified from Alberts B, Bray D, Lewis J, et al. *Molecular Biology of the Cell,* 2nd ed. New York: Garland Publishing, 1989.)

plasma membrane. This exocytosis requires a specific chemical signal; therefore, it is referred to as *regulated exocytosis*. Exocytosis that does not require a chemical signal and, therefore, is continuous, is referred to as *constitutive exocytosis*.

The Endoplasmic Reticulum Is a Network of Single Membrane-Enclosed Cisternae, With (Rough Endoplasmic Reticulum) or Without (Smooth Endoplasmic Reticulum) Associated Ribosomes

The basic morphologic difference between rough endoplasmic reticulum (RER) and smooth endoplasmic reticulum (SER) is the association of ribosomes to the cytoplasmic surface of RER. The ER within animal cells can constitute as much as 50% of the total cellular membrane. It consists of a single membrane-enclosed network of flattened sacs or cisternae that enclose a continuous internal compartment called the *lumen*. The RER is continuous with the outer membrane of the nuclear envelope, which itself has associated ribosomes (Fig. 4–16). The SER in some cells (i.e., skeletal muscle) is clearly distinct from the RER, whereas in other cells the SER is continuous with the RER. In the latter case, this ribosome-free region of the ER is referred to as the *transitional element* because it is the region of the ER that buds off, creating vesicles that carry newly synthesized protein and lipid to the Golgi apparatus. The RER is the site of protein synthesis for ER, Golgi, lysosomal, endosomal, plasma membrane, and secretory proteins. The

synthesis of membrane lipids (phospholipids, steroids, triglycerides) occurs within both the RER and SER. In skeletal and cardiac muscle cells (fibers), the smooth ER is referred to as the *sarcoplasmic reticulum* (SR). The SR has the specialized function of serving as a reservoir for Ca^{2+}, which is released to initiate contraction and is pumped back into the SR to cause relaxation.

The SER is also involved in the detoxification of drugs by the cytochrome P_{450} electron transfer chain and the formation of unsaturated fatty acids by the cytochrome b_5 electron transport chain. In response to drugs such as phenobarbital, liver hepatocytes produce high concentrations of cytochrome P_{450} and expand the surface area of the SER. When the drug is effectively removed, the SER is reduced in surface area by removal via lysosomal autophagocytosis. This process will be described later.

Signal Peptides, Signal Recognition Particles, and the Signal Recognition Particle Receptor Play Essential Roles in the Endoplasmic Reticulum's Association of Ribosomes and Cotranslational Insertion of Protein Into the Endoplasmic Reticulum's Lumen

How do ribosomes involved in organelle protein synthesis, membrane protein synthesis, or secreted protein synthesis become associated with the ER? The answer is that such proteins have an NH_2-terminal signal sequence of 16–30 amino acids, or leader sequence, which is recognized by a signal recognition particle as it is being

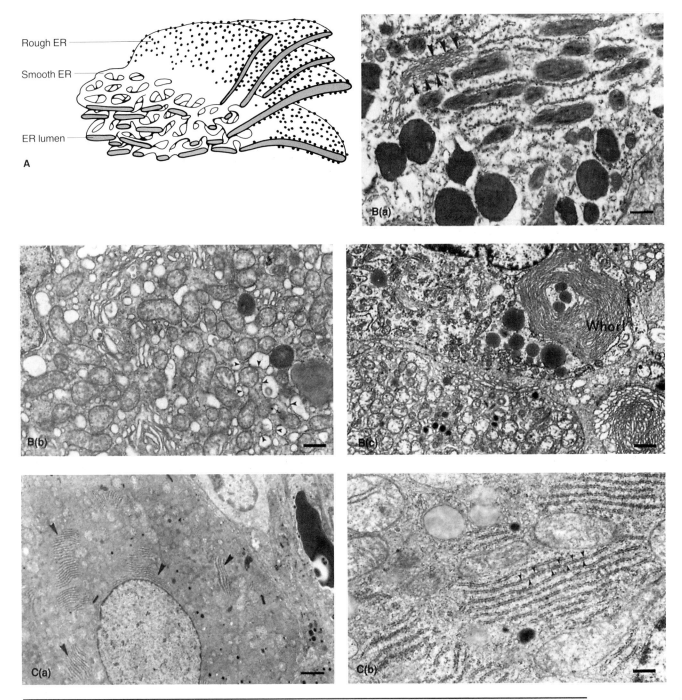

Figure 4–16. Structure of rough and smooth endoplasmic reticulum. **A:** The rough endoplasmic reticulum (RER) consists of oriented stacks of flattened cisternae studded with ribosomes on its cytoplasmic surface. The luminal space is 20–30 nm. The smooth endoplasmic reticulum (SER) is sometimes connected to the RER, and appears as 30- to 60-nm diameter membranous tubes. The SER has no associated ribosomes. **B:** Electron micrographs of SER from a mammalian luteal cell. The SER can be found in various forms. At least three types are presented in the following micrographs. **(a)** Lamellar stacks of SER are observed (*between the arrowheads*) (small luteal cell from a pregnant cow corpus luteum). **(b)** The cytoplasm of the cell may be filled with SER that appears as empty vesicles (*small arrowheads inside the vesicles*). This is characteristic of active steroid-secreting cells (large luteal cell from a pregnant cow corpus luteum). **(c)** Smooth endoplasmic reticulum may be observed in spirally arranged cisternae (*whorl*) (small luteal cell from a pregnant cow corpus luteum) (Bars: **a,** 0.39 μm; **b,** 0.39 μm; **c,** 1.31 μm). **C:** Electron micrographs of RER from a mammalian large luteal cell (21-day pregnant rat). **(a)** Stacks of RER (*arrowheads*) are observed throughout the cytoplasm of the cell. **(b)** At a higher magnification, ribosomes (*small arrowheads*) can be observed lining both membrane surfaces of the RER (Bars: **a,** 1.6 μm; **b,** 0.27 μm). (**A,** modified from Alberts B, Bray D, Lewis J, et al. *Molecular Biology of the Cell*, 2nd ed. New York: Garland Publishing, 1989; **B, C,** electron micrographs courtesy of Dr. Phillip Fields.)

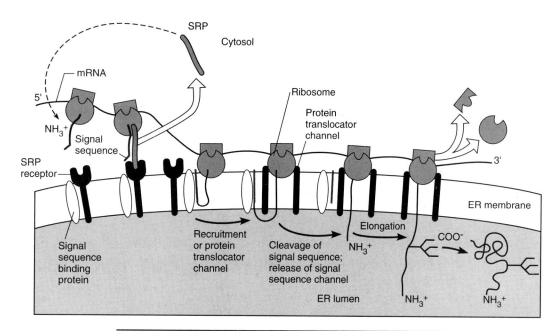

Figure 4–17. Synthesis of secretory proteins and insertion into the ER lumen. This diagram presents the role of the signal sequence, signal recognition particle (SRP), and SRP receptor in insertion of secretory proteins into the ER lumen. It also demonstrates the cleavage of the signal sequence; elongation and glycosylation of the nascent peptide; and release of the newly synthesized protein into the lumen. Details of these steps are discussed in the text. (Modified from Darnell J, Lodish H, Baltimore D. *Molecular Cell Biology*, 2nd ed. New York: Scientific American Books, 1990.)

translated from the 5′ end of mRNA. This signal recognition particle (SRP) contains six protein subunits in combination with a 7S RNA. The SRP binds both to the ribosome and the signal peptide and, in doing so, causes a brief pause in protein translation, or translation arrest. During this translation arrest, the SRP, with bound ribosome and newly translated NH_2-terminus of a nascent polypeptide, becomes bound to an ER integral membrane-docking protein called the *SRP receptor*, an event which causes translation to continue. The signal peptides typically contain one or more positively charged amino acids followed by a stretch of 6–12 nonpolar amino acids. The signal peptide is inserted and oriented within the membrane by attachment to the signal sequence binding protein. The remainder of the translated protein is then inserted through a protein translocator channel that forms around the growing nascent peptide (Fig. 4–17). Because the entry into the lumen is occurring as the protein is being synthesized, the process is referred to as *cotranslational insertion*. This differs from proteins that are destined for the mitochondria and peroxisomes that are inserted posttranslationally.

Protein insertion requires ATP hydrolysis, which supplies energy either for a protein pump that drives insertion or for protein unfolding as the nascent protein emerges on the luminal side of the ER membrane. If the newly synthesized protein is not a transmembrane protein, its NH_2-terminal signal peptide is cleaved by a signal peptidase located on the luminal side of the ER membrane, and the protein is released into the lumen.

Once within the ER lumen, these unfolded proteins become associated with a binding protein (BIP), which is probably a catalyst of protein folding. The BIP is related in sequence to the family of heat-shock proteins (hsp) hsp 70, thought to be ATP-dependent reconstruction enzymes for damaged proteins. Another protein involved in creating the proper conformation of proteins found in the ER lumen is protein disulfide isomerase, a protein that is associated with the luminal surface of the ER membrane. This enzyme helps the proper disulfide (S-S) bonds to form between cysteine residues within the newly synthesized protein.

The Insertion of Integral Membrane Proteins Into the Endoplasmic Reticulum Requires Carefully Placed Start-Transfer Signals and Stop-Transfer Signals

Once one understands the mechanism by which a soluble protein is transferred to the lumen of the ER, it becomes easier to understand how membrane proteins are inserted into the ER membrane. The simplest case is a single-pass integral protein with its NH₂-terminus facing the cytoplasmic compartment. In this case, the signal peptide is internal and is inserted into the membrane by the mechanisms described above. If the hydrophobic signal peptide has more positively charged amino acids immediately flanking its NH₂-terminal side, it will be oriented by the signal sequence binding protein, as shown in Fig. 4–18A. The internal hydrophobic signal sequence contains 20–30 nonpolar amino acids and it is not cleaved by the signal peptidase. The result is a single-pass transmembrane protein that has its NH₂-terminus on the cytosolic side of the ER membrane, then has the hydrophobic signal peptide making a single α-helical pass through the bilayer, followed by the COOH-terminus on the luminal side of the ER membrane.

A slightly more complicated case (Fig. 4–18B,C) is a single-pass transmembrane protein in which the COOH-terminal end of the protein is in the cytoplasm. This orientation can be obtained by either of two mechanism. In Fig. 4–18B, a NH₂-terminal start-transfer signal peptide is followed by a hydrophobic stop-transfer sequence that terminates the insertion. The start-transfer NH₂-terminal signal peptide is then cleaved, leaving behind a single-pass integral membrane protein with its COOH-terminus in the cytoplasm and NH₂-terminus in the ER lumen.

The second mechanism for obtaining the same final topography is shown in Fig. 4–17C. In this case, an internal signal peptide containing a hydrophobic stretch of 20–30 amino acids is flanked by more positively charged amino acids on its COOH-terminal side than on its NH₂-terminal side. In this case, the signal peptide would be inserted and oriented in the membrane with its COOH-terminus facing the cytosol. The internal signal peptide is not cleaved by the signal peptidase, and the result is a single-pass transmembrane protein with the orientation shown in Fig. 4–18C.

It now becomes easy to understand how multipass integral membrane proteins can obtain such varied topographies. The multipass protein is topographically placed within the bilayer by a series of specifically placed alternating start-and stop-transfer signals (Fig.

4–18D). The start-and-stop-transfer signals are hydrophobic stretches of 20–30 amino acids that form an α-helix within the bilayer.

The asymmetry of membrane proteins that we discussed in Chapter 2 is created in the ER. This asymmetry is maintained as proteins are transferred to the Golgi complex and then to the plasma membrane. This is because these membrane proteins are transferred by the budding off of vesicles, then fusion of these vesicles with the target membrane. During these fusion events, protein domains facing the cytoplasm continue facing the cytoplasm, and domains facing the ER lumen continue to face the Golgi lumen, which then becomes the external face of the plasma membrane.

Glycosylation of Secretory and Membrane Proteins Begins Within the Endoplasmic Reticulum

Most proteins synthesized on the ER become glycosylated on entry into the ER lumen, whereas proteins translated on free ribosomes are, in general, not glycosylated. This suggests that the enzymes required for the initial steps in glycosylation must be located on the luminal side of the ER, and this is indeed correct. In human and animal cells, sugar residues are either N-linked (to the NH₂ amide nitrogen of asparagine) or O-linked (to the OH hydroxyl group of serine, threonine, or hydroxylysine) to the glycoprotein. The initial steps of glycosylation of N-linked glycoproteins occur within the lumen of the ER as the protein is being translated and inserted. The precursor of the N-linked oligosaccharide is a 14-sugar branched-chain oligosaccharide that contains two N-acetylglucosamine residues linked by a high-energy pyrophosphate bond to a membrane-anchored lipid called *dolichol* (long-chain unsaturated hydrocarbon), nine mannose residues, and three glucose residues (Fig. 4–19). This branched oligosaccharide, which is located on the luminal side of the ER, is transferred as a unit to asparagine residues of the nascent polypeptide by an enzyme called *oligosaccharide transferase*, which is also located on the luminal

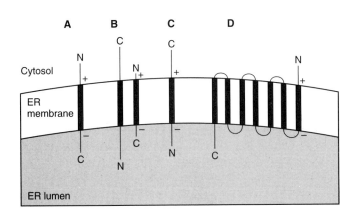

Figure 4–18. Insertion of integral membrane proteins into the ER membrane. Three examples of integral membrane proteins are demonstrated. **A:** Single-pass with a cytoplasmic NH₂-terminus. **B, C:** single-pass with a luminal NH₂-terminus. **D:** A multipass integral membrane protein. The steps by which these integral membrane proteins obtain their final configuration within the membrane are discussed in the text. In the diagram presented in this figure start-transfer sequences are shown (*red*), and stop-transfer sequences are identified (*black*).

Figure 4–19. Early steps in *N*-linked glycosylation of nascent proteins require a dolichol carrier. As soon as an asparagine in an Asn-X-Ser or Asn-X-Thr recognition sequence is transferred onto the luminal side of the ER membrane, it immediately receives a 14-sugar-branched chain oligosaccharide from a dolichol carrier. The details are discussed in the text. (Modified from Darnell J, Lodish H, Baltimore D. *Molecular Cell Biology*, 2nd ed. New York: Scientific American Books, 1990.)

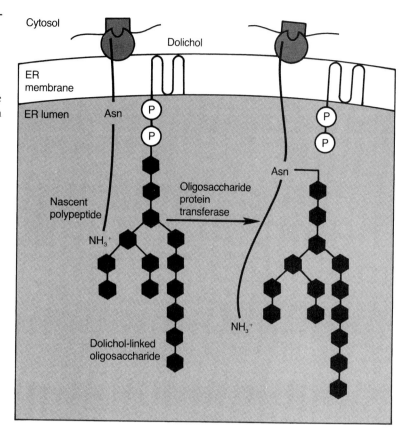

surface of the ER membrane. Only asparagines that occur in the tripeptide recognition sequence Asn-X-Ser or Asn-X-Thr (X can be any amino acid except proline) are transferred.

Once the 14-sugar branched oligosaccharide is transferred to the nascent protein, the initial trimming and processing steps also occur within the ER. The three glucose residues and one mannose residue are trimmed off in three enzymatic steps (Fig. 4–20). The remaining step in the processing of *N*-linked oligosaccharides occurs within the Golgi apparatus and will be discussed later.

Glycosylphosphatidylinositol Anchors Are Attached to Some Proteins Within the ER Lumen

We now know that many integral membrane proteins are anchored to the membrane, not by a stretch of hydrophobic amino acids, but by covalent linkage to lipids. Some membrane proteins are associated with the membrane via a glycosylphosphatidylinositol (GPI) anchor that forms in the ER lumen by the mechanism displayed in Fig. 4–21. The precursor protein is associated with the ER membrane by a hydrophobic stretch of approximately 20 amino acids. Shortly after insertion is complete, an enzyme within the ER lumen causes proteolytic cleavage, releasing the bulk of the protein from its COOH-terminal membrane spanning domain. Concur-

rently the new COOH-terminus of the released protein is attached to the amino group of the ethanolamine moiety of a preassembled GPI intermediate. The result is a membrane-bound protein with its amino acids found within the lumen of the ER.

Membrane Lipids Are Synthesized Within the Endoplasmic Reticulum and Then Moved to the Recipient Organelles by Transport Vesicles or Phospholipid-Exchange Proteins

In human and other animal cells, the synthesis of membrane phospholipids, cholesterol, and ceramide occurs within the ER, usually the SER. The synthesis of phospholipids [e.g., phosphatidylcholine (PC); Fig. 4–22] occurs on the cytoplasmic leaflet of the ER membrane. The enzymes involved are associated with the cytoplasmic leaflet, and the substrates are found in the cytosol. The steps are similar for all phospholipids. First, acyltransferase adds two fatty acids to glycerol phosphate. The result of this enzymatic reaction is the formation of phosphatidic acid, which is incorporated into the cytoplasmic leaflet of the ER membrane. The next step is the removal of the phosphate from phosphatidic acid, forming diacylglycerol. This is followed by the transfer of the phosphorylated head group from cytidine diphosphocholine (CDP)-choline, CDP-serine, or CDP-ethanolamine to diacylglycerol by a phosphotransferase.

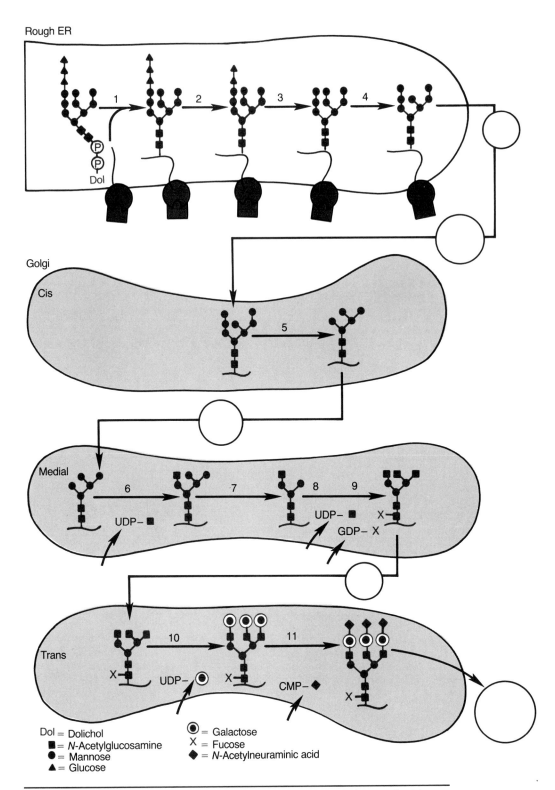

Dol = Dolichol
■ = *N*-Acetylglucosamine
● = Mannose
▲ = Glucose
◉ = Galactose
X = Fucose
◆ = *N*-Acetylneuraminic acid

Figure 4–20. Formation of a mature glycoprotein with *N*-linked oligosaccharide occurs in a series of stages occurring in the RER and *cis-, medial-, and trans-*Golgi. The known stages occur in 11 enzymatic steps, beginning with the precursor *N*-linked oligosaccharide shown (*step 1*). The steps with their locations are shown in the diagram. First, three glucose residues are removed (*steps 2 and 3*), then four mannose residues are cleaved (*steps 4 and 5*), one *N*-acetylglucosamine is added (*step 6*), two mannose residues are removed (*step 7*), one fucose and two *N*-acetyl-glucosamine residues are added (*steps 8 and 9*), three galactose residues are added (*step 10*), and then three *N*-acetylneuraminic acid residues are added (*step 11*). (Modified from Darnell J, Lodish H, Baltimore D. *Molecular Cell Biology.* 2nd ed. New York: Scientific American Books, 1990.)

Although the phospholipids are synthesized on the cytoplasmic leaflet, some must be moved to the luminal leaflet of the ER; otherwise, a monolayer, rather than a bilayer, would be formed. This is accomplished by phospholipid-translocating enzymes called *flippases*. Flippases transfer phospholipids from the cytosolic to the luminal leaflet of the ER membrane. The most active flippase in the ER membrane specifically moves choline-containing phospholipids, but not phosphatidylethanolamine (PE), phosphatidylserine (PS), or phosphatidylinositol (PI). Therefore, membrane phospholipid asymmetry is established within the ER

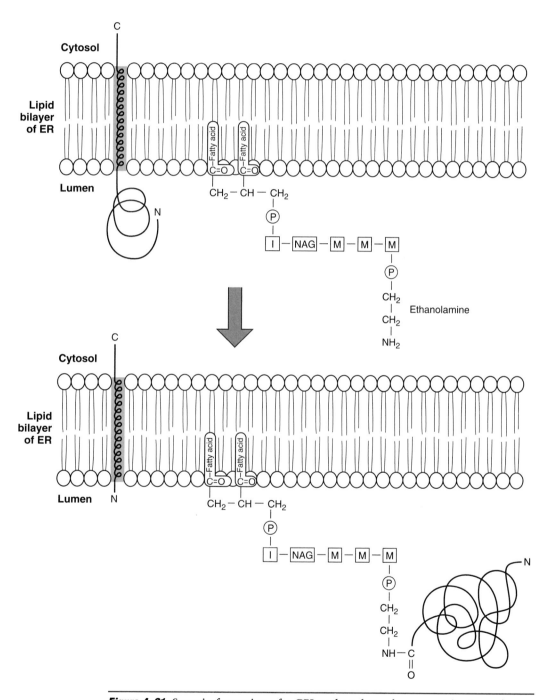

Figure 4–21. Steps in formation of a GPI-anchored membrane protein. An endoprotease within the lumen of the ER cleaves the protein away from its COOH-terminal membrane-spanning domain, then attaches the new COOH-terminus of the released protein to the amine within the ethanolamine moiety of the glycosylphosphatidylinositol (GPI) anchor.

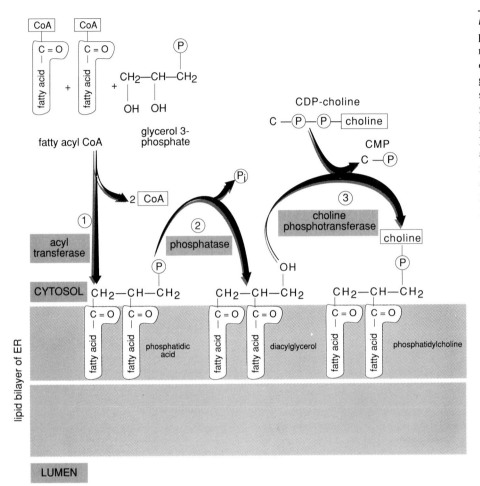

Figure 4-22. The synthesis of phospholipids involves three integral membrane proteins, the active sites of which face the cytosol. The diagram presents the pathway for synthesis of phosphatidylcholine from fatty acyl-CoA, glycerol-3-phosphate, and cytidine-diphosphocholine (CDP-choline). Details are discussed in the text. (Modified from Alberts B, Bray D, Lewis J, et al. *Molecular Biology of the Cell*, 2nd ed. New York: Garland Publishing, 1989.)

because of the presence of specific flippases within this organelle (Fig. 4-23).

The phospholipids synthesized within the ER are transferred to the Golgi, lysosomes, and plasma membrane by the budding off of phospholipid-containing transport vesicles from the ER and their fusion with the target organelle. Because the Golgi, lysosomes, and plasma membrane contain little flippase activity, the asymmetry of phospholipids established within the ER remains essentially unchanged. The mitochondria and peroxisomes, however, do not receive their membrane phospholipid by this mechanism of membrane budding and fusion. Mitochondria synthesize their own cardiolipin (found in the inner membrane) and phosphatidylglycerol. However, mitochondria and peroxisomes both appear to obtain their PC, PE, PS, and PI from the ER by a mechanism that uses water-soluble phospholipid-exchange proteins. Phospholipid-exchange proteins have the ability to pluck a specific phospholipid from the cytoplasmic leaflet of the ER membrane, carry the phospholipid to the target membrane, and insert the phospholipid into the cytoplasmic leaflet of the mitochondrial or peroxisomal membrane. How lipid asymmetry is established within the mitochondrial and peroxisomal membranes is unknown.

Cholesterol and ceramide are also synthesized within the ER membrane. Ceramide is further processed in the Golgi complex to form glycosphingolipids and sphingomyelin. The enzymes involved in the synthesis of glycosphingolipids and sphingomyelin from ceramide are found on the luminal leaflet of the Golgi membrane. Therefore, these lipids are found on the noncytoplasmic leaflet of cellular membranes.

The Golgi Apparatus: A Series of Cisternal Compartments Involved in Posttranslational Modification and Sorting of Newly Synthesized Proteins and Lipids

The Golgi apparatus is a series of flattened sacs or cisternae that function in the posttranslational modification and sorting of protein and lipids synthesized within the ER. The Golgi apparatus appears in the electron microscope like a stack of plates located close to the nucleus, near the centrosome (Fig. 4-24). The cisternae enclose a lumen of approximately 1 μm diameter. At the edges, the cisternae are dilated. The Golgi cisternae can be separated into five functional compartments. The *cis*-compartment is close to the ER, and vesicles that have

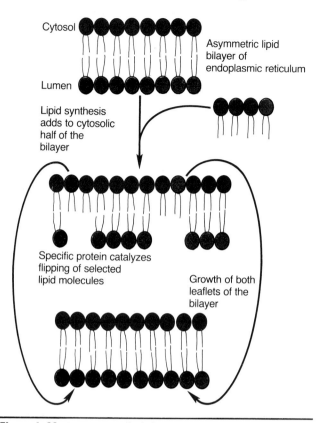

ferent transport vesicles destined for the lysosomes, plasma membrane, or secretion (Fig. 4–25).

These Golgi compartments have their own specific functions. In the discussion of the initial steps of *N*-linked glycosylation that occurs within the ER, it was noted that a 14-sugar, branched-chain oligosaccharide

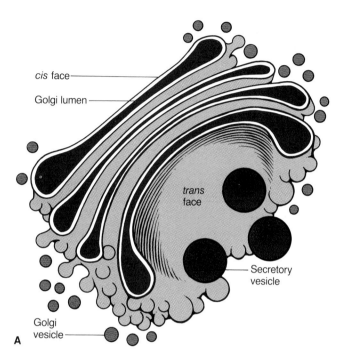

Figure 4–23. Enzymes called *flippases* catalyze transbilayer movement of ER phospholipids. Because phospholipids are synthesized on the cytoplasmic side of the ER membrane, there must be a mechanism to move selected phospholipids to the luminal leaflet. Enzymes that specifically catalyze the transbilayer movement of selected phospholipids are present in the ER membrane. These enzymes are called *flippases*. (Modified from Alberts B, Bray D, Lewis J, et al. *Molecular Biology of the Cell*, 2nd ed. New York: Garland Publishing, 1989.)

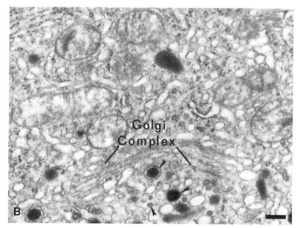

budded off the transitional element of the ER fuse with a specialized tubular region of the *cis* face called the *cis-Golgi network*. Only properly folded proteins are transported, via transport vesicles, from the ER to the Golgi Complex. If a protein with an ER retention signal (KDEL) reaches the *cis-Golgi* network, it binds to a specific receptor and is repackaged into vesicles that return it to the ER. Once the newly synthesized proteins and lipids have been modified within the *cis*-cisternae, 50-nm-diameter coated vesicles bud off the dilated ends of this compartment and fuse with the medial cisternae. After the proteins and lipids are acted on by the enzymes of the medial cisternae, the process continues in an assembly-line fashion by coated vesicles budding from the terminal dilations and fusing with the *trans*-cisternae. From the *trans*-cisternae, the proteins and lipids that will exit the Golgi are passed into the *trans*-Golgi network (TGN), where they are sorted into dif-

Figure 4–24. Structure of the Golgi apparatus. **A:** The Golgi contains stacks of flattened cisternae. These cisternae have dilated edges from which small vesicles bud. Large secretory vesicles bud from the *trans*-Golgi. **B:** Electron microscopy of the Golgi apparatus of a mammalian luteal cell. Numerous secretory granules and vesicles of various sizes (*arrowheads*) can be observed budding from the *trans*-cisternal face of the Golgi complex (large luteal cell from a day-20 pregnant rat corpus luteum; bar, 0.27 μm). (**A,** Modified from Alberts B, Bray D, Lewis J, et al. *Molecular Biology of the Cell*, 2nd ed. New York: Garland Publishing, 1989; **B,** courtesy of Dr. Phillip Fields.)

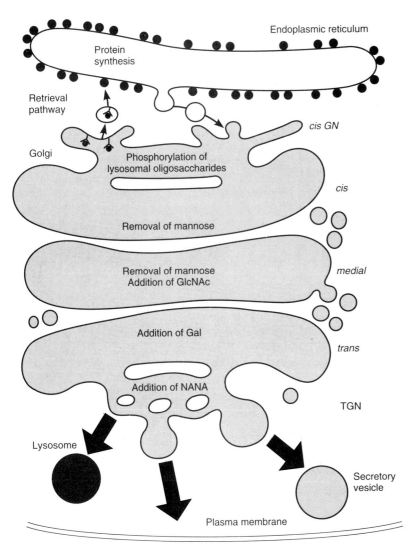

Figure 4–25. Compartmentalized functions of the Golgi. This diagram summarizes the compartmental localization of posttranslational modifications of proteins within the Golgi apparatus. Proteins that contain KDEL ER retention signal and are, therefore, returned from the *cis*-Golgi network to the ER are represented (*red spheres*). (Modified from Alberts B, Bray D, Lewis J, et al. *Molecular Biology of the Cell*, 2nd ed. New York: Garland Publishing, 1989.)

is transferred as a unit to appropriate asparagines within nascent polypeptides. Once transferred, three glucose residues and one mannose residue were trimmed off within the ER (see Fig. 4–20). In the *cis*-Golgi compartment, further trimming of mannoses occurs by the enzyme α-mannosidase I (see Fig. 4–20), which leads to the high mannose oligosaccharides (Man)$_8$(GlcNAc)$_2$ and (Man)$_5$(GlcNAc)$_2$ that are common final *N*-linked oligosaccharides. In the *medial-* and *trans*-Golgi, a series of glycosyltransferases found on the luminal surface of the Golgi membrane use sugar nucleotides as substrates to sequentially add *N*-acetyl-glucosamine, galactose, fucose, and sialic acid residues to form the complex *N*-linked oligosaccharides typical of many complete glycoproteins and glycolipids (see Fig. 4–19). All of the essential enzymes involved in glycosylation are found on the luminal side of the ER and Golgi membranes, which is why all plasma membrane, lysosomal membrane, ER membrane, Golgi and endosomal membrane, glycoproteins, and glycolipids have

their sugar residues on the noncytoplasmic side of the membrane. The terminal sialic acid residues that indicate the completion of glycosylation are incorporated via CMP-sialic acid in the *trans*-Golgi.

In addition to *N*-linked glycosylation, some proteins have sugar residues attached via the hydroxyl group of selected serine or threonine residues. This type of glycosylation, called *O-linked glycosylation*, takes place within the Golgi complex. O-linked glycosylation is catalyzed by glycosyl-transferase enzymes that are resident within the lumen of the Golgi apparatus and utilizes sugar nucleotides to add sugar residues to the protein, one by one.

Several other important modifications occur specifically within Golgi compartments. The phosphorylation of mannose residues of proteins destined for the lysosomes occurs within the *cis*-Golgi network. Sulfation of proteins on selected tyrosine residues occurs within the Golgi. An important modification of several secreted proteins is specific proteolytic cleavage

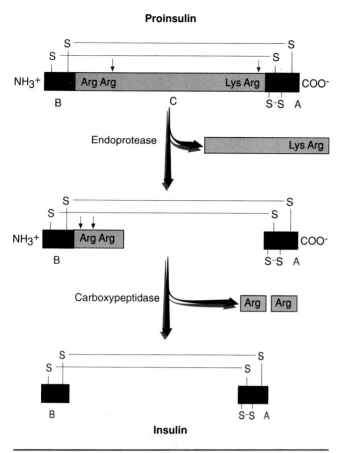

Proinsulin

Endoprotease

Lys Arg

Carboxypeptidase

Arg   Arg

Insulin

Figure 4–26. Proteolytic cleavage of proinsulin occurs in the TGN and secretory vesicles. The proteolytic processing of proinsulin is presented. Details are given in the text. (Modified from Darnell J, Lodish H, Baltimore D. *Molecular Cell Biology*, 2nd ed. New York: Scientific American Books, 1990.)

of a proenzyme peptide within the TGN and resulting secretory vesicles. For example, insulin is initially synthesized as a preproinsulin. The preinsulin peptide is the signal peptide that is cleaved off cotranslationally in the ER. The proinsulin is further processed (Fig. 4–26) within the TGN and secretory granule by removal of an internal C peptide. The remaining A chain (COOH-terminal) and B chain (NH$_2$-terminal) are linked together by a disulfide bond to form insulin. This proteolytic processing also occurs during the maturation of several other important hormones and neurotransmitters.

Targeting of Acid Hydrolases to the Lysosome Requires Mannose Phosphorylation and Recognition by a Mannose-6-Phosphate Receptor

Lysosomes are scavenger organelles that are 0.2–0.5 μm in diameter and are enclosed by their own single membrane. The lysosomes contain a diverse array of acid hydrolases, including proteases, nucleases, glycosidases, lipases, phospholipases, sulfatases, and phosphatases. The lysosomal membrane keeps these acid hydrolases compartmentalized away from cytoplasmic macromolecules. Furthermore, the acid hydrolases are active at pH 5.0, which is the intralysosomal pH but inactive at the neutral pH of the cytoplasm. Finally, the acid hydrolases are translated within the ER as inactive proenzymes and cleaved to the active form at pH 5.5 within the late endosome. The acidic pH of the inner lysosome is maintained by an ATP-dependent proton pump in the lysosomal membrane.

Materials that reach lysosomes for destruction come by four different routes (Fig. 4–27).

1. Macrophage and neutrophils are capable of engulfing large foreign particles, such as bacteria, by the process of phagocytosis (described in detail later). The bacteria are brought into the cell within a membrane-encapsulated phagosome, which, in turn, fuses with a late endosome (immature lysosome) or lysosome.

2. In receptor-mediated endocytosis, specific ligands are brought into the cell at clathrin-coated pits (discussed in detail later). The resulting coated vesicles lose their coats, forming early endosomes. Some of the endocytosed molecules are retrieved from the early endosomes and returned to the plasma membrane. Other molecules are sent by transport vesicles to the late endosomes, which contain acid hydrolases that initiate digestion. The late endosomes, by mechanisms that are unknown, mature into lysosomes.

3. Obsolete organelles (such as the mitochondria shown in Fig. 4–27) are degraded by lysosomes through a process called *autophagy*. The spent mitochondria is first wrapped in ER-derived membranes, forming an autophagosome. The autophagosome then fuses with a late endosome or lysosome.

4. Cytosolic proteins that contain the amino acid sequence KFERQ on their surface are selectively taken up by lysosomes. The mechanism by which these proteins are internalized by the lysosome remains to be determined.

How are proteins that are synthesized within the ER and modified and sorted within the Golgi targeted to the lysosome? A partial answer (for the acid hydrolases) has come from the study of a rare recessive lysosomal storage disease called *inclusion cell disease* (I cell disease). In this disorder, extracellular and intracellular glycoproteins and glycolipids, which would normally be digested by lysosomes, accumulate as undigested substrates within large cytoplasmic inclusions in fibroblasts and macrophages; however, hepatocytes from I cell subjects had no such accumulation. Study of

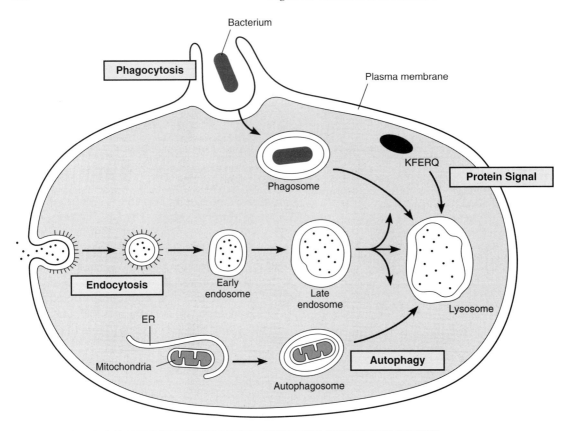

Figure 4–27. **Four pathways by which materials are delivered to the lysosome.** The pathways, which are described in the text, are (1) phagocytosis, (2) endocytosis, (3) autophagy, and (4) KFERQ protein signal.

fibroblasts from patients with I cell disorder indicated that they lack an enzyme called *N-acetylglucosamine-phosphotransferase*. All *N*-linked oligosaccharides have been processed to a similar form that contains two *N*-acetylglucosamine residues and five or eight mannose residues by the time they reach the *cis*-Golgi. Within the *cis*-Golgi lumen, the enzyme N-acetylcosamine-phosphotransferase is responsible for transferring *N*-acetylglucosaminephosphate to multiple mannose residues of selected glycoproteins. That only selected mannose-containing proteins, such as the acid hydrolases, are acted on by the phosphotransferase indicates that a signal patch within the protein sequence must act as a recognition site for the phosphotransferase. The acid hydrolases are then acted on by a second enzyme called *phosphoglycosidase*, which removes *N*-acetylglucosamine, leaving mannose 6-phosphate (M-6-P) on multiple mannose residues (Fig. 4–28). Therefore, in I cell disease, the acid hydrolases within the fibroblasts or macrophages do not have phosphorylated mannose residues, and, instead of being found in the lysosomes, the acid hydrolases are found within the patient's blood. Why?

The answer to the mistargeting of acid hydrolases in I cell disease can be found by inspecting Fig. 4–29. Acid hydrolases, with associated M-6-P, within the *cis*-Golgi are transferred from the *medial-* to the *trans*-Golgi, and finally to the site of sorting within the TGN. Within the TGN, there is a transmembrane glycoprotein called the *mannose 6-phosphate receptor* that binds acid hydrolases by their M-6-P residues on the luminal side of the membrane. Once the acid hydrolases are bound to the M-6-P receptor, the receptor–ligand complexes are segregated to regions of the TGN that are coated with the protein clathrin. Clathrin-coated vesicles containing the receptor-bound acid hydrolases bud off the TGN, and these transport vesicles rapidly lose their clathrin coat. The transport vesicles now fuse with the lysosome, and the receptor–acid hydrolase complex is internalized within the lysosome. Although the M-6-P receptor can bind the M-6-P-containing acid hydrolase at neutral pH, this complex becomes dissociated at pH 5.0 within the lysosome. The free acid hydrolase is then acted on by a phosphatase within the lysosome, assuring that it cannot reassociate with its receptor. The vacant M-6-P receptor is now recycled by budding off

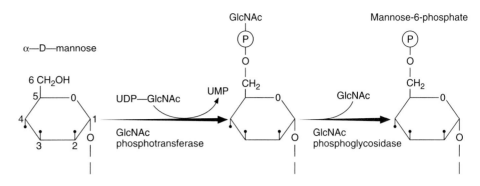

Figure 4–28. Synthesis of mannose-6-phosphate on proteins targeted for the lysosomes. Within the *cis*-Golgi network compartment, an enzyme called *N*-acetylglucosamine-phosphotransferase transfers P-GlcNac residues to the 6 position of several mannose residues on the *N*-linked oligosaccharides of a lysosomal precursor glycoprotein. Next, a phosphoglycosidase removes *N*-acetylglucosamine, leaving mannose-6-phosphate (M-6-P).

of the lysosomal membrane within a coated vesicle, rapidly losing the coat, which then allows the vesicle to fuse with the Golgi TGN.

Although I cell disease is a problem in targeting of acid hydrolases to lysosomes within fibroblasts and macrophages, hepatocytes are not affected. This suggested that there must be a second mechanism for lysosomal targeting that is independent of M-6-P in hepatocytes (and perhaps other cells). Other lysosomal storage diseases are based on a genetic defect within a specific subset of acid hydrolases. For example, Hunter and Hurler syndromes result from defective hydrolases involved in the catabolism of sulfated mucopolysaccharides, and Tay-Sachs disease, prevalent among the Jewish population, is caused by a defect in β-*N*-hexosaminidase A.

Proteolytic Pathways in the Cytoplasm Are Responsible for the Programmed Turnover of Cytoplasmic Proteins As Well As Destruction of Damaged Protein

In addition to the degradation of proteins within the lysosomes, the cytoplasm contains several proteolytic enzyme systems that are responsible for the programmed turnover of cytoplasmic proteins and destruction of damaged proteins. A simple mechanism is in place in eukaryotic cells for such protein turnover. Cytoplasmic proteins that contain Met, Ser, Thr, Ala, Val, Cys, Pro, or Gly as their NH_2-terminal amino acid are stabilized against proteolysis. Those proteins that contain any of the other 12 amino acids at their NH_2-terminus are destabilized, and these proteins become post-translationally modified by the covalent linkage of ubiquitin, a small protein, at their NH_2-terminus. Ubiq-

uitin is linked to a lysine residue at the NH_2-terminus of target proteins, followed by the addition of multiple ubiquitins to the original one. The multiubiquitin complexes then serve to identify the target protein as a substrate for an ATP-dependent protease. Misfolded or denatured proteins are also substrates for the ubiquitin-linked protease system, independently of their NH_2-terminal amino acid.

This ubiquitin-linked protease system is not found in the lumen of the ER or Golgi complex. Therefore, proteins packaged within secretory vesicles are protected from the cytoplasmic proteases.

Proteins and Lipids Destined for the Plasma Membrane, and Proteins and Other Biomolecules Destined for Secretion Are Transferred to the Cell Surface by Secretory Vesicles

The Golgi complex has its own array of membrane proteins and enzymes, and, indeed, the Golgi compartments—CGN, *cis*-, *medial*-, *trans*-, and TGN are functionally distinct. Therefore, membrane proteins that are needed within individual Golgi compartments must have compartment-specific peptide retention signals. The Golgi retention signal is found within the membrane-spanning α-helical segment of transmembrane proteins.

In the last section, we discussed *N*-linked M-6-P as the signal that targets proteins from the Golgi to the lysosome. Now we need to turn our attention to those proteins and lipids that are destined either for the plasma membrane or secretion.

Membrane proteins and lipids that reach the TGN and contain no retention signal for the Golgi or targeting signals for either lysosomes or regulated secretion

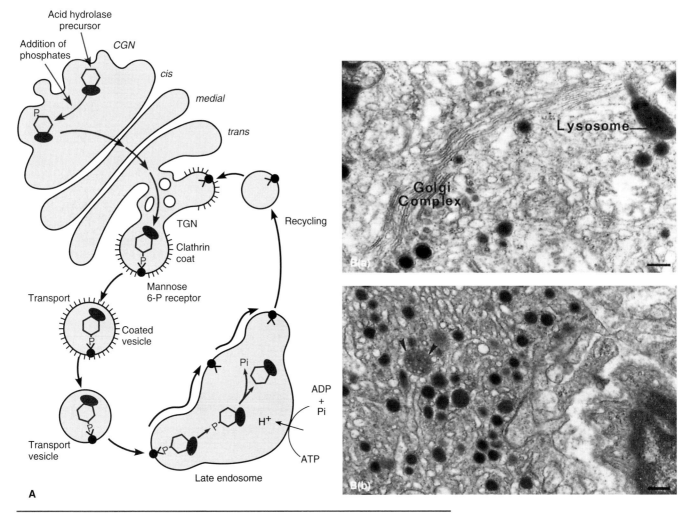

Figure 4–29. Targeting of proteins to the lysosomes and recycling of mannose-6-phosphate receptors. **A:** Precursors to lysosomal hydrolases are identified for lysosomal targeting with M-6-P groups. Within the *trans*-Golgi, the acid hydrolases are concentrated and segregated by binding to M-6-P receptors, which are clustered within clathrin pits. Clathrin-coated vesicles bud off the TGN, carrying the acid hydrolase bound to the M-6-P receptor. The transport vesicle loses its clathrin coat and fuses with late endosomes. At the acid pH of the late endosome, the acid hydrolase dissociates from the M-6-P receptor. A phosphatase within the late endosome removes the phosphate from the M-6-P groups attached to the acid hydrolases, eliminating their ability to rebind to the receptors. The receptors are then recycled to the *trans*-Golgi network, where they are reused. **B:** Lysosome structure. (a) A lysosome is observed budding from the Golgi complex. This is a primary lysosome or one not yet involved in degradative processes (large luteal cell from a day-20 pregnant rat). (b) A secondary lysosome or residual body (*arrowheads*) is a lysosome that is involved in degradative processes. They contain material inside their limiting membrane (large luteal cell from a day-21 pregnant rat) (bars: B(a), 0.22 μm; B(b), 0.31 μm). (Electron micrographs courtesy of Dr. Phillip Fields.)

vesicles bud off from the TGN. These secretory vesicles, which have a nonclathrin coat, continuously bud from the TGN and fuse with the plasma membrane, releasing their contents by exocytosis. Because the process is continuous (requiring no regulatory signal), it is called *constitutive secretion* or *exocytosis*. Most new membrane proteins and lipids are incorporated into the plasma membrane, and many proteins are secreted (e.g., extracellular matrix proteins) by this default pathway (Fig. 4–30). However, some important hormones and neurotransmitters are packaged within a different form of secretory vesicles. For example, proinsulin within pancreatic β-cells is concentrated within clathrin-coated vesicles that bud from the TGN (see Fig. 4–30). Proinsulin and other hormones appear to be concentrated within the TGN by binding to a 25-kDa M_r receptor protein that has the ability to bind aggregates or multiple copies of the hormone and cluster them within the TGN membrane. Once these clathrin-coated vesicles bud off the TGN, they rapidly lose their clathrin coat, and internal pH drops to 5.5 because of an ATP-driven proton pump. The result of this acidification is that proinsulin and other prohormones are released from the 25-kDa receptor protein and are then cleaved to the active hormone.

The secretion of hormones, such as insulin, and of neurotransmitters, such as acetylcholine or glutamate, requires a chemical signal (such as a rise in cytosolic Ca^{2+}) before the secretory vesicle or granule will move to the plasma membrane, then fuse and release its contents. This form of secretion, which requires a chemical signal, is referred to as *regulated exocytosis*. Why does this form of exocytosis require a chemical signal? In synaptic vesicles that carry the neurotransmitters acetylcholine and glutamate, recent studies have indicated that they are restrained by attachment to a cytoskeletal network of actin filaments and nonerythroid spectrin (spectrin II). The rise in cytosolic Ca^{2+} within the presynaptic terminal leads to a series of events that probably includes (a) release of the vesicles from cytosolic actin filaments; (b) translocation of the vesicles toward the presynaptic plasma membrane; (c) binding of vesicles to spectrin II on the cytoplasmic surface of the presynaptic plasma membrane; and (d) fusion of the synaptic vesicles with the plasma membrane, releasing neurotransmitter. It is quite likely that other forms of regulated exocytosis (e.g., insulin secretion) are based on similar mechanisms. If so, a major difference between constitutive secretory vesicles and regulated secretory vesicles would be their ability to bind to the cytoskeleton in a Ca^{2+}-dependent manner.

For nonpolarized cells, constitutive and regulated secretion occurs, and new membrane proteins are incorporated into the plasma membrane. The story becomes a bit more complex for polarized cells that have plasma membrane domains with quite different protein and lipid compositions. For example, in epithelial cells, the apical and basolateral segments of the plasma membrane, separated by tight junctions between cells, have a very different set of membrane-associated enzymes and transport proteins. Although the mechanisms for targeting newly synthesized proteins to the apical and basolateral membranes have not been clearly established, there is some evidence for two possible mechanisms. Targeting of some plasma membrane proteins within polarized cells occurs within the TGN. For example, all proteins with a GPI membrane anchor are sorted from the TGN to the apical surface of the plasma membrane. In other cases, the proteins can be inserted anywhere within the plasma membrane and then selectively sorted by endocytosis, incorporation into endosomes, and transcytosis until they are locked in place by interaction with a polarized membrane skeleton (ankyrin, spectrin II) on the cytoplasmic surface of the plasma membrane (Fig. 4–31).

Without a mechanism for recycling membrane, the events described in this section would create an ever-expanding plasma membrane. This does not occur because membrane is retrieved and reemployed by the process of endocytosis, which is described next.

Macromolecules and Biological Particles Are Internalized by Endocytosis and Phagocytosis With Concurrent Retrieval of Membrane

Exocytosis adds new membrane to the plasma membrane of eukaryotic cells. This process typically does not add to the surface area of the plasma membrane because of concurrent internalization of segments of this membrane by the process of endocytosis. Indeed, this endocytotic-exocytotic cycle must be under tight control because of the rapidity of the events. For example, a macrophage ingests 3% of its plasma membrane per minute.

Classically, endocytosis has been divided into two categories: *pinocytosis*, which means "cell drinking," and *phagocytosis*, which means "cell eating." These two processes are quite different (Fig. 4–32). Phagocytosis is carried out by two types of cells in mammals—macrophages and neutrophils. The functions of these cells are to rid the body of senescent or damaged cells and to protect us against pathogenic microorganisms. Macrophages and neutrophils carry out this function by engulfing the cell or particles with their plasma membrane, followed by internalization of the membrane-enclosed *phagosome*. Phagosomes are the size of the particle they are ingesting, and tend to be >1.0 μm in diameter. Once ingested by the cell, the phagosome, with entrapped particle, fuses with the cell's lysosome, forming a phagolysosome in which the ingested mater-

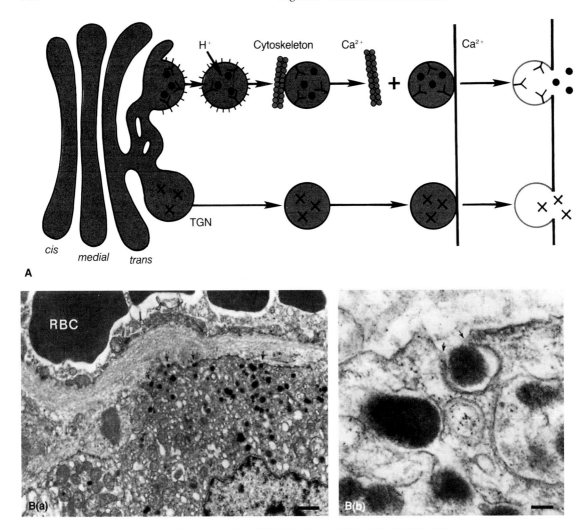

Figure 4–30. Constitutive versus regulated secretion. **A:** Although vesicles involved in both constitutive and regulated secretion bud from the *trans*-Golgi network, they appear to carry out their functions with the following differences: (1) vesicles involved in regulated secretion bud from clathrin-coated pits, whereas constitutive vesicles have coatomer coats; (2) the coated pits contain a receptor that concentrates the proteins secreted by the regulated pathway; (3) the regulated vesicles have a low internal pH, thereby releasing the secretory protein from its receptor; (4) the regulated secretory vesicles bind to the actin-based cytoskeleton and are released by an increase in intracellular Ca^+; and (5) the fusion event with the plasma membrane requires Ca^+. **B:** Endocrine cells contain membrane-bound secretory granules that fuse with the cell membrane and release their content into the extracellular space. The product enters the bloodstream through capillary beds and is then transported to a target tissue. (**a**) Large luteal cells of the cow corpus luteum contain numerous secretory granules (150 nm diameter) that are readily observed being exocytosed (*arrows*) into the extracellular space adjacent to a blood vessel containing red blood cells (RBC). (**b**) A higher-power photomicrograph of (**a**) showing the fusion of the granule membrane with the cell membrane (*arrows*). The dense product is being released into the extracellular space (bars: **B(a)**, 0.74 μm; **B(b)**, 0.05 μm). (Electron micrographs courtesy of Dr. Phillip Fields.)

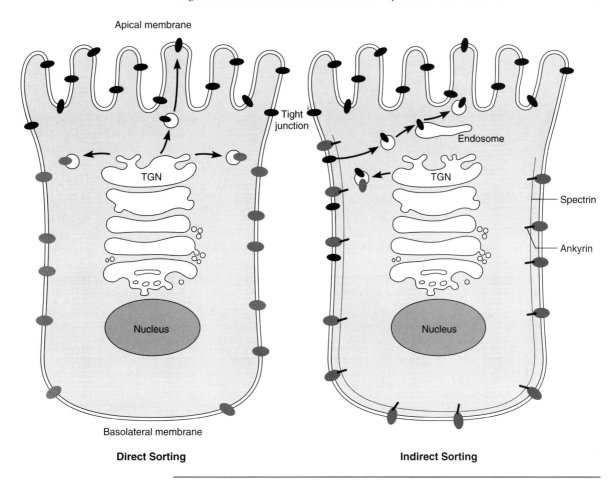

Figure 4–31. Sorting of membrane proteins into the apical and basolateral surface of epithelial cells. Proteins are trafficked to the apical and basolateral surfaces of these polarized cells, either by *direct sorting* based on signals that direct the proteins into secretory vesicles targeted for one surface or the other, or, alternatively, due to *indirect sorting*. In indirect sorting, membrane proteins are delivered from the TGN to the basolateral membrane. Those that are linked to the spectrin membrane skeleton via ankyrin remain in the basolateral membrane. Those that are not linked to the spectrin membrane skeleton are endocytosed, fuse with endosomes, then bud from the endosome and, by transcytosis, fuse with the apical membrane.

ial is digested by acid hydrolases, leaving behind indigestible substances that form residual bodies within the lysosome. The mechanism by which particles are engulfed by the plasma membrane of macrophages and neutrophils is shown in Fig. 4–33. Pathogenic bacteria and damaged cells are seen by the immune system as foreign material and, therefore, are coated on their surfaces by IgG antibodies. The Fc region of the IgG molecule is recognized by Fc receptors on the surface of the macrophage or neutrophil. If the IgG molecules completely surround the particle, the plasma membrane of the phagocyte will completely wrap around and engulf the foreign particle by a membrane-zippering mechanism (see Fig. 4–33). If the IgG molecules are localized to one region of the particle, the phagocyte will bind the

particle through the IgG–Fc receptor interaction but phagocytosis will not occur.

True endocytosis (see Fig. 4–32) is quite different from phagocytosis, as just described. In endocytosis, the plasma membrane invaginates instead of evaginating around a foreign particle. The invagination leads to adherence and fusion of the plasma membrane, creating an endocytic vesicle that is typically 0.1–0.2 μm in diameter (see Fig. 4–32). There are two forms of invaginating endocytosis fluid-phase endocytosis (pinocytosis) and receptor-mediated endocytosis. Most eukaryotic cells undergo both fluid-phase endocytosis and receptor-mediated endocytosis, primarily at specialized plasma membrane sites called *clathrin-coated pits*. In fluid-phase and receptor-mediated endocytosis, the

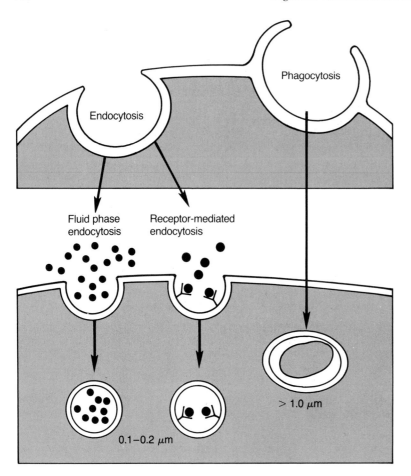

Figure 4–32. Endocytosis and phagocytosis. In phagocytosis, membrane evaginations engulf a large particle, followed by internalization of the membrane-enclosed phagosome. The phagosome is >1.0 μm in diameter. For endocytosis, the membrane invaginates and fuses, internalizing small vesicles of 0.1–0.2 μm diameter. In receptor-mediated endocytosis, the molecule being internalized first binds to its receptor on the plasma membrane and is then captured within the budding vesicle. This is a concentrating step because these receptors are enriched at the site of internalization (often a coated pit). In pinocytosis or fluid-phase endocytosis, there is no such receptor-stimulated concentrating step.

endocytic vesicles are constantly forming, and all molecules entrapped within the invagination of the plasma membrane are carried into the cell.

Receptor-mediated endocytosis involves the binding of a molecule or small particle to a receptor. The receptor is usually clustered within a clathrin-coated pit, and the resulting endocytic vesicles that are formed are referred to as *coated vesicles*. Therefore, receptor-mediated endocytosis differs from fluid-phase endocytosis in the concentrating of ligand within the endocytic vesicles. The polyhedral coat that is found on the cytoplasmic surface of the plasma membrane and endocytic vesicle is composed primarily of the protein clathrin. Clathrin takes the form of a three-legged triskelion in which each leg is formed from one heavy chain (M_r 180 kDa) and one light chain (35–40 kDa) (Fig. 4–34). The triskelions assemble to form the polyhedral coat, which is a mixture of hexagons and pentagons, and this process drives the invagination of the membrane. Because the receptors are clustered within the coated pit, they tend to concentrate their specific ligand within the coated vesicle. A general scheme of the events following receptor-mediated endocytosis is shown in Figure 4–35.

Once the coated vesicle enters the cortical cytoplasm, clathrin is rapidly removed by an uncoating ATPase, and the clathrin returns to the plasma membrane to form new coated pits. The uncoated endocytic vesicle fuses with an early endosome. The resulting endosome contains an ATP-dependent proton pump that maintains its acidic internal pH at 6.0. Many ligand–receptor complexes formed at neutral pH become dissociated at pH 6.0. For those complexes that dissociate within the endosome, the receptors are recycled to the plasma membrane by receptor-enriched vesicles budding off the early endosome and returning to the plasma membrane. The ligands still enclosed within the early endosome are released within budding transport vesicles, which are carried to late endosomes, probably via microtubule-based translocation. Transport vesicles from the Golgi that are carrying newly made acid hydrolases fuse with the late endosomes. Late endosomes mature into lysosomes, although the mechanism underlying this conversion remains to be determined. The ligands within the lysosomes are degraded and their building blocks reused or excreted.

It would be appropriate to discuss two physiologically important examples of receptor-mediated endocy-

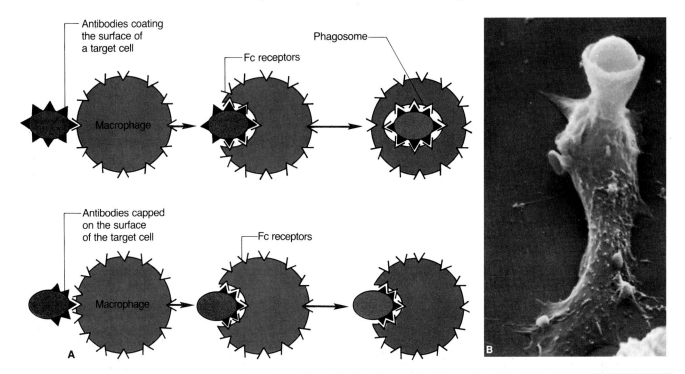

Figure 4–33. Membrane-zippering mechanism for phagocytosis. **A:** Drawing of the experimental evidence that suggests a membrane-zippering mechanism for phagocytosis. Details are discussed in the text. **B:** A macrophage shown ingesting an aged red blood cell. (**A,** Modified from Alberts B, Bray D, Lewis J, et al. *Molecular Biology of the Cell,* 2nd ed. New York: Garland Publishing, 1989; **B,** courtesy of Dr. Marguerite Kay.)

tosis. The first relates to the Nobel prize-winning studies of Brown and Goldstein on the inherited disorder, familial hypercholesterolemia. This disorder is characterized by high levels of cholesterol in the blood, leading to atherosclerotic plaques and premature heart attacks. We must first understand that cholesterol does not circulate in its free form, but instead as a component of 20-nm diameter, spherical LDL particles (Fig. 4–36). The LDL particles have an outer monolayer composed of phospholipids and unesterified cholesterol, and a single large (4563-amino acid) integral protein called *apo-B*. Within the LDL particle is a hydrophobic core of cholesterol that is esterified primarily to linoleic acid through a hydroxyl group. Most cells contain an LDL receptor in their plasma membrane that binds LDL and causes its uptake by receptor-mediated endocytosis in coated vesicles. The LDL receptor is a homodimer consisting of two single-pass integral membrane subunits of 839 amino acids, of which 767 are on the exoplasmic side, 22 amino acids form the transmembrane domain, and 50 amino acids form the cytoplasmic domain (see Fig. 4–36). In subjects with familial hypercholesterolemia, the LDL receptors are not produced, or they have defective binding sites for

LDL. Either of these alterations leads to defective uptake of cholesterol-containing LDL. In some subjects, the LDL receptor lacks the COOH-terminal domain that localizes it within the coated pits. Here, LDL can bind to the LDL receptor throughout the plasma membrane, but it is not internalized within the coated vesicles. In normal LDL endocytosis, the ligand–receptor complex is dissociated at pH 6.0 within the early endosome, and the receptor is recycled to the plasma membrane. The LDL particles are transferred to the late endosome and then to the lysosomes, where the cholesterol esters are hydrolyzed to free cholesterol and free fatty acid, and apo-B is proteolyzed to amino acids. The cholesterol and fatty acids can exit the lysosome and be incorporated into membranes or reused for phospholipid synthesis, respectively. Cells can regulate their cholesterol levels in several ways. When the cell has an excess of cholesterol, it can turn off cholesterol synthesis and downregulate the synthesis of the LDL receptor by preventing the transcription of its mRNA.

A second example of receptor-mediated endocytosis deals with transferrin, a serum glycoprotein involved in transport of iron from the liver and intestine to other tissues. Apotransferrin is capable of binding two ferrous

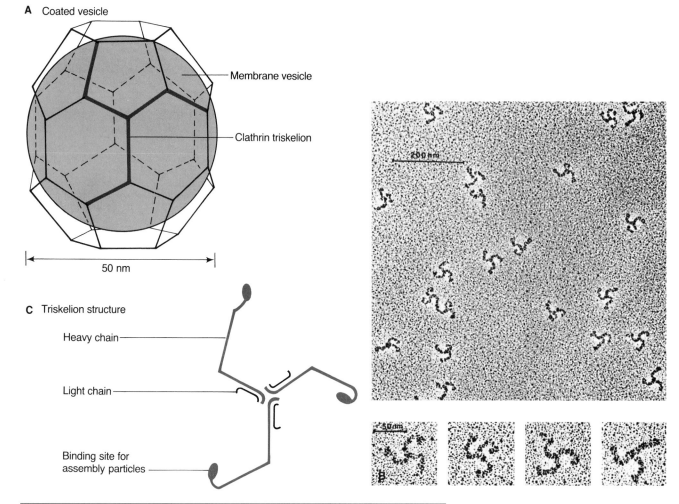

Figure 4–34. Structure and assembly of the "coat" of a coated vesicle. **A:** The basketlike appearance of the coat is created by 36 clathrin triskelions (*red*). **B:** Electron micrograph of rotary shadowing and electron microscopy of clathrin triskelions. **C:** Subunit interactions within the triskelion. (**B,** courtesy of Dr. D. Branton.)

ions (Fe^{2+}), forming ferrotransferrin. All cells contain ferrotransferrin receptors, which avidly bind ferrotransferrin, but not apoferritin, at the neutral pH of the extracellular interstitial fluid. The binding of ferrotransferrin triggers receptor-mediated endocytosis by coated vesicles. Within the acidic environment of the endosome, the two ferrous ions dissociate from transferrin, but the apotransferrin–receptor complex remains intact. The free ferrous ions are transported into the cytoplasm, and the apotransferrin–receptor complex is recycled to the plasma membrane. However, once the complex reaches the extracellular interstitial fluid, the apotransferrin is released at neutral pH and can be reused to transport more iron.

We have discussed examples for which the receptor in receptor-mediated endocytosis is reincorporated into the plasma membrane from which it came. This, how-

ever, is not always true (Fig. 4–37). In polarized cells such as epithelial cells, endocytic vesicles from the apical plasma membrane can be reincorporated into the basolateral plasma membrane. The signals that direct these vesicles to specific membrane domains are unknown. Furthermore, occasionally, when the low pH of the endosome does not dissociate the ligand–receptor complex, the receptor is degraded in the lysosome along with the ligands. An example is epidermal growth factor (EGF), which binds an EGF receptor that is then clustered within a coated pit. The EGF–receptor complex does not dissociate at pH 6.0 within the endosome; consequently, the EGF receptor and its ligand are both transferred to the lysosome. This process is referred to as *receptor down-regulation*, because the binding of EGF to the plasma membrane results in a decrease in the number of EGF receptors on the plasma membrane (see Fig. 4–37).

Figure 4–35. **Steps by which the contents of receptor-mediated endocytotic vesicles reach the lysosomes and the receptors are recycled to the plasma membrane.** These steps are explained in the text. An endocytosed protein (*red spheres*), the destination of which is within a lysosome, is represented, as are acid hydrolases (*black spheres*).

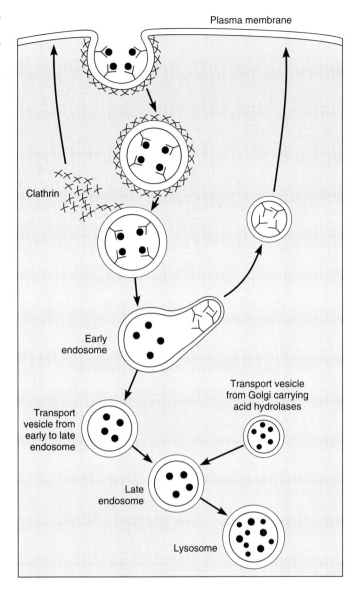

Clathrin-Coated Vesicles Mediate Selective Transport of Membrane Proteins, While Coatomer-Coated Vesicles Are Involved in Nonselective Vesicular Transport

Clathrin-coated vesicles mediate the movement of selected transmembrane proteins from the TGN to the late endosomes (e.g., M-6-P receptor) or into regulated secretory vesicles on the way to fusion with the plasma membrane. The clathrin-coated vesicles are also involved in receptor-mediated endocytosis in which specific receptors are internalized in clathrin-coated vesicles, which lose their coat prior to fusing with early endosomes. However, transport vesicles that move between the ER and Golgi, between Golgi compartments, or between the Golgi and plasma membrane via constitutive secretory vesicles contain a *coatomer* coat.

The assembly of the clathrin triskelions into a network of twelve pentagons and six hexagons drives the curvature of the membrane and the budding process that leads to a coated vesicle. In addition to pulling membrane into the bud of what will become a coated vesicle, the clathrin coat also is involved in capture of transmembrane receptor proteins and their ligands into selective transport vesicles. Key to this process is a family of multisubunit transmembrane proteins called *adaptins*. As shown in Figure 4–38, the adaptins link the cytoplasmic domain of the transmembrane receptors to the clathrin coat. Any bound ligand, e.g., the LDL particle attached to the LDL receptor at the cell surface, will become incorporated into the lumen of the coated vesicle that forms in response to clathrin assembly. A stretch of four amino acids—F-R-X-Y—in the COOH-terminal cytoplasmic domain of cell surface receptors serves as an endocytosis signal for

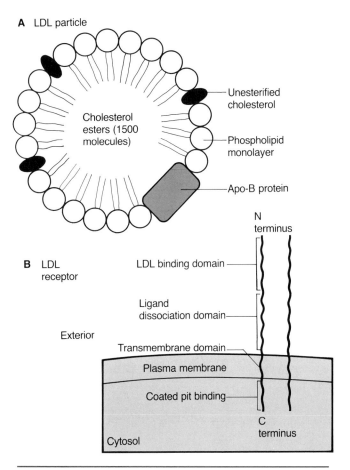

A LDL particle

Unesterified cholesterol

Cholesterol esters (1500 molecules)

Phospholipid monolayer

Apo-B protein

B LDL receptor

N terminus

LDL binding domain

Ligand dissociation domain

Exterior

Transmembrane domain

Plasma membrane

Coated pit binding

Cytosol

C terminus

Figure 4–36. Structure of the LDL particle and LDL receptor. Diagrammatic presentation of (**A**) the structure of the LDL particle and (**B**) the LDL receptor. The LDL receptor is a dimer of two identical 839-amino acid polypeptides. Receptors lacking the ligand-dissociation domain can bind LDL, but cannot release it.

docked at its target site. A monomeric G protein, called *ARF*, with a covalently linked fatty acid chain plays an essential role in coatomer assembly and disassembly. When it has guanosine diphosphate (GDP) bound, it is soluble within the cytosol and inactive. A guanine nucleotide releasing protein (GNRP) that is associated with a donor membrane that produces coatomer-coated vesicles causes release of the bound GDP and binding of GTP. As a result, the activated ARF-GTP changes conformationally in a manner in which its fatty acid chain can insert into the donor membrane. The ARF-GTP then binds the COP subunits, forming the coatomer coat. Assembly of the coatomer coat causes the underlying membrane to bud, then fuse, pinching off a coatomer-coated vesicle. Disassembly of the coatomer coat also involves the

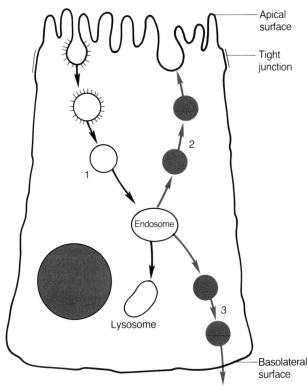

Apical surface

Tight junction

Endosome

Lysosome

Basolateral surface

Figure 4–37. Fates of endocytosed receptors in a polarized cell. **Pathway 1:** The receptor is not dissociated from the ligand at the acid pH of the endosome. Therefore, both the ligand and receptor proceed to the lysosome. **Pathway 2:** The receptor has dissociated from the ligand within the endosome and is recycled to the apical plasma membrane. **Pathway 3:** An example of the recycled receptor being directed to the basolateral membrane (transcytosis). Pathway 1 requires no special signal; therefore, it is the default pathway (*black*). Pathways 2 and 3 must require a signal on the surface of the vesicles that bleb off of the endosome, directing the vesicles to the apical or basolateral plasma membrane; therefore, these pathways are differentiated (*red*).

attachment to the adaptins. Likewise, a stretch of phosphorylated amino acids at the COOH-terminus of the M-6-P receptor signals its attachment to adaptins within the TGN. Once a clathrin-coated vesicle forms, the clathrin coat must be removed before the transport vesicle can fuse with its selected target. The coat is removed by an uncoating ATPase that is a chaperon protein of the hsp 70 family. This protein utilizes the energy of ATP hydrolysis to disassemble and remove the clathrin from the cytoplasmic surface of the coated vesicle.

The coatomer vesicles that are involved in nonselective vesicular transport are composed of seven coat protein subunits termed *COPs*. One of the COPs, (β COP) has sequence homology to the adaptins. The formation of the coatomer complex requires energy in the form of ATP hydrolysis. Furthermore, the coatomer does not disassemble into COPs until the vesicle is

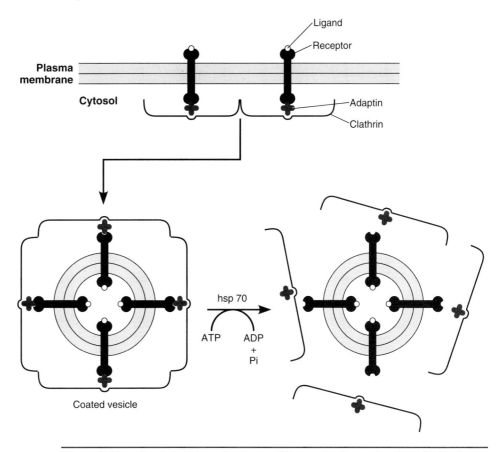

Figure 4–38. Role of clathrin, adaptins, and hsp 70 in the production of selective transport vesicles.

ARF protein. As the coatomer-coated vesicle reaches the target membrane, a GTPase-activating protein (GAP) associated with the target membrane causes the GTP bound to ARF to be hydrolyzed to GDP+P_i. As a result, ARF is released from the membrane, and the coatomer coat disassembles. A summary of these events is found in Fig. 4–39.

There Appear To Be Common Pathways for Vesicular Docking and Fusion

As a transport vesicle approaches a potential target membrane, it must identify whether that is the appropriate target membrane and it must penetrate the proteinaceous meshwork of the membrane skeleton in order for the fusion complexes to form. We will go through the steps using the fusion of synaptic vesicles with the active zone of the presynaptic terminal membrane as an example. Common pathways are found for all eukaryotic vesicular docking and fusion events.

First, as the synaptic vesicle approaches the plasma membrane, it encounters the spectrin membrane skele-ton. The available evidence suggests that the synaptic vesicle binds to the tails of spectrin tetramers (described in Chapter 3) via synapsin, causing the tail to dissociate from membrane-bound actin filaments. The result is that the synaptic vesicle is initially docked to the membrane via this attachment with spectrin (Fig. 4–40). With this initial attachment in place, VAMP [a vesicular-SNARE (v-SNARE) protein] and synaptotagmin (a Ca^{2+} sensor protein) on the donor vesicle associate with two t-SNAREs (syntaxin and SNAP-25) on the target membrane. This binding allows a monomeric G-protein on the donor membrane to cleave bound GTP to GDP and P_i, locking the vesicle onto the target membrane. In the next step, a soluble NSF attachment protein (SNAP) binds to the v-SNARE–t-SNARE complex, causing release of synaptotagmin from the complex. An N-ethylmaleimide-sensitive fusion (NSF) protein binds to SNAP, completing the fusion complex; NSF is an ATPase that hydrolyzes ATP, causing release of SNAP and NSF, and fusion of the membranes. Finally, in a Ca^{2+}-dependent step that follows the NSF-dependent ATP hydrolysis, the membranes become one, and neurotransmitter is released.

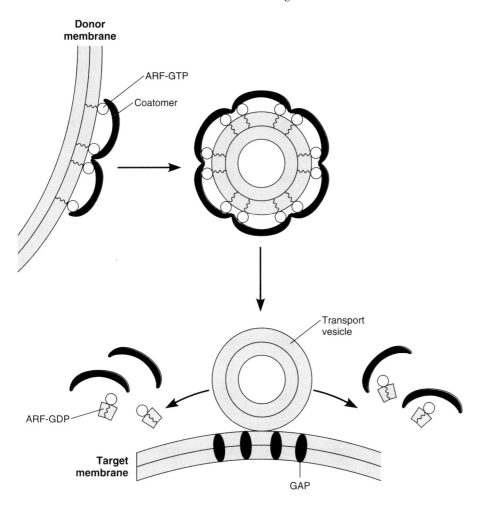

Cytoplasmic Organelles that Import Most of their Proteins from the Cytosol

Mitochondria Use the Proton Motive Force Created by Electron Transfer from NADH to Oxygen to Drive ATP Production

Mitochondria are ideally configured to carry out their function of producing ATP by oxidative phosphorylation. Electron microscopy of mitochondria indicate a roughly ellipsoid organelle, with a length of 1–2 μm and a width of 0.1–0.5 μm (Fig. 4–41). This static view, however, is quite misleading because mitochondria viewed within a living cell, with the fluorescent dye rhodamine 123, are constantly changing shape, fusing, dividing, and moving. The average mitochondrion will double its mass and divide in two during each cell cycle; however, some individual mitochondria will divide more rapidly, whereas others not at all during this fixed time. Mitochondria have their own autonomous double-stranded circular DNA. The human mitochondrial genome has been completely sequenced and contains 16,569 nucleotides, which is less than 10^{-5} times the size of the human nuclear genome. There are five to ten mitochondrial genomes per human mitochondria and multiple mitochondria per cell. Therefore, this DNA represents less than 1% of the total cellular DNA.

The mitochondrion has two membranes: outer and inner. These membranes create two compartments: an intermembrane space that separates the inner and outer membranes, and a matrix that is the compartment enclosed by the inner membrane (see Fig. 4–41). The composition and functions of the individual mitochondrial compartments have been extensively studied. The outer membrane contains a major integral protein, called *porin*. Porin can form channels within the outer membrane, through which molecules that are less than 10,000 M_r can freely pass. Therefore, the intermembrane space and the cytoplasm can be considered continuous compartments for molecules that are smaller than 10 kDa. The inner membrane has a very large surface area, owing to its characteristic infolds, called *cristae*. This membrane is composed of 75% protein and is extremely active functionally. The inner membrane contains proteins involved in the respiratory chain, ATP

Figure 4–40. The role of v-SNAREs, t-SNAREs, SNAPs, NSF, spectrin, and rabs in vesicle docking, activation, and fusion.

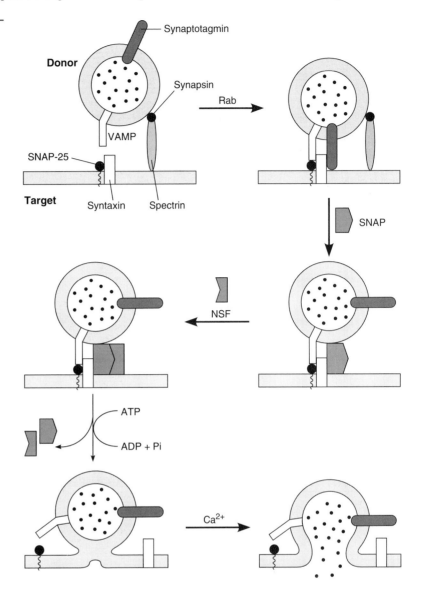

synthesis, and the transport of substrates and products of oxidative phosphorylation into and out of the mitochondria. The inner membrane has a high concentration of cardiolipin (diphosphatidylglycerol), a phospholipid thought to decrease the permeability of this bilayer to small ions. The matrix contains enzymes involved in the oxidation of pyruvate and fatty acids, as well as most of the enzymes of the citric acid cycle. In addition, the mitochondrial genome and mitochondrial ribosomes are located in the matrix. The matrix protein composition is approximately 500 mg/ml or a 50% protein solution; therefore, it has a viscous gel-like consistency.

In 1961, Mitchell proposed that the production of ATP within mitochondria was powered by a mechanism that he called *chemiosmotic coupling*. To understand chemiosmotic coupling, we will begin in the cytoplasm with the formation of pyruvate and fatty acyl-CoA. Glucose is converted to pyruvate in the cytoplasm by

the Embden-Meyerhof glycolytic pathway (Fig. 4–42). The net reaction is

$$\text{Glucose} + 2\text{NAD}^+ + 2\text{ADP}^{2-} + 2\text{P}_i \rightarrow$$
$$2 \text{ Pyruvate} + 2\text{NADH} + 2\text{ATP}^{4-} + 2\text{H}^+$$

The two reduced nicotinamide adenine dinucleotide (NADH) molecules can reduce NAD^+ molecules in the mitochondria that will serve as electron donors for the reduction of oxygen. Pyruvate also enters the mitochondrial matrix, where it is immediately converted to acetyl-CoA, the key substrate for the citric acid cycle. Another source of mitochondrial acetyl-CoA is the oxidation of fatty acids. Hormone stimulation causes free fatty acids, which have been stored primarily in adipose tissue as triacylglycerol, to be released into the bloodstream. Free fatty acids can cross the plasma membrane of cells and are converted in the cytoplasm to fatty acyl-

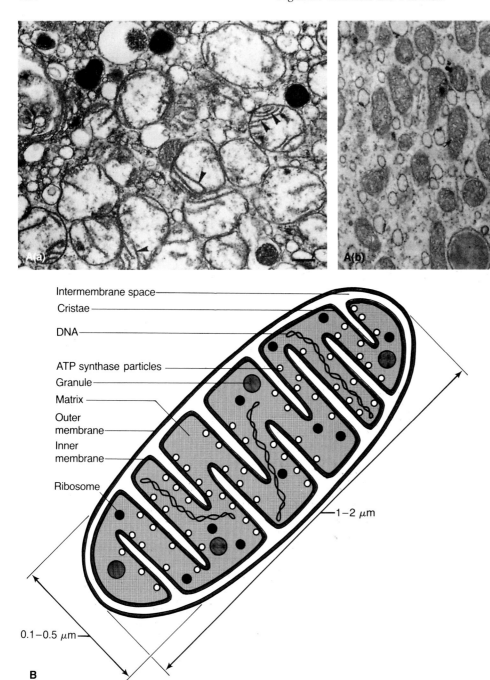

B

Intermembrane space
Cristae
DNA
ATP synthase particles
Granule
Matrix
Outer membrane
Inner membrane
Ribosome

1–2 μm

0.1–0.5 μm

Figure 4–41. **Structure of the mitochondrion. A:** Mitochondria exhibit numerous types of internal architecture in the inner membrane. **(a)** Mitochondria with lamellar or foliate cristae (*arrowheads*) in the large luteal cell of a late-pregnant cow. **(b)** Mitochondria with tubular cristae (*arrows*). This is the preponderant type found in active steroid-secreting cells, such as the corpus luteum of this early-pregnant cow (bars, **A(a),** 0.11 μm; **A(b),** 0.35 μm). **B:** Diagrammatic presentation of mitochondrial compartments. (**A,** courtesy of Dr. P. Fields; **B,** modified from Darnell J, Lodish H, Baltimore D. *Molecular Cell Biology,* 2nd ed. New York: Scientific American Books, 1990.)

Figure 4–42. The glycolytic (Embden-Meyerhof) pathway.

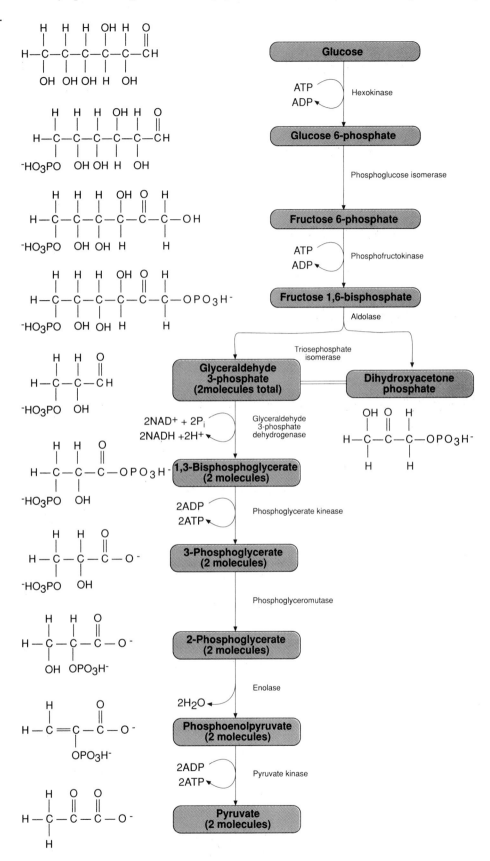

CoA, which can enter the mitochondrion. Within the mitochondrial matrix, free fatty acyl-CoA is broken down by a cycle of reactions (Fig. 4–43) that removes two carboxyl carbons and produces one acetyl-CoA molecule per cycle.

The acetyl-CoA formed by the oxidation of pyruvate and fatty acyl-CoA fuels the citric acid cycle (Fig. 4–44). During the citric acid cycle, acetyl-CoA is oxidized to two molecules of CO_2 (which are released from the cell), and the released electrons are transferred to NAD and flavin adenine dinucleotide (FAD). The net reaction for the citric acid cycle is

$$Acetyl\text{-}CoA + 3\ NAD^+ + FAD + GDP^{3-}$$
$$+ P_i^{2-} + 2H_2O \rightarrow 2CO_2 + 3NADH + FADH_2$$
$$+ GTP^{4-} + 2H^+ + HSCoA$$

The $3NADH+FADH_2$ molecules formed by the citric acid cycle and the NADH molecules generated during glycolysis can now transfer electron pairs to acceptor molecules on the inner mitochondrial membrane, eventually leading to the reduction of oxygen and formation of water.

This brings us to the steps of chemiosmotic coupling (Fig. 4–45). When NADH (or $FADH_2$) is oxidized, it releases a hydride ion (H^-). The hydride ion is immediately converted to a proton (H^+), and two high-energy electrons ($2e^-$). It is these two high-energy electrons that reduce molecular oxygen to water during mitochondrial respiration. During this process, there is generation of approximately 53 kcal/mol of free energy, which would be lost as heat if the transfer was direct. Instead, the transfer of electrons from NADH (or $FADH_2$) to oxygen is catalyzed by a series of electron carriers associated with four multiprotein complexes. Three of these multiprotein complexes are presented schematically in Fig. 4–45. The two high-energy electrons from NADH are initially transferred to the NADH dehydrogenase complex. The NADH dehydrogenase complex has an M_r of approximately 800,000 and consists of at least 22 polypeptides. The high-energy electrons are transferred by a flavin and several iron–sulfur prosthetic groups attached to the protein complex to ubiquinone. Ubiquinone (Fig. 4–46), or coenzyme Q, can accept the two electrons in two steps, in which it is first converted to ubisemiquinone and then ubiquinol. The hydrophobic ubiquinone can then move laterally in the membrane and transfer the electrons to the b–c_1 complex. The b–c_1 complex contains eight polypeptides that form two linked proteins with three cytochromes and one iron–sulfur group attached per protein. The cytochromes contain a bound heme group with a Fe^{3+} (ferric) iron atom that converts to a Fe^{2+} (ferrous) state when it accepts an electron. The cytochromes and iron–sulfur complex transfer the pair of electrons from ubiquinone to cytochrome c, a peripheral membrane protein.

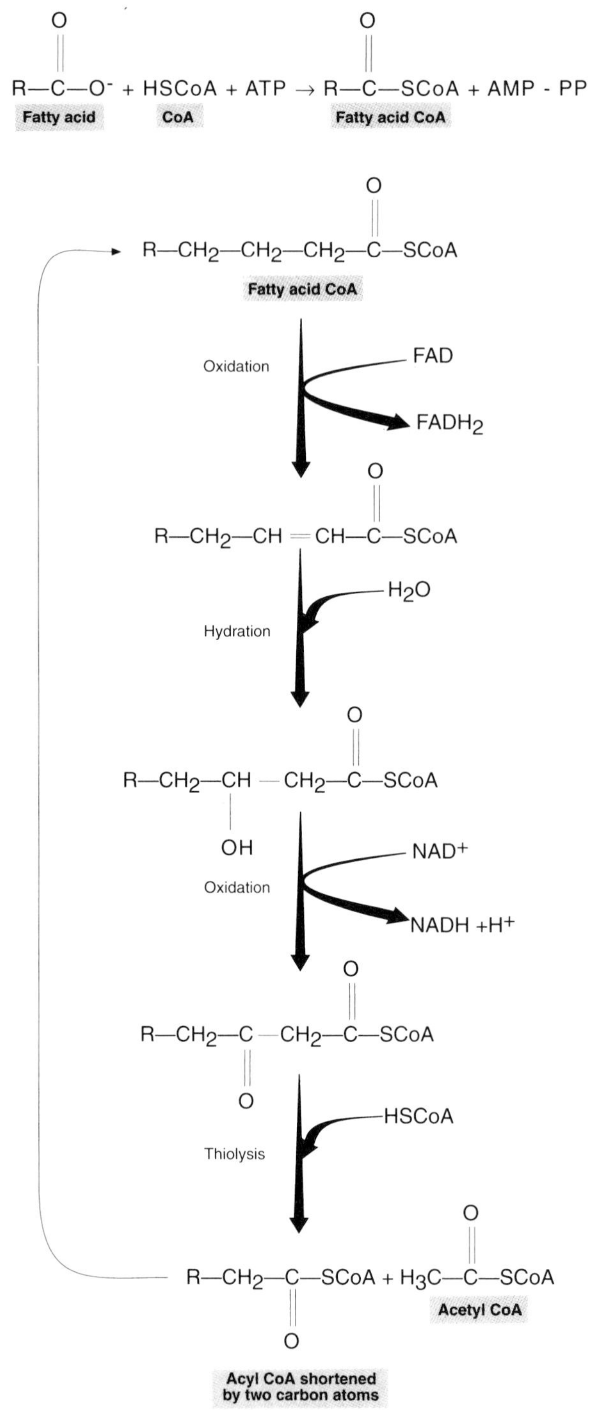

Figure 4–43. Pathway for oxidation of fatty acids.

Cytochrome c, which is attached to the inner mitochondrial membrane facing the intermembrane space, transfers the electrons to cytochrome oxidase. Cytochrome oxidase contains at least eight polypeptides that associate to form two attached proteins. The complex has an M_r of approximately 300 kDa. Each protein monomer contains two cytochromes and two copper atom carri-

Figure 4–44. The citric acid cycle.

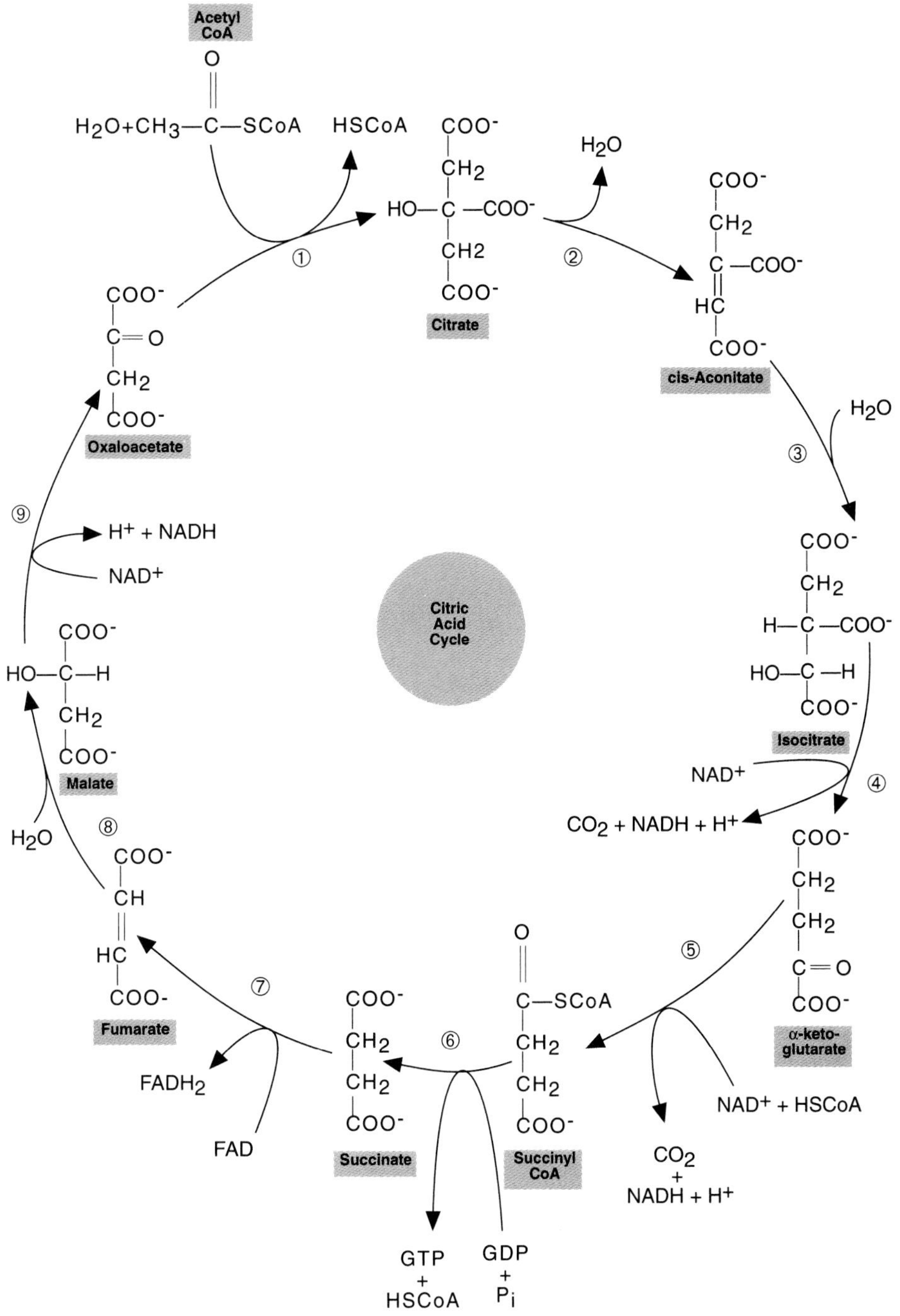

ers, which accept the electrons from cytochrome c and transfer them to oxygen. Not shown in Figure 4–45 is the succinate–ubiquinone–reductase complex that transfers high-energy electrons created during the conversion of succinate to fumarate from $FADH_2$ to an iron sulfur center to ubiquinone. Ubiquinone can then transfer these high-energy electrons to the b–c_1 complex.

As the high-energy electron pair is transferred to each of the three protein complexes (see Fig. 4–45), there is an allosteric conformational change in these complexes that moves protons from the matrix to the intermembrane space. The inner membrane is impermeable to ions such as H^+ but, because of the porin channels, once in the intermembrane space, the H^+ can freely pass into the cytoplasm. This movement of protons out of the matrix has two effects: the membrane potential of the inner membrane of the mitochondria is 160 mV and is negative on the matrix side; the pH within the cytoplasm and intermembrane space (pH $\cong$ 7) is 1 pH unit lower than in the matrix (pH $\cong$ 8). Hence, the first aspect of the chemiosmotic coupling theory is that the transfer of electrons from NADH down the electron transport chain causes protons to be pumped out of the matrix. This pumping of protons causes a proton motive force (PMF)

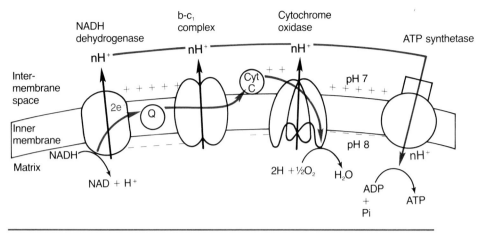

Figure 4–45. Steps in chemiosmotic coupling.

to form, which is the additive force placed on a proton in the intermembrane space to move down its electrochemical gradient (proton concentration gradient+Δ membrane electric potential). Put in mathematical terms:

$$PMF = MP - \frac{2.303(RT)}{F} \times \Delta pH$$

where

MP = membrane potential = 160 mV across inner membrane
R = gas constant = 1.987 cal/degree · mol
T = temperature (°K)
F = Faraday constant = 23.062 cal · mV^{-1}

∴ at 37°C

$$PMF = 160 - \frac{2.303(1.987)(310°)}{23.062} \times (-1)$$
$$PMF = 160 + 61.5 = 221.5 \text{ mV}$$

Therefore, protons have a proton motive force of 221.5 mV at 37°C, pulling them across the inner membrane and back toward the matrix. How can this proton motive force drive ATP synthesis? The inner membrane contains multiple lollipop-like protrusions (termed *elementary particles* in electron micrographs) associated with its inner surface (facing the matrix) that are respon-

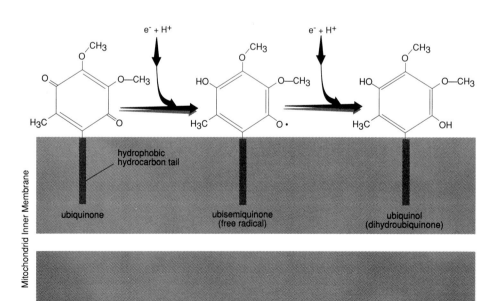

Figure 4–46. Ubiquinone as an electron carrier. Ubiquinone, or coenzyme Q, accepts two electrons in a two-step process discussed in the text.

sible for ATP production. Submitochondrial particles containing these protrusions can be isolated from the inner membrane as sealed inside-out vesicles (Fig. 4–47). This means that the protein-containing spheres, attached by stalks to the inner membrane, are facing out. These submitochondrial particles can use the electron pair from NADH to reduce oxygen and convert ADP+phosphate to ATP in the test tube (see Fig. 4–47). If the submitochondrial particles are mechanically disrupted so that the protein spheres are separated from the vesicles, the vesicles can still transfer electrons to oxygen but cannot form ATP; and the heads can form ATP, but cannot transfer electrons. If the heads and vesicles are reconstituted, both functions return.

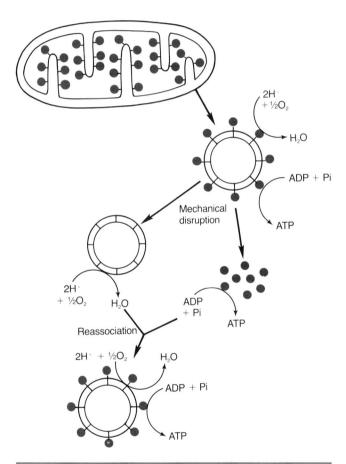

Figure 4–47. **Submitochondrial particles can be used to study oxidative phosphorylation.** Submitochondrial particles can be isolated by differential centrifugation of homogenized mitochondria. These submitochondrial particles contain lollipop-like protrusions on their outer surface that are involved in ATP synthesis from ADP+P_i. If the submitochondrial particles are mechanically disrupted such that they lose the protein spheres, they also lose the ability to produce ATP from ADP+P_i. However, the sheared particles can still transfer an electron pair from NADH to oxygen to form H_2O. If the heads and particles are reconstituted, they can again carry out both functions.

This important experiment indicates that ATP synthetase activity resides within the heads of the elementary particles associated with the mitochondrial inner membrane. This ATP synthetase contains two segments (Fig. 4–48): one portion is buried in the bilayer called F_0 and a peripheral portion (which is attached to the bilayer by F_0) called F_1. F_0 contains five to eight subunits, including a and b proton conduction subunits, c proteolipid, and two to five accessory proteins. F_1 has nine subunits with an $\alpha_3\beta_3\gamma\delta\epsilon$ oligomeric structure. The α-subunit is regulatory and senses the surrounding ADP–ATP levels. The β-subunit catalyzes the ADP+P_i→ATP conversion. The γ and δ subunits bind F_0, and the function of ϵ is unknown. This ATP synthetase, therefore, converts ADP+P_i to ATP, and is powered by protons moving down their electrochemical gradient into the matrix, powered by the proton motive force. The ATP synthetase is reversible, and, were it not for the large proton motive force, it could theoretically hydrolyze ATP and use the free energy of hydrolysis to move protons out of the matrix. Therefore, the second point of the chemiosmotic coupling theory is that the proton motive force causes movement of protons down their electrochemical gradient by ATP synthetase, and this movement drives ATP synthesis.

The movement of protons down their electrochemical gradient is also coupled to the movement of inorganic phosphate, Ca^{2+}, and pyruvate into the mitochondrial matrix by symport mechanisms (see Chapter 2). The mechanism for getting the newly synthesized ATP out of the mitochondria and into the cytoplasm is accomplished by an ADP–ATP antiport system by which the movement of ATP down its electrochemical gradient (from matrix to intermembrane space and into the cytoplasm) is coupled to the movement of ADP in the reverse direction. Figure 4–49 summarizes this section.

The Mitochrondrion Must Import Most of Its Protein from the Cytosol

The entire human mitochondrial genome has been sequenced, and its circular organization is shown in Figure 4–50. There are several fascinating features of this mitochondrial genome. First, almost the entire genome is coding sequence. The 16,569 nucleotides encode 13 proteins, many of which are subunits of protein complexes involved in electron transport and oxidative phosphorylation, 2 rRNA genes and 22 tRNA genes. That mitochondrial protein synthesis requires only 22 tRNAs, whereas there are 31 tRNAs involved in cytosolic protein synthesis, leads to the second distinctive feature of the mitochondrial genome. Most of the mitochondrial tRNAs recognize any one of four nucleotides in the third position of codons. Therefore, there is far greater "wobble" in the third codon position within mitochondrial mRNA, leading to "two-out-of-three" pairing of tRNAs. Lastly, the genetic code is

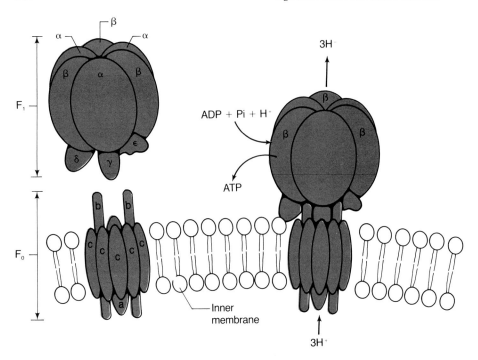

Figure 4–48. Structure of the ATP synthetase. This diagrammatic presentation of the structure of ATP synthetase makes several important points. The protein is composed of two large oligomeric complexes. F_0 is an integral membrane protein containing a, b, and c subunits. F_1 is an peripheral protein that associates with the membrane through F_0. F_1 contains α, β, δ, and ϵ subunits. The β-subunit catalyzes the conversion of ADP+P_i to ATP, with energy supplied by protons moving down their electrochemical gradient, from the intermembrane space to the matrix.

altered so that in human mitochondria, three codons have different meanings than they would have within cytoplasmic mRNA (Table 4–2).

Because the mitochondrial genome can encode only 13 protein subunits, it is clear that the mitochondria must import almost all its protein from the cytosol.

Indeed, most of the 13 protein subunits encoded by the mitochondrial genome become associated with the inner membrane, complexed to subunits encoded by the nuclear genome. We will break our conversation of mitochondrial import into two questions: How are proteins imported from the cytosol to the mitochondrial

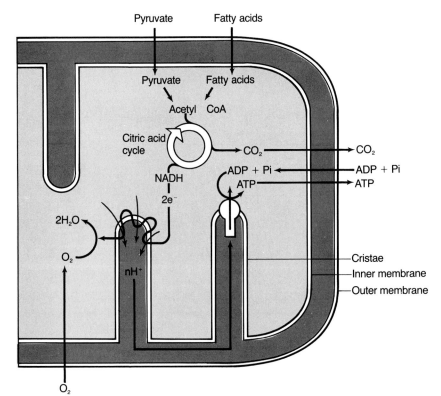

Figure 4–49. Summary of mitochondrial function. Pyruvate and fatty acids enter the mitochondrion and are metabolized by the citric acid cycle. NADH is produced by the citric acid cycle, and electrons produced from the hydride ion of NADH are passed to oxygen by the electron transport chain on the inner mitochondrial membrane. A proton motive force is created during this electron transfer, and protons, moving back down their electrochemical gradient into the matrix, power ATP synthetase to produce ATP. ATP produced in the mitochondria is transported to the cytoplasm by an ADP–ATP antiport system.

Figure 4–50. Organization of the human mitochondrial genome.

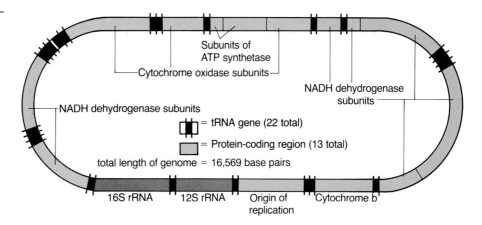

Subunits of ATP synthetase

Cytochrome oxidase subunits

NADH dehydrogenase subunits

NADH dehydrogenase subunits

= tRNA gene (22 total)

= Protein-coding region (13 total)

total length of genome = 16,569 base pairs

16S rRNA   12S rRNA   Origin of replication   Cytochrome b

matrix? and, How do these imported proteins reach their final destination within the four compartments of the mitochondria? The answer to the first question is summarized in Fig. 4–51.

Proteins that are synthesized on free ribosomes, but are destined for the mitochondria have a mitochondrial signal peptide. The mitochondrial signal peptides are stretches of 20–80 amino acids, at the NH_2-terminus of the protein, that have the characteristic of forming an amphipathic α-helix where one side of the helix is nonpolar and the other side contains positively charged amino acids. The protein is maintained in a nonfolded condition by the attachment of a cytosolic chaperon protein that is a member of the hsp 70 family of heat shock proteins. The signal peptide binds to a putative signal receptor on the outer membrane, and the complex moves laterally across the outer membrane until it reaches a contact site at which the inner and outer membranes are joined. The amphipathic signal peptide crosses both membranes at the contact site, using the membrane potential difference of the inner membrane as the energy source (remember that the inner membrane has a membrane potential of 160 mV, with the matrix side negative). In order for the protein to pass through the contact site, the cytoso-

lic hsp 70 proteins must be released. This step requires ATP hydrolysis. As the protein emerges on the matrix side of the contact site, there is binding of a mitochondrial hsp 70 protein. The release of the mitochondrial hsp 70 protein also requires ATP hydrolysis. The release of hsp 70 allows the attachment of the mitochondrial hsp 60 protein, which uses the energy of ATP hydrolysis to catalyze the folding of the protein. Once inside the matrix, the signal peptide is cleaved by a signal peptidase.

How proteins reach final destinations within the inner membrane, intermembrane space, and outer membrane is just a variation on the foregoing theme (Fig. 4–52). Proteins that will remain in the outer membrane have a hydrophobic stop-transfer signal that halts the transfer as the protein is being inserted at the contact site. The protein is then thought to move laterally from the contact site into the outer membrane. Proteins that will be incorporated into the mitochondrial inner membrane have a start-transfer signal that follows shortly after the NH_2-terminal signal peptide (which is cleaved in the matrix). Therefore, after the protein enters the matrix, it is immediately threaded through the inner membrane, beginning with this new start-transfer signal. The mechanism is identical with that for insertion of integral proteins into the ER. Therefore, multipass proteins would be inserted by a series of start-and-stop transfer signals. Finally, those proteins designated for the intermembrane space initially enter the matrix (see Fig. 4–52). After removal of the NH_2-terminal signal peptide in the matrix, they are threaded through the inner membrane, beginning at the new NH_2-terminal start-transfer signal. However, they have no stop-transfer signal, so they completely enter the intermembrane space. They then are released into the intermembrane space by a start-transfer signal peptidase located within this compartment. These mechanisms are very similar to those used within the ER.

Table 4–2
Mitochondrial and "Universal" Genetic Codes

Codon	Universal Code	Human Mitochondrial Code
UGA	Stop	Trp
AUA	Ile	Met
AGA	Arg	Stop
AGG		

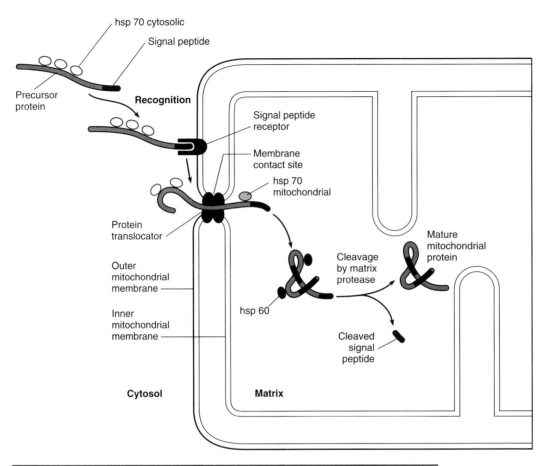

Figure 4–51. Import of protein into the mitochondrion. An NH_2-terminal signal peptide is recognized by a signal receptor on the outer mitochondrial membrane. The complex moves within the membrane to a contact site. The protein is translocated through the contact site, powered by the negative electrochemical gradient across the membrane (the signal peptide is positively charged). Once within the matrix, a specific protease cleaves the signal peptide, and the remainder of the protein enters, after unfolding by a heat-shock protein. The signal peptide is cleaved, then the protein folds into its native structure. (Modified from Alberts B, Bray D, Lewis J, et al. *Molecular Biology of the Cell*, 2nd ed. New York: Garland Publishing, 1989.)

Peroxisomes Import All of Their Proteins from the Cytosol

Peroxisomes are small, spherical organelles with a single membrane and a diameter that ranges from 0.15 to 0.5 μm (Fig. 4–53). This heterogeneous group of organelles contains several enzymes that can remove hydrogen atoms from organic substrates, transferring the hydrogens to O_2. This reaction, which produces hydrogen peroxide, is

$$RH_2 + O_2 \rightarrow R + H_2O_2$$

Catalase, which can represent as much as 40% of the total peroxisomal protein, uses the hydrogen peroxide that is formed to oxidize other substrates, such as alcohols, formaldehydes, and formic acid. The general reaction is

$$H_2O_2 + R'H_2 \rightarrow R' + 2H_2O$$

This is an important reaction because about half of consumed alcohol is detoxified in the liver by this mechanism. When excess hydrogen peroxide is formed in the peroxisome, catalase can convert it to H_2O.

$$2H_2O_2 \rightarrow 2H_2O + O_2$$

New peroxisomes are formed by the growth and fission of existing peroxisomes. The peroxisome does not have its own genome or ribosomes; therefore, all of its protein (and membrane lipids) must be imported.

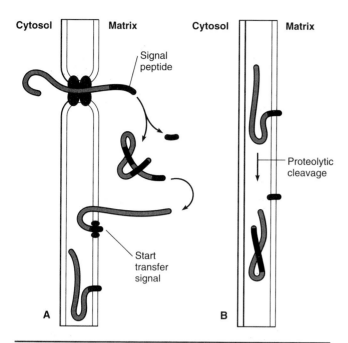

Figure 4–52. How proteins reach their final destination within the mitochondrion. Diagrammatic presentation of how proteins synthesized on cytoplasmic ribosomes become integral membrane proteins within (**A**) the inner mitochondrial membrane or (**B**) soluble proteins within the intermembrane space. See text for details. (Modified from Alberts B, Bray D, Lewis J, et al. *Molecular Biology of the Cell*, 2nd ed. New York: Garland Publishing, 1989.)

Catalase contains a signal of three amino acids at its COOH-terminus that directs it to the peroxisome. Presumably, there is a signal peptide receptor on the surface of the peroxisomal membrane that binds the signal peptide and begins the insertion into the mem-

brane. Zellweger syndrome is a fatal genetic disorder where the defect lies in an inability to import protein into the peroxisome. The defective gene product appears to be a membrane import receptor protein. Both the peroxisome and mitochondrion must import their phospholipids, which are synthesized within the ER. Phospholipid exchange proteins may be responsible for carrying phosphatidylserine and phosphatidylcholine to peroxisomes and mitochondria. Mitochondrial membranes contain enzymes that can convert phosphatidylserine to phosphatidylethanolamine and can produce cardiolipin from imported phospholipids. Similar mechanisms are probably involved in the formation of phosphatidylethanolamine within the peroxisome membrane.

Clinical Case Discussion

In this chapter, we discussed an important example of receptor-mediated endocytosis, uptake of LDL into various cells. Lack of a LDL receptor or mutations in the LDL receptor can lead to the disorder familial hypercholesterolemia.

Familial hypercholesterolemia, familial hypertriglyceridemia, and familial combined hyperlipidemia are the most common disorders of lipid metabolism; about 1 in 160 individuals carry a gene for one of the disorders. Familial hypercholesterolemia (FH) is recognizable in infancy and early childhood by demonstrating a two- to threefold increase in plasma concentrations of total cholesterol and LDL cholesterol, with decreased HDL cholesterol. Homozygotes are rare, but 1 in 500 in the general population are heterozygous for the gene. Clinical symptoms include premature coronary artery disease due to atherosclerosis, tendon xan-

Figure 4–53. Electron microscopy of a peroxisome. Electron micrograph of a large luteal cell from a day-17 pregnant rat. This cell type contains two sizes of membrane-bound granules. The smaller (150 nm diameter), or secretory granules (*small arrows*) contain the hormone relaxin. The larger (250 nm diameter) are peroxisomes (*arrowheads*) and lysosomes. Except for their size, these granules cannot be distinguished by morphologic criteria. Both types have a limiting membrane. Special staining techniques are needed to demonstrate the granule content and thus type (bar, 0.26 μm). (Courtesy of Dr. Phillip Fields.)

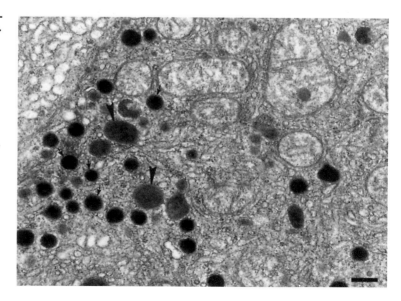

thomas, xanthelasmas, and arcus corneas in young adult life. The mean age for manifestation of coronary artery disease is 43.8 years. The abnormality in FH is a cell membrane defect affecting the receptor for LDL, either a lack of functional LDL receptor or an alteration in the LDL receptor, leading to defective uptake of cholesterol into various cells. Over 140 different mutations of the LDL-receptor gene located on chromosome 19 have been characterized at the DNA level.

Screening children and adolescents if a parent or grandparent has been diagnosed with premature cardiovascular disease or if a parent has an elevated total cholesterol level (>240 mg/dl) is recommended. Treatment of FH consists of a diet low in cholesterol and high in polyunsaturated fats. In addition, exchange resins that bind intestinal bile acids such as cholestyramine or colestipol can be administered to patients who do not respond to diet alone. Liver transplantation offers a cure for a limited number of patients. Newer approaches to treatment of FH include transduction of primary hepatocytes with recombinant retroviruses carrying the LDL gene, followed by transplantation into the liver via a portal venous catheter. This approach has been successful in humans.

Suggested Readings

The Cell Nucleus

Adolph KW. Adolph KW, ed. In *Chromosomes and Chromatin*, vols 1–3. Boca Raton: CRC Press, 1988.

Widnell C, Pfenninger K. *Essential Cell Biology*. Baltimore: Williams & Wilkins, 1990.

Organelles Involved in Macromolecular Synthesis and Turnover

Bishop WR, Bell RM. Assembly of phospholipids into cellular membranes: biosynthesis, transmembrane movement and intracellular translocation. *Annu Rev Cell Biol* 1988;4:579.

Kornfield S, Mellman I. The biogenesis of lysosomes. *Annu Rev Cell Biol* 1989;5:483.

Rose JK, Doms RW. Regulation of protein export from the endoplasmic reticulum. *Annu Rev Cell Biol* 1988;4:257.

Cytoplasmic Organelles that Import Most of Their Proteins from the Cytosol

Attardi G, Schatz G. Biogenesis of mitochondria. *Annu Rev Cell Biol* 1988;4:289.

deDuvc C. Microbodies in the living cell. *Sci Am* 1983;248:74.

Hartl FU, Pfanner N, Nicholson DW, et al. Mitochondrial protein import. *Biochim Biophys Acta* 1989;988:1.

Review Questions

1. Which of the following classification of proteins was *not* synthesized on a "free" ribosome:
   a. Cytoskeletal protein
   b. Peripheral plasma membrane protein
   c. Integral transmembrane plasma membrane protein
   d. Peroxisomal protein
   e. Nuclear matrix protein

2. Liver hepatocytes respond to the presence of drugs such as phenobarbital by expanding the surface area of the following membranous organelle:
   a. Mitochondria
   b. Peroxisome
   c. Lysosome
   d. RER
   e. SER

3. A protein synthesized on the ER has an internal signal peptide flanked on its *N*-terminal side with 4 aspartic acid residues and on its *C*-terminal side with 3 lysine residues. It contains no subsequent stop transfer signal. Its final topography within the ER membrane will be as follows:
   a. Multipass transmembrane protein with *N*-terminus facing the cytosol
   b. Multipass transmembrane protein within *C*-terminus facing the cytosol
   c. Single-pass transmembrane protein with *N*-terminus facing the cytosol
   d. Single-pass transmembrane protein with *C*-terminus facing the cytosol
   e. A peripheral membrane protein

4. A protein synthesized on the RER, containing a KDEL signal reaches the *cis*-Golgi network. It then binds a KDEL receptor and is packaged into transport vesicles, which are then trafficked to:
   a. The ER
   b. The *medial*-Golgi compartment
   c. The *trans*-Golgi compartment
   d. The lysosomes
   e. The plasma membrane

5. During eukaryotic vesicular fusion events a fusion complex forms that contains all of the following protein **EXCEPT**:
   a. v-SNAREs (VAMP)
   b. t-SNAREs (syntaxin and SNAP-25)
   c. SNAP
   d. NSF
   e. Kinesin

6. Protons moving through the following mitochondrial protein complex drives the formation of ATP from ADP and P_i:
   a. ATP synthetase

b. Cytochrome oxidase

c. Cytochrome C

d. NADH dehydrogenase

e. Succinate dehydrogenase

7. Familial hypercholesterolemia is a common disorder leading to premature coronary heart disease and death. All of the following statements are true regarding familial hypercholesterolemia **EXCEPT**:

a. The basic abnormality is a defect in the receptor for LDL cholesterol.

b. An elevation of LDL cholesterol usually occurs.

c. Treatment consist of dietary modifications and drugs.

d. The mutations for the LDL receptor is located on chromosome 19.

e. Reduction in HDL cholesterol levels is desirable.

8. The reformation of the nuclear envelope during mitosis is primarily dependent upon:

a. Lamin A

b. Lamin B

c. Lamin C

d. Nuclear spokes

e. Nucleoplasmin

9. The DNA protein complex, thought to be transcriptionally active is referred to as:

a. Heterochromatin

b. Euchromatin

c. Histochromatin

d. Actochromatin

e. Fibrochromatin

10. All of the following concerning the transport of proteins from the cytoplasm to the nucleus is true **EXCEPT**:

a. Protein to be transported to the nucleus contain a short segment of amino acids that act as a nuclear localization signal

b. Proteins to be transported to the nucleus bind with receptors found on the nuclear spokes

c. Protein transport to the nucleus requires the participation of several proteins, some of which enter the nucleus with the transported protein

d. Protein transport to the nucleus requires an intact nuclear pore complex

e. Protein transport to the nucleus requires the hydrolysis of GTP to GDP

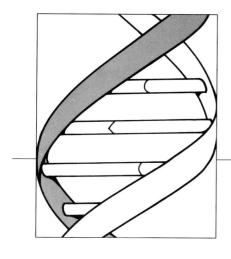

Chapter 5
Regulation of Gene Expression

Clinical Case

A 2-month-old male, Chip Smithers, was seen by his pediatrician for well child care. His birth weight was 6 pounds, 9 ounces. Mom states the baby eats well but has intermittent diarrhea that responds temporarily to a change in diet. He weighs 7 pounds and has a Candida diaper rash. The remainder of his physical examination was unremarkable. Dietary counseling was given, and an antifungal cream was prescribed to treat the rash. Over the next few months, the infant developed two episodes of otitis media and severe pneumonia that required extensive intravenous antibiotics to clear. At his 6 months well child visit, he weighed 10 pounds and had very little increase in his length or head circumference. The baby's physical examination was remarkable for a thin, undergrown infant with a temperature of 101°F, a severe Candida diaper rash, and yellow nasal discharge. On chest x-ray, the costochondral junctions showed cupping and flaring. The lung fields were clear. A complete blood count revealed profound lymphopenia; serum immunoglobulin levels were decreased. Candida and tuberculin skin tests were nonreactive.

Gene Cloning and Sequencing

The Study of Gene Expression Has Been Facilitated by Recombinant DNA Technology

A major focus of modern cell biology is to understand how a cell works in molecular detail. Although classic biochemical approaches have made possible the purification and examination of many cellular components, until recently, the only way to investigate the informational content of the cellular genome was the examination of phenotype of mutant organisms to deduce gene function. This approach remains as an important investigative and diagnostic tool (karyotyping analyses, as discussed in Chapter 4); however, we now possess the ability to directly examine a specific gene and how it functions in normal and pathologic circumstances through the application of techniques referred to as *recombinant DNA technology*. Although new technical advances occur rapidly in this field of study, the key techniques that constitute the basis of recombinant DNA technology are:

1. The cleavage of DNA at specific locations by restriction endonucleases, which facilitate the identification and manipulation of individual gene sequences.
2. The propagation of eukaryotic DNA fragments in bacterial cells by gene cloning, which allows the isolation of large quantities of a specific DNA.

3. The determination of the order of nucleotides contained in a purified DNA by DNA sequencing, which allows the examination of gene structure and the amino acid sequence it encodes.

4. The direct amplification of a DNA sequence by the polymerase chain reaction (PCR), which enhances the ability to quickly examine specific regions of the genome for genetic defects.

Because recombinant DNA techniques are becoming increasingly important as clinical diagnostic tools, we will devote the initial sections of this chapter to explain these techniques, after which we will examine the concepts of gene expression, leading to a discussion of the importance of these concepts toward establishing viable genetic therapies.

Restriction Nucleases: Enzymes That Cleave DNA at Specific Nucleotide Sequences

One of the most important developments of recombinant DNA technology was the discovery of enzymes that catalyze the double-stranded cleavage of DNA at specific nucleotide sequences, the restriction endonucleases. This discovery came from an understanding of the defense mechanism used by bacteria to protect themselves from foreign DNA molecules carried into the cell. The genome of a bacteria contains a host-specific pattern of DNA methylation, and when a DNA that does not contain this pattern is encountered by the cell (e.g., carried in by bacteriophage), it is degraded, thereby protecting the bacteria from the foreign DNA. In 1970, Smith and his colleagues were the first to purify an enzyme from *Escherichia coli* responsible for

this degradation. Remarkably, this enzyme catalyzed the double-stranded cleavage of DNA within a specific short nucleotide sequence (Fig. 5–1). Hundreds of enzymes capable of cleaving DNA at specific nucleotide sequences have been isolated from different species of bacteria, providing powerful tools for the characterization of DNA molecules.

With these enzymes, the DNA isolated from a particular cell can be cleaved into a series of discrete fragments called *restriction fragments*. The size of DNA fragments produced by an enzyme digest can be analyzed by resolving the cleaved DNA on an electrophoretic gel (see Chapter 1). Thus, by examining the sizes of restriction fragments produced from a particular gene region after treatment with combinations of different restriction endonucleases, a map of that region can be drawn that shows the location of each restriction site relative to adjacent restriction sites (Fig. 5–2). Because restriction endonucleases cleave DNA at positions of specific nucleotide sequences, a restriction map reflects the arrangement of these sequences within a given fragment of DNA. This is useful to characterize similarities and differences between isolated, homogeneous DNA fragments (e.g., cloned DNA) (see Fig. 5–2).

The DNA fragments containing a specific sequence can be identified within the thousands of fragments produced when a population of DNA molecules, such as the total genome of an organism, are cleaved with restriction enzymes by using a technique referred to as *Southern hybridization* (Fig. 5–3). This is a very powerful way to examine the organization of specific genetic loci among individual members of a family or population. A difference in restriction maps between two individuals is called a *restriction fragment length*

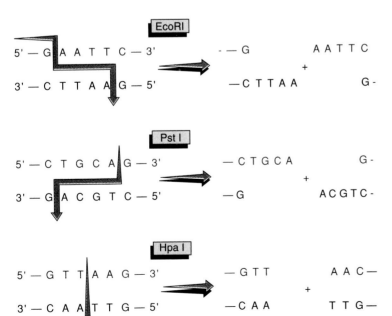

Figure 5–1. Restriction endonuclease recognition and digestion. The recognition sequences and patterns of double-stranded DNA cleavage are illustrated for three different restriction enzymes. Notice that the nucleotide sequence recognized is the same in the two strands of the DNA. This arrangement of nucleotides is referred to as a *palindromic DNA sequence*. The cleavage of the DNA helix often occurs in an offset fashion, such as that shown for *Eco* RI and *Pst* I, which creates a staggered or cohesive end on the DNA fragment. However, some enzymes illustrated by the *Hpa* I cutting site leave blunted ends. That is, the DNA is cleaved such that the ends of the resultant fragments do not contain short single-stranded regions.

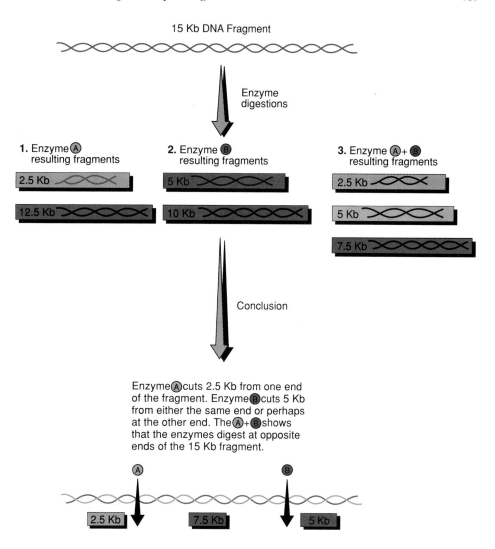

Figure 5-2. Example of restriction mapping. Restriction enzymes are useful tools to characterize segments of DNA. Here, we illustrate an experiment that shows how the cleavage sites for different restriction endonucleases are positioned relative to each other to create a restriction map of a DNA fragment. *Kb* is an abbreviation that means 1,000 nucleotides or 1,000 nucleotide pairs.

polymorphism (RFLP). This analysis has become an important technique to identify loci that are close to, or contain, a defective gene associated with a genetic disease (Fig. 5–4). For example, the molecular basis for Duchenne's muscular dystrophy (the dystrophin gene) and cystic fibrosis (the cystic fibrosis transporter gene) have been recently elucidated using RFLP technology.

Gene Cloning Can Produce Large Quantities of Any DNA Sequence

Many restriction enzymes cleave DNA in a staggered or offset manner that produces short single-stranded regions of DNA at the ends of the restriction fragment (see Fig. 5–1). In 1973, Boyer and Cohen recognized that because of the exact nature of restriction endonuclease cleavage, the staggered ends of a DNA restriction fragment produced by an enzyme are complementary to the ends of any DNA fragment produced from treatment with the same enzyme (Fig. 5–5). Therefore, DNA molecules from any source are capable of being joined covalently by DNA ligase, an enzyme that links

DNA molecules together (see Fig. 5–5). In their pioneering experiments, Cohen and Boyer linked a fragment of eukaryotic DNA to a DNA molecule isolated from bacteria that was capable of directing its own replication in bacterial cells. This hybrid DNA formed *in vitro* is called a *recombinant DNA molecule* because it is formed from the end-to-end joining of two different DNAs. Moreover, because the bacterial DNA fragment replicates when the bacterium grows, the eukaryotic DNA fragment of the recombinant molecule is also replicated.

This experiment is an example of *DNA cloning*. Successful cloning requires two basic elements. The first is a suitable host bacterial strain. Most, if not all, bacteria contain a restriction-modification system. However, there are many bacterial strains that have been modified or engineered by classic genetic selection, such that they no longer contain this defense mechanism. Because these bacterial strains can undergo genetic transformation (the uptake of DNA) with a foreign DNA molecule with the DNA becoming resident within the recipient cell, they provide excellent host for cloning experiments.

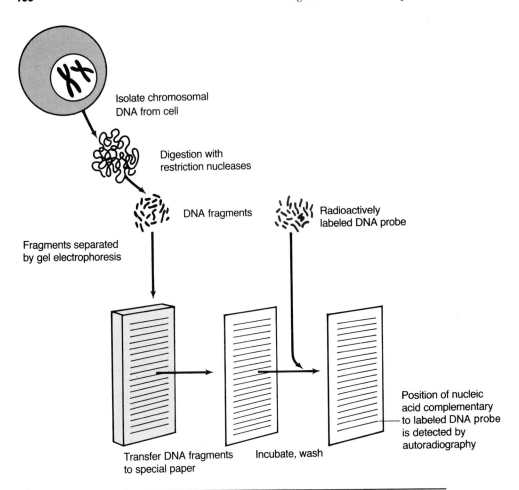

Figure 5–3. **Southern-blotting analysis.** The size or migration of specific DNA fragments within a mixture of DNA can be examined by the Southern hybridization technique. After treatment with restriction endonucleases, the DNA fragments are resolved on electrophoretic gels, and the fragments contained within the gel are then transferred to a nitrocellulose paper by blotting. The paper is incubated with a radioactive DNA fragment under conditions that permit this DNA probe to bind with complementary molecules on the paper sheet (hybridization). Following hybridization, the sheet is washed free of nonspecific probe binding, and the immobilized molecules complementary to the probe are visualized as radioactive bands on x-ray films placed next to the nitrocellulose paper.

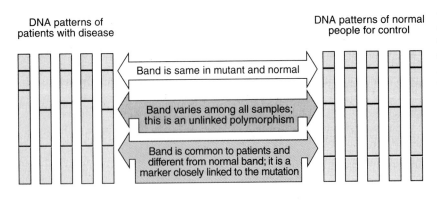

Figure 5–4. **An example of restriction fragment length polymorphism analysis.** Southern-blotting examination of the chromosomal DNA from patients who have a disease trait and comparison with the DNA from unaffected individuals has been instrumental in elucidating the molecular basis for many genetic disorders. (Modified from Lewin B. *Genes*, 4th ed. New York: Oxford University Press, 1991.)

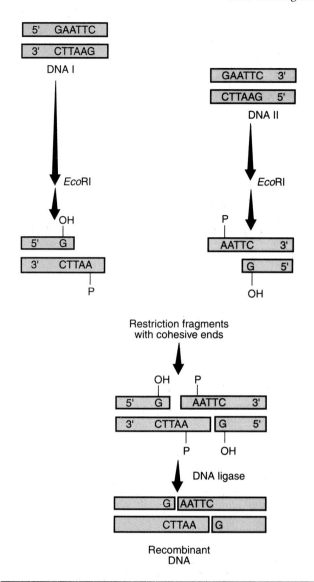

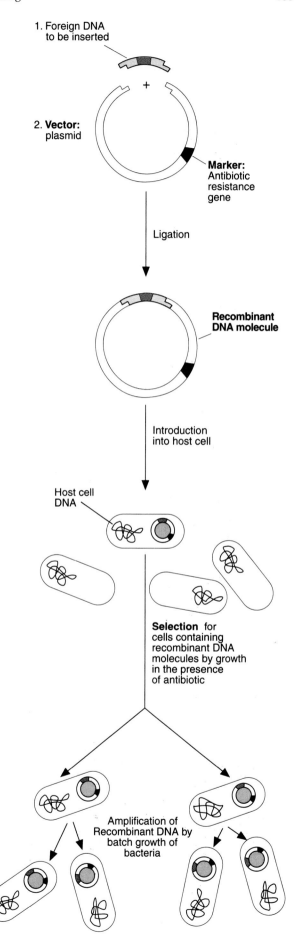

Figure 5–5. Formation of recombinant DNA molecules. Recombinant DNA molecules are formed from the covalent joining of two DNA fragments. Restriction fragments derived from treatment with the same restriction enzyme will have cohesive ends that are complementary, and the two molecules can be joined together by the enzyme DNA ligase.

Figure 5–6. Cloning DNA fragments. Two DNA fragments can be covalently joined together, forming a recombinant DNA molecule (see Fig. 5–5). If one of the DNAs contains a bacterial origin of replication, a vector (such as a plasmid, as shown here), then the products of the *in vitro* ligation reaction can be placed into a host bacteria. Plasmid vector molecules also contain genes that confer resistance to antibiotics, allowing the selection of bacteria that take up the recombinant molecule.

A vector, or DNA molecule to which the fragment of interest (e.g., eukaryotic DNA) can be covalently linked, is necessary. A vector can be any DNA molecule that contains an origin of replication and is capable of replicating after entry into a suitable host cell (Fig. 5–6). The most common vectors are small circular DNAs, called *plasmids*, because they can replicate independently of the bacterial host genome, allowing the amplification of the recombinant DNA molecule. That is, under appropriate conditions, the foreign DNA molecule can replicate many times, compared with the host genome, effectively increasing the concentration of the recombinant DNA molecule within the cell. The bacterial cell then becomes a factory to produce large quantities of the recombinant DNA.

If we begin with a heterogeneous mixture of DNA fragments, each connected with a vector DNA, and transform bacteria with this mixture (see Fig. 5–6), we can construct a library of DNA fragments. Each bacterial colony observed will contain one of the target DNAs from the starting mixture linked to the vector molecule. Thus, each colony represents a clone housing a unique recombinant DNA molecule, and the aggregate of all of the colonies (the library) will reflect the mixture of heterogeneous DNAs in the starting material. There are two types of libraries commonly used in molecular biology experiments: genomic DNA and complementary DNA (cDNA) libraries. A genomic DNA library contains fragments representing the entire genome of an organism (e.g., DNA fragments formed from cleavage of chromosomal DNA isolated from the nucleus). A cDNA library contains fragments that represent the population of messenger RNA (mRNA) purified from a particular cell or tissue. This involves using the purified mRNA as a template for reverse transcriptase, an enzyme from certain viruses that synthesizes DNA by copying an RNA sequence. From this can be produced a complementary DNA copy of each mRNA present. The cDNAs can then be joined with an appropriate vector for bacterial transformation. There are important distinctions between genomic and cDNA clones or libraries. Only a cDNA clone will contain an uninterrupted form of the sequences that code for proteins, in particular, because genes in higher eukaryotes contain extra nucleotides that are removed to form mRNAs (discussed in detail later in this chapter).

How are libraries made from such heterogeneous DNA fragments useful? Once the DNAs, ligated to a vector DNA, have been introduced into bacterial cells, the transformed bacteria can easily be spread onto agar plates. Each colony that grows on the plates represents a clone (that is, a single target DNA), and all that is necessary is to find the one that contains the DNA of interest. Although this may seem like finding a needle in the haystack, the task is easier if hybridization techniques similar to those for Southern blotting are used (Fig.

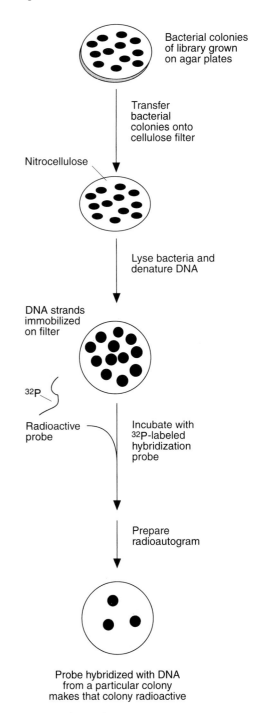

Probe hybridized with DNA from a particular colony makes that colony radioactive

Figure 5–7. Selection of bacterial colonies in a library that contains a specific DNA fragment. The bacterial colonies that contain a DNA fragment of interest (e.g., a specific gene or cDNA) can be identified using a procedure similar to that of Southern hybridization (see Fig. 5–3). The colonies are transferred to a nitrocellulose paper and, following immobilization of the DNA onto the paper, the colonies of interest are located by hybridization with a radioactive DNA probe.

5–7). After we have identified the colony containing the DNA fragment of interest, we can grow large quantities of the bacteria in a liquid culture. Because the vector contains an origin of replication that directs the replication of the recombinant DNA molecule as the bacteria grow, there is now an unlimited supply of the target DNA molecule identified from the heterogeneous DNA fraction of the library.

The Primary Structure of a Gene Can Be Rapidly Determined by DNA Sequencing

Many human diseases are the consequence of single-base changes in genes, causing abnormal proteins to be formed, or perhaps abnormal function of the gene. Sickle-cell anemia, for example, is the result of a single-base change that ultimately replaces a glutamic acid at residue 6 with a valine in the β-chain of the hemoglobin molecule. There are other diseases of hemoglobin caused by single-nucleotide changes; however, the changed base is not within the protein-coding sequence of the gene. For example, a particular class of disease in which hemoglobin is not produced (the β-thalassemias) is caused by a single-nucleotide change of the β-globin gene that creates an improper processing of the pre-mRNA to the mature mRNA capable of directing the synthesis of the protein. The basis of these thalassemic diseases remained a mystery until the genetic material from affected persons was cloned and the primary structure of the genes was determined by DNA sequencing.

DNA sequencing is simply the determination of the order of molecules in a particular DNA fragment. There are two principal methods of DNA sequence analysis. The first method, pioneered by Maxam and Gilbert, employs the chemistry of nucleic acids. In this method, a radioactive tag (usually a [^{32}P]phosphate) is placed on one end of a purified DNA fragment, and the labeled fragment is then subjected to chemical modifications that cleave the DNA at specific nucleotide residues (e.g., reactions for each of the four nucleotide bases). The cleaved DNA fragments are resolved on electrophoretic gels, with the resultant cleavage pattern indicative of the nucleotide sequence (Fig. 5–8).

The second method of DNA sequence determination is referred to as the *dideoxynucleotide chain termination technique.* Generally, DNA fragments to be sequenced are cloned into a vector DNA that is capable of synthesizing a single-stranded copy of the target molecule. The bacteriophage M13 is the most commonly used vector for this purpose (Fig. 5–9). The single-stranded recombinant M13 DNA is then used to synthesize DNA *in vitro* by using the enzyme DNA polymerase in the presence of a nucleotide analogue that contains a hydrogen at the 2 and 3 positions of the ribose sugar (2′, 3′ dideoxynucleotides). When the dideoxynucleotide is incorporated into the newly syn-

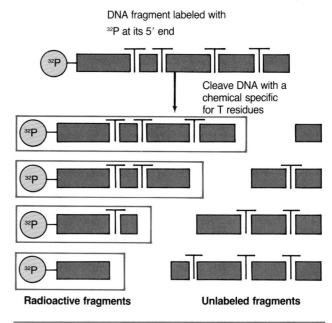

DNA fragment labeled with ^{32}P at its 5′ end

Cleave DNA with a chemical specific for T residues

Radioactive fragments   **Unlabeled fragments**

Figure 5–8. **DNA sequencing by chemical cleavage.** The method of DNA sequencing described by Maxam and Gilbert takes advantage of the chemical properties of the DNA. One end of the DNA fragment is first radioactively labeled, then the fragment is subjected to random cleavage by chemicals that recognize a specific nucleotide (in this example, cleavage at *T* residues). The resulting DNA fragments are separated by size on an electrophoretic gel. Because only one end of the DNA contains a radioactive tag, the size of a fragment observed from autoradiography of the gel reflects the distance, in bases, from the radioactive tag to the cleavage site within the DNA helix. The order of nucleotides within a DNA fragment can be obtained by examining the banding pattern of the same DNA subjected to chemical cleavage specific for each base (A, G, C, T) and resolved on the same gel.

thesized DNA, the chain cannot be elongated further and, under the right conditions, a buildup of DNA fragments that stops specifically at each point where the dideoxynucleotide can be incorporated will be obtained. Therefore, similar to the chemical cleavage method, the sequence from resolving the *in vitro* synthesized DNA fragments on an electrophoretic gel can be determined (see Fig. 5–9).

Both methods of DNA sequence determination are rapid and reliable. It is now more common to determine the sequence of a protein by deducing the amino acids from a nucleotide sequence; there are only four bases in nucleic acids and 20 different amino acids, making DNA a chemically simpler molecule. Moreover, the importance of knowing and understanding the primary structure of genes as related to human disease can be inferred from the current scientific effort to sequence the entire approximately 10^9 nucleotides of the human genome.

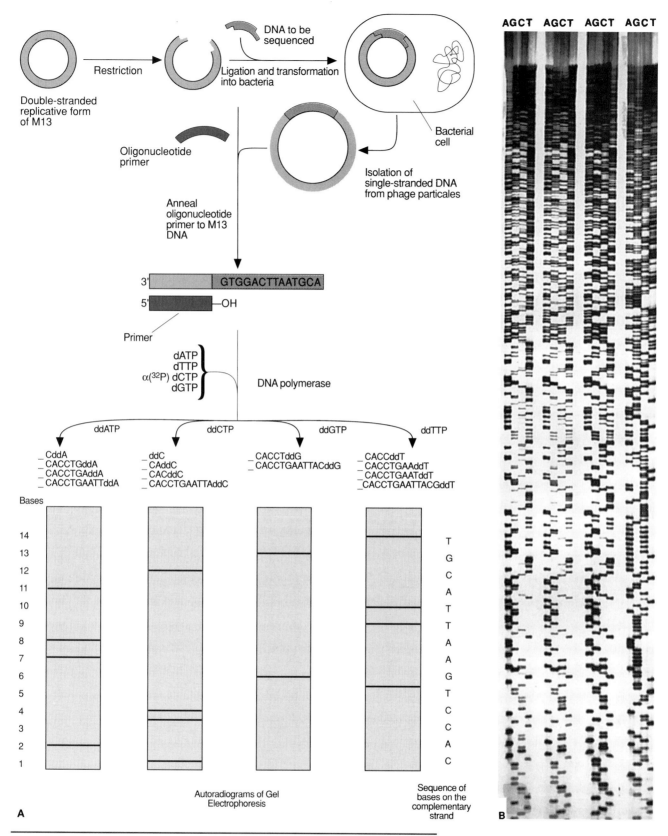

Figure 5–9. Scheme of dideoxynucleotide DNA sequencing analysis. **A:** The DNA fragment to be sequenced is cloned into a vector that is capable of synthesizing a single-stranded version of the recombinant DNA molecule. The single-stranded DNA is then used as a template for *in vitro* DNA synthesis in the presence of a dideoxynucleotide. The resultant DNA fragments are resolved on an electrophoretic gel, which allows determination of the DNA sequence. **B:** An example autoradiogram from a dideoxynucleotide chain termination sequence experiment. (Courtesy of Dr. Adrienne Kovacs.)

Specific Regions of the Genome Can Be Amplified with the Polymerase Chain Reaction

The techniques discussed thus far have revolutionized our understanding of how cells work. However, they are somewhat laborious and time-consuming. In 1987, a novel technology was introduced that allows amplification of a nucleic acid sequence without the need to clone it. This technique, called the *polymerase chain reaction*, does require a knowledge of the sequences that surround the region to be amplified. Synthetic oligonucleotides complementary to these sequences are used as primers in a series of reactions that use a special thermostable DNA polymerase isolated from a bacterial species that lives at high temperatures (Fig. 5–10). The reactions involve a cycle of steps that first denatures the DNA duplex at high temperature, then allows binding of the oligonucleotides by cooling to a lower temperature and extension of the oligonucleotide primers by DNA polymerase. Because the reagents are not inactivated by high temperatures, they can be added to a single tube, and the cycles of denaturation, annealing, and extension can be repeated multiple times. Each cycle increases the concentration of the duplex DNA, bound by the oligonucleotide primers (see Fig. 5–10), such that microgram quantities of DNA can be obtained from analysis of DNA isolated from the nucleus of a single cell (a specific sequence is amplified exponentially; for example, 30 cycles would represent a 2^{28}, or 27 millionfold, amplification). Moreover, this can be accomplished in hours, compared with weeks or months needed for conventional cloning techniques.

Although PCR is a relatively new technique, it is rapidly becoming an important diagnostic tool. Because of its speed and sensitivity, PCR is becoming a vital technique for the clinical diagnosis of infectious diseases such as acquired immune deficiency syndrome (AIDS). The PCR-based technologies can detect the presence of the viral genome much earlier than can antibody-based tests that require months (and even years) of infection to detect the presence of a viral protein. Additionally, PCR tests have been developed for prenatal diagnosis of a variety of genetic diseases. Finally, because PCR is such a sensitive technique, it is possible to analyze DNA from a tissue sample as small as a human hair. Thus, PCR is an ideal technology for use in forensic medicine.

With this foundation for how genes are studied, attention can be turned to how genes function.

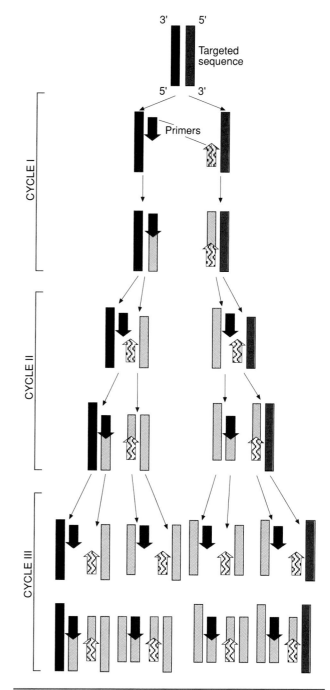

Figure 5–10. The polymerase chain reaction. This illustrates the amplification of a DNA fragment through three cycles of a polymerase chain reaction (PCR) experiment. In each cycle, the DNA strands are denatured, permitting annealing of synthetic oligonucleotides and the synthesis of the complementary DNA strand. Notice that there is an exponential increase in the amount of the target DNA at the end of each cycle, resulting in an amplification of the DNA segment.

Gene Expression: The Transfer of Information from DNA to Protein

According to most standard college dictionaries, the word *gene* is defined as "a complex molecule associated with the chromosomes and acting as a unit or in various biochemically determined combinations in the transmission of specific hereditary characteristics." The word *expression* is defined as an "outward indication or manifestation of some feeling, condition, or quality." Thus, the phrase *gene expression* is an outward indication or manifestation of the complex molecules associated with the chromosomes.

What is a measure of this outward manifestation? The most obvious answer is that the appearance of a functional protein within a given cell would be the outward indication of the expression of a gene. The aggregate functional capabilities of the collective set of proteins expressed in a particular cell is what specifies the biochemical and phenotypical properties of the cell. For example, skeletal muscle is composed of elongated, multinucleated cells that are able to contract when stimulated by neuronal input because of the collective set of proteins expressed in these cells. This does not mean that all of the proteins found in a skeletal muscle cell are expressed exclusively in this cell type; in fact, many of these proteins are expressed in a variety of cells. How-

ever, there are proteins unique to skeletal muscle, implying that there must be mechanisms that stringently govern their appearance. The appearance of proteins in a cell occurs by a transfer of the information housed within the genetic material to the machinery responsible for protein synthesis. Thus, the mechanisms that govern the expression of proteins must, in some way, act on this information transfer.

The Basic Steps of Information Transfer Are Transcription and Translation

The process of information transfer from DNA to protein involves two major steps: transcription and translation (Fig. 5–11). Transcription occurs in the nucleus and is the synthesis of a single-stranded RNA copy of the

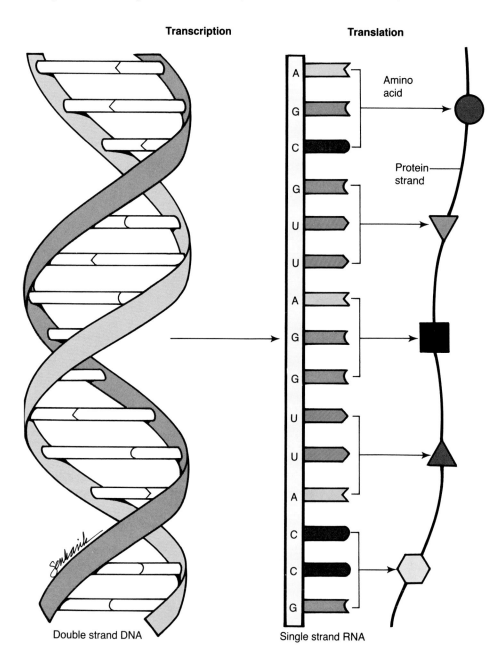

Figure 5–11. Two major steps involved in information transfer from DNA to protein. The first step involves the synthesis of a single-stranded RNA molecule from a double-stranded DNA template. This process, catalyzed by the enzyme RNA polymerase, is called *transcription*. The product of transcription by eukaryotic RNA polymerase II is an RNA that directs the synthesis of a protein molecule by the translation machinery, the *ribosome*. The genetic code of three nucleotide units, the codons, stored in the double-stranded DNA, is passed to the single-stranded RNA molecules (the mRNA) which are the "words" that specify a particular amino acid building block of the protein sequence.

information stored in the double-stranded DNA molecule. Synthesis of the RNA is catalyzed by the enzyme RNA polymerase, and the synthesis of RNA from DNA is asymmetric; that is, only one strand of the DNA is transcribed to make the RNA copy. In eukaryotes, there are three types of RNA molecules transcribed: transfer RNA (tRNA), ribosomal RNA (rRNA), and messenger (mRNA), each of which are synthesized by a different RNA polymerase. Ribosomal RNA (transcribed by RNA polymerase I) and the tRNA, (transcribed by RNA polymerase III) are often considered to be structural RNAs because they make up the integral components of the translational machinery, the ribosome. The RNA molecules that are synthesized by RNA polymerase II have a special role; they carry the information needed to code a protein sequence from the DNA to the ribosome. The information is stored in this RNA as discrete "words" made up of three nucleotides, referred to as *codons*; and because this RNA provides the link between the genome and the protein synthesis machinery, it is called *messenger RNA*.

Regardless of the RNA that is synthesized, the process of transcription begins with binding of RNA polymerase to specific sequences of DNA in or near the gene, called *promoter DNA elements* (Fig. 5–12). Synthesis of the RNA then occurs by the progressive addition of ribonucleotides to form an RNA chain with the polarity of synthesis being 5' to 3'. Therefore, tran-

scription is a multistep process, and the interruption or enhancement at any of the steps necessary for transcription would be important factors for the regulation of expression. Transcriptional controls are thought to be paramount regulatory mechanisms because the synthesis of RNAs from the DNA template constitutes the primary step of gene expression.

The second basic step of information transfer is the translation of codons encrypted in the nucleic acid of the mRNA into a chain of amino acids. Protein synthesis is carried out by the ribosomes, and in eukaryotic cells, this synthesis occurs in the cytoplasmic compartment of the cell. Ribosomes are large complexes of RNA and protein composed of one large subunit and one small subunit that come together to form the protein synthesis machinery. The basic function of the small subunit is binding the mRNA and tRNA, whereas the large subunit is required for catalyzing the peptide bonds of the growing protein chain. Thus, the synthesis of a protein requires all three of the RNA classes synthesized in the nucleus working in concert in the cytoplasm.

Translation occurs through a series of distinct steps (Fig. 5–13). The first step is binding of the small ribosomal subunit with an mRNA molecule. The small ribosomal subunit is prepared for binding of a mRNA by its association with proteins, referred to as *initiation factors*, and its binding of a special tRNA, called the *initiation tRNA*, which contains a methionine amino acid

Figure 5–12. **Transcription by the three eukaryotic RNA polymerases.** The synthesis of single-stranded RNAs from a double-stranded DNA template occurs through the RNA polymerase enzymes. In eukaryotic cells, there are three distinct RNA polymerase molecules, each synthesizing a class of RNAs. The process of transcription always begins with binding of the RNA polymerase to specific regions of DNA, called *promoter elements* (shown by the *shaded DNA*). Promoter DNA elements may be binding sites for proteins, in addition to polymerases, that are referred to as *transcription factors*. Also notice that promoter DNA elements may occur near the gene sequences (which are indicated by the *+1 initiation start site* and continues in the direction of the *arrows*) as in polymerases I and II complexes or within a gene sequence, as in RNA polymerase III complexes.

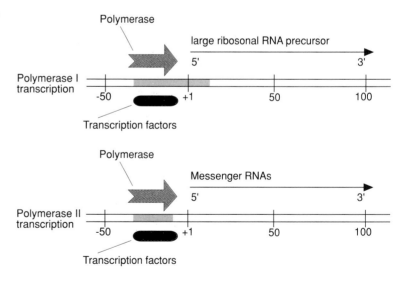

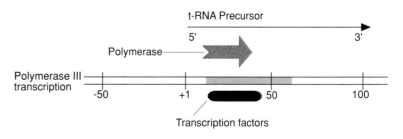

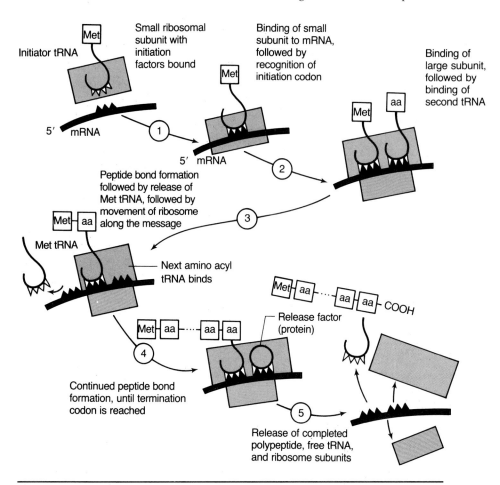

Figure 5–13. Scheme of events for protein translation from an mRNA molecule. The process of translation begins with binding of activated small ribosomal subunits with the mRNA. On identification of an initiation codon (AUG), the large ribosomal subunit binds to form the translation machinery that recruits tRNAs specified by the three nucleotide codons of the mRNA. This enables the ribosome to join the amino acids of the polypeptide chain. When a termination codon is encountered, the protein chain is released, and the translation machinery breaks up, which can then reform a translation complex upon reactivation of the small ribosomal subunit. (Modified from Widnell C, Pfenninger K. *Essential Cell Biology*. Baltimore: Williams & Wilkins, 1990.)

residue. The binding of the small ribosomal subunit occurs at or near the 5′ end of the mRNA; after which the activated ribosomal subunit scans the messenger for an AUG initiation codon for methionine. This results in the formation of an initiation complex.

Following the formation of the initiation complex, the large ribosomal subunit binds to the small-subunit mRNA structure, permitting progressive "reading" of subsequent codons. The reading of mRNA involves the recognition of the three-nucleotide codon, with recruitment and binding of the appropriate tRNA–amino acid complex to the ribosome. Once the tRNA is bound, the amino acid it is carrying is linked to the growing protein chain by peptide bond formation, catalyzed by the large ribosomal subunit. The formation of the peptide bond causes a release of the tRNA from the previous cycle, a

shift of the tRNA polypeptide chain within the ribosomal complex, and the subsequent reading of the next codon sequence. This cycling continues until the ribosome encounters a codon that specifies the end of the protein, called the *termination* or *stop codon*, after which the mature protein is released from the ribosome. The ribosome–mRNA complex dissociates into separate components, which, after the appropriate charging of the small ribosomal subunit, can reassemble to make more protein.

Because mRNAs are composed of four nucleotides (adenosine, cytosine, guanosine, and uridine), there are 64 possible combinations available to form the three nucleotide codon sequences. Three of these nucleotide combinations, UAA, UAG, and UGA, do not specify an amino acid, but rather instruct the ribosomal machinery to end the protein synthesis and are referred to as *stop codons*. This leaves 61 possible combinations to encode only 20 amino acids; the amino acids being specified by binding of complementary nucleotides carried in specific tRNA molecules. Two amino acids, methionine and tryptophan, are encoded by a single codon, and the other 18 amino acids are encoded by multiple codons. Thus, the genetic code is said to be degenerate, meaning that most amino acids are specified by more than one triplet sequence.

In prokaryotic cells, mRNAs are available to the translational machinery as soon as it is transcribed. Translation is often observed to begin before the transcription process is finished. This is not true in eukaryotes, in which there is a separation of the transcriptional and translational machinery. In eukaryotic cells, transcription occurs in the nucleus, and the transcribed product is often larger than the final, mature mRNA. The primary RNA transcript is modified or processed to form the mature RNA, which is then transported from the nucleus to the cytoplasm to participate in protein synthesis. Thus, gene expression in eukaryotes is more complex than that of prokaryotes; it requires additional steps, and each step in the process provides a potential point of regulation.

Each Cell Type of Multicellular Organisms Contains a Complete Complement of Genes

Before a discussion of strategies of gene control is begun, there are two important concepts that will be addressed: (a) every cell of a multicellular organism contains a complete complement of genetic material; and (b) the molecular definition of a gene as an independent unit of information. There are two central observations indicating that the entire blueprint, or genetic plan, for an organism is contained within the nuclei of all different cell types. First, examination of the DNA content of different cell types within an organism demonstrated that all somatic cells had approximately equal amounts of DNA, and this DNA content was twice that found in the gametic cells. For example, physical, chemical, and kinetic experiments showed that a liver cell contained the same DNA content as a brain cell. This is referred to as the *constancy of DNA*. However, it is difficult to imagine that cells that differ so dramatically in morphologic structure and function do not suffer some type of irreversible change in genetic material (e.g., loss of nonessential genes) during the progressive specialization of the cells during development. Because only a fraction of the total genetic material is expressed in a given cell, it was reasoned that a loss of genetic material might be so minimal that it had escaped detection when examining gross DNA content. This was shown to be incorrect by an elegant set of experiments by Gurdon and colleagues, who examined differentiation in frogs. The

seminal experiment from these studies is diagrammed in Figure 5–14. When a nucleus of a fully differentiated frog cell (e.g., intestinal epithelium cells) was injected into a frog egg from which nuclei had been removed, the injected donor nucleus was capable of programming the recipient egg to produce normal, viable tadpoles. Because a tadpole contains the full range of differentiated cell types found in adult frogs, it was concluded that the nucleus of the original differentiated donor cell contained all the necessary information to specify the frog's many different cell types. There was no irreversible loss of important DNA sequences in the differentiated cell nucleus; the development of an organism requires the expression of a specific set of sequences at the appropriate time.

The primary function of the genome is to produce RNA molecules, and only specific regions of the DNA sequence are copied or transcribed into a corresponding RNA nucleotide sequence that can function to encode a protein (mRNA) or as structural molecules (tRNAs and rRNAs). Therefore, every segment of the DNA molecule that produces a functional RNA would constitute a gene.

Until recently, genes were defined primarily by their abilities to confer a biochemical or phenotypic trait on a cell. The application of molecular-cloning technologies has allowed a refinement of the definition of a gene based on the structural organization of the DNA nucleotide sequence. One of the most notable findings from these analyses is that many of the DNA sequences transcribed by RNA polymerase II produce functional mRNAs that represent more nucleotides than are found in the cytoplasmic messenger.

One of the first eukaryotic genes in which extra DNA sequences were identified was the chicken ovalbumin gene (Fig. 5–15). The extra length of nucleotides consists of long stretches of noncoding DNA that interrupt the segments of informational DNA. The informational or coding sequences are called *exons*, and the interrupting stretches of noncoding DNA are referred to as *introns*. Thus, the noninformational segments of the RNA molecule as synthesized from the DNA (called the *RNA primary transcript*) must be removed and segments of informational content joined together to form the mature mRNA. These events occur in the nucleus and are termed *RNA splicing*. Therefore, in eukaryotes a gene does not necessarily reflect, nucleotide for nucleotide, the functional RNA that it encodes.

A second refinement of the definition of a gene has evolved from a correlation of structural organization of the DNA sequence with the ability to visualize regions of chromatin supporting active transcription. Electron micrographs of transcribed DNA segments (genes) demonstrate the transcribed segment as an expanded, bead-on-a-string region of chromatin (e.g.,

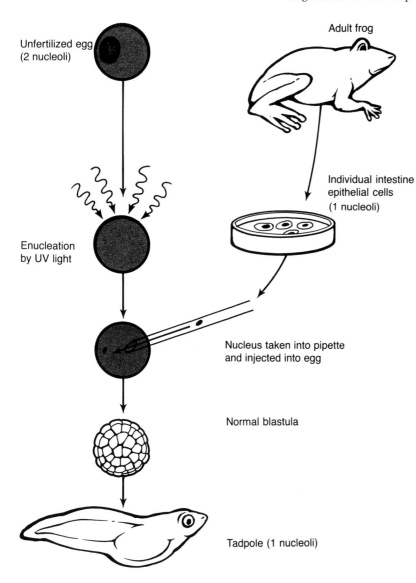

Unfertilized egg
(2 nucleoli)

Adult frog

Enucleation
by UV light

Individual intestine
epithelial cells
(1 nucleoli)

Nucleus taken into pipette
and injected into egg

Normal blastula

Tadpole (1 nucleoli)

Figure 5–14. Nuclear transplantation experiments to examine the constancy of DNA in eukaryotic cells. The diagram outlines an experiment that examined the capacity of nuclei from an adult differentiated tissue to express specific gene sequences. Nuclei from the intestinal epithelium of an adult frog were injected into an oocyte that has had its genetic material inactivated by ultraviolet light. Genetic markers for the two kinds of nuclei were derived from the difference in the number of nucleoli expressed in the cells; the donor nucleus had one nucleolus, and the acceptor oocyte had two nucleoli. The injected nuclei have the capacity to express all the genes necessary to make a tadpole and, subsequently, an adult frog, demonstrating that differentiated cells contain an entire complement of genes to specify the many differentiated cells of an adult organism. (Modified from Gurdon JB, *Sci Am* 1968;219(6):24–35.)

~11-nm chromatin fiber) on which RNA polymerase molecules appear as globular particles, with a single RNA molecule trailing the polymerase particle. Active RNA polymerase II molecules are often observed to be single units, indicating an infrequent transcription of many gene segments. However, on occasion, many polymerase particles and associated transcripts are observed on a single gene, indicative of high-frequency transcription. Here, the length of the associated RNA molecules are observed to increase progressively in the direction of transcription. Micrographs of such gene regions demonstrate a characteristic "Christmas tree" pattern. More importantly, these experiments demonstrated that transcription of a DNA segment begins and ends at discrete sites, defining a transcription unit (Fig. 5–16).

Because transcription begins with the binding of a polymerase molecule to specific DNA segments, an expanded definition of a gene as a unit of transcription must then take into account the segments of DNA that

are associated with the transcribed segments. These associated DNA regions are said to direct or promote transcription. Comparisons of DNA sequences from several genes show that certain of these elements are conserved in sequence and position relative to the transcribed sequences.

In summary, nuclei of somatic cells from multicellular organisms contain a complete complement of genetic material. This constancy of DNA implies that the drastic physiologic and biochemical differences among differentiated cell types in these organisms arise by the specific activation of individual genes, allowing their expression within a given cell. This activation process includes sequences that are copied into a RNA molecule as well as adjacent DNA sequences that are required for appropriate transcription, giving rise to transcription units. Moreover, the primary transcripts synthesized from eukaryotic genes contain informational segments (exons) and noninformation segments (introns) and must be significantly modified to form a functional RNA sequence. Accord-

Figure 5–15. The chicken ovalbumin gene makes an RNA that is larger than the translated mRNA. Transcription of the ovalbumin gene leads to the formation of an RNA that is larger than the mRNA found in the cytoplasm for translation. This larger RNA, called the *primary RNA transcript*, contains segments of noninformational content—introns—that are found in the double-stranded DNA of the gene. The intron segments are removed, and the segments containing informational content are joined together at specific sites, called *donor* (*D*) and *acceptor* (*A*) sequences, to form the mature mRNA by a mechanism of RNA processing.

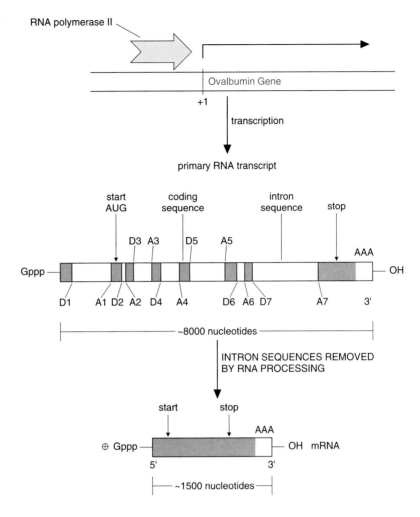

ingly, the pathway of information transfer from DNA to protein in eukaryotic cells involves a complex set of steps; altering the pathway at any of these steps can be a point of regulatory control for gene expression. A cell may exert control of the proteins it expresses by:

1. Transcription control specifying when and how often a gene sequence is copied into RNA;
2. Processing of the primary transcript altering the modifications of the synthesized RNA molecule to form a functional mRNA;
3. RNA transport selection of which mRNAs are exported from the nucleus to the cytoplasm;
4. mRNA stability selectively degrading the mRNA molecules in the cytoplasm;
5. Translation control selection of which mRNAs in the cytoplasm are translated; and
6. Protein posttranslational control activating, inactivating, or compartmentalizing specific polypeptide chains after they have been translated (Fig. 5–17).

Although there is evidence to show that each of these steps may function as points of gene expression control, the expression of any given gene is governed by the collective set of interactions along the pathway of information transfer.

Transcription Control Requires Two Basic Steps: Activation and Modulation of Gene Sequences

The control of gene transcription occurs at two levels: activation, the conversion of compacted chromatin to an extended structure; and modulation, the fine-tuning of transcription mediated by DNA-binding proteins called *transcription factors* (Fig. 5–18). When chromatin is in its most compacted state, such as at metaphase of the cell cycle, there is no RNA transcription. This inhibition of transcription occurs because the DNA is packaged so tightly to form the metaphase chromosomes that it becomes inaccessible to the nonhistone proteins responsible for transcription (RNA polymerases, transcription factors, and others). To be transcribed, a gene sequence must first be made available to the RNA polymerases and other proteins (activated state) so that the regulatory proteins can provide their functions of influencing the rate that a gene is transcribed (modulation).

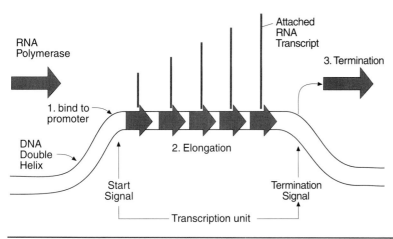

Figure 5-16. A model of a transcription unit, derived from electron microscopy experiments, visualizing transcribed DNA. A transcribed segment of DNA appears as an unfolded 11-nm chromatin fiber that is thought to be looped from the chromosome. Transcription begins with the binding of an RNA polymerase to a promoter DNA element that specifies the starting place for polymerization of the RNA molecule. The growing RNA chain remains linked to the polymerase as transcription proceeds, forming the characteristic "Christmas tree" pattern observed in the electron microscope. Transcription terminates by the release of the primary RNA transcript, which undergoes RNA-processing reactions. Thus, the transcription unit is defined by discrete start and termination sites of the transcription process. (Modified from Alberts B, Bray D, Lewis J, et al. *Molecular Biology of the Cell*, 2nd ed. New York: Garland Publishing, 1989.)

In electron micrographs of transcription units (a DNA segment that is being transcribed), the chromatin appears as an 11-nm fiber. There are nucleosomes present; however, the transcribed segment of DNA is in an extended fiber arrangement of chromatin. These observations correlate well with experiments showing that transcriptionally active gene segments are arranged in a chromatin structure that is biochemically distinct from inactive genes. For example, if a gene is expressed in skeletal muscle but not in liver cells, the gene sequence is more accessible to probes, such as nucleases, in nuclei isolated from skeletal muscle than in those isolated from liver cells. Chromatin in this accessible state is referred to as *active chromatin*, and its nucleosomes are thought to be altered in such a way that their packing is less condensed.

The mechanisms that form active chromatin are not understood. It is thought that active chromatin acquires an extended loop structure that emerges from the surrounding highly condensed chromatin. This model implies that there must be some way to recognize the gene sequences that are to be expressed, such as the possibility that they are converted by a specific protein-DNA interaction to an active, less-condensed form. Evidence now indicates that the conversion of chromatin from an inactive to an active form requires DNA synthesis and that the DNA near active gene sequences contains fewer modified bases, in particular, methylated cytosine residues. Additionally, there is less histone H1

and more nonhistone proteins associated with active chromatin. It is unknown whether these changes cause the activation of gene sequences or whether they occur subsequent to the conversion of the chromatin to an active state.

The conversion of chromatin to an activated state does appear to be irreversible; that is, once the chromatin containing an expressed gene has been changed to an activated state, it retains a structure that is different from highly condensed inactive chromatin, even in the absence of apparent transcriptional activity. Therefore, gene sequences that have adopted an activated chromatin structure have the potential for transcription, with the actual rate that a sequence is transcribed being governed by the interaction of activated DNA sequences with nuclear gene regulatory proteins.

Eukaryotic cells contain a variety of sequence-specific DNA-binding proteins that function to modulate gene expression by turning transcription on or off. Collectively, these proteins are known as *gene regulatory proteins*. These proteins contain structural domains that can "read" the DNA, allowing their binding to specific sequences. DNA binding domains are conserved, allowing the broad classification of regulatory proteins as helix-loop-helix, homeodomain, zinc finger, or leucine zipper proteins. Gene regulatory proteins are generally present in very few copies in the individual cells and perform their function by binding to a specific DNA nucleotide sequence that is generally 8–10 nucleotides

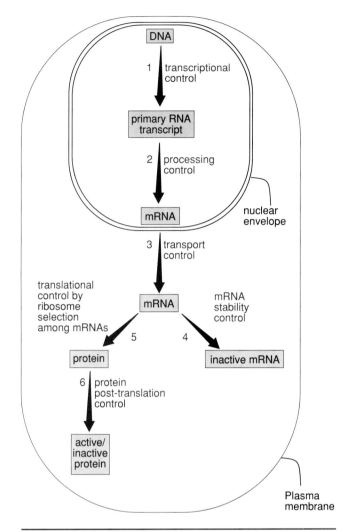

Figure 5–17. Six steps of information transfer in eukaryotes that constitute potential regulatory points of gene expression. Three steps of potential regulation occur in the nucleus of the cell: (*1*) transcription, (*2*) RNA processing, (*3*) transport of the mRNA from the nucleus to the cytoplasm. Once in the cytoplasm, an RNA is subjected to degradation: (*4*) mRNA stability control or selective translation and (*5*) translation control by the pool of ribosomes in the cytoplasm. The translated protein may be modified to form an active protein, to inactivate the protein, or (*6*) to compartmentalize the protein by posttranslational controls.

long. The DNA sequences recognized by these proteins can be classified into two broad categories: the core or basal promoter sequences and the enhancer sequences (Fig. 5–19).

Core promoter sequences are generally located very close to the transcribed portion of DNA and function to specify the exact point of RNA chain initiation. These sequences are often enriched in adenine and thymidine bases located approximately 20–30 bases (called *TATA sequence*) and 70–80 bases (called *CAAT sequence*) to the 5′ side of the transcribed DNA segment. Because they appear to function in all cell types, these sequences

are thought to promote basal transcriptional activity. The second class of DNA elements, the enhancer sequences, are regulatory DNA sequences that activate or enhance transcription from a core promoter, with RNA synthesis beginning at the site specified by the core promoter, and appear to function regardless of their sequence polarity and location relative to the transcribed DNA. The enhancer sequences are variable and capable of providing their function even when located at relatively long distances away from a transcribed gene. In some cases, enhancer sequences have been found in DNA that follows (to the 3′ side) the transcribed gene. Each of the different enhancer elements appears to bind a specific, distinct protein factor, allowing the specificity of its function to be based on the appearance or absence of the protein factor within the nucleus of a given cell.

RNA polymerase catalyzes the synthesis of RNA from a DNA template at a singular rate of about 30–40 nucleotides per second. Because the rate of synthesis (transcription) is constant, the absolute rate that a gene is transcribed is effectively regulated by the number of polymerases that are synthesizing RNA from the gene sequence. Thus, the principal role of the gene regulatory proteins is to modulate the number of polymerase molecules actively synthesizing RNA from a given segment of DNA. That is, an increase in transcription requires an increase in active RNA polymerase molecules, and reduction in transcription is accomplished by fewer active RNA polymerase molecules. There are two ways that binding of a protein to a specific DNA sequence can affect gene transcription (see Fig. 5–19).

First, binding of a regulatory protein may affect the conformation of surrounding DNA. The binding may unmask sequences, allowing an optimized presentation of polymerase-binding sequences to increase transcription or, conversely, the binding may present a block to transcription by tightening DNA conformation or by occupation of polymerase-binding elements (or both).

Second, a regulatory factor may interact directly with the polymerase. The binding of a protein to a specific sequence near the gene could create a complex that attracts polymerase molecules by an optimal binding complex formed between the polymerase and the regulatory factor. Experiments have shown that both mechanisms are used (often in combination) and support the conclusion that the major function of regulatory factors is the recruitment of RNA polymerase molecules to a particular gene locus.

It would seem that a way to have tissue- and cell-specific transcription of a gene would be simply to have a single regulatory region for each gene. Transcription could then be governed by a single protein–DNA interaction. However, most eukaryotic genes contain multiple DNA elements that are able to bind different regulatory factors. The synthetic rate of a gene is then governed by

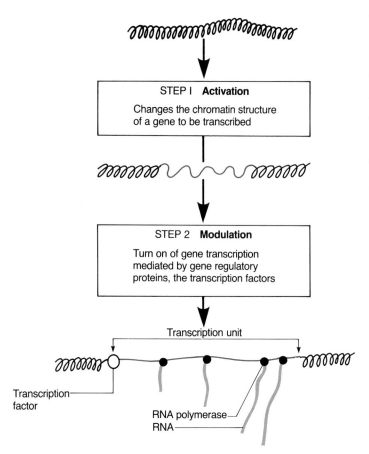

Figure 5–18. Steps involved in the transcriptional control of gene sequences. The transcriptional control of gene expression occurs in two discrete stages. The first step, *activation*, involves changes in the structure of chromatin containing the DNA sequences to be transcribed. The molecular interactions that cause this change in chromatin structure are unknown, but they result in the formation of a region of DNA-protein complex that is biochemically distinct from that of inactive (condensed) chromatin. The second step of transcriptional control, *modulation*, involves the binding of gene regulatory proteins to the activated DNA sequences that operate to fine-tune the transcription of a specific gene.

the sum total effects derived from multiple protein–DNA interactions. The best example of this regulatory scheme is the expression of the β-globin gene. The DNA surrounding the chicken β-globin gene contains 13 distinct sites (7 that are 5′ and 6 that are 3′ to the gene) capable of binding 8 different protein factors (Fig. 5–20). In early development, the chromatin containing β-globin gene is converted to an active state, and nine of the binding sites for regulatory factors are occupied, even though the gene is not transcribed. Later, all but one of the sites have bound their regulatory factors allowing transcription; after which the change in binding of proteins at the 5′ region of the gene again inhibits the transcriptional capacity of the β-globin gene. Even though this represents a singular example, most eukaryotic genes appear to be subject to transcriptional regulation derived from the sum of activities of multiple regulatory proteins. The cellular specificity of gene transcription is thereby governed by a delicate balance of regulatory protein activities.

Recently, a class of regulatory factors has been defined that is able to regulate multiple genes specifying a particular cell type (Fig. 5–21). MyoD1, a nuclear protein found in skeletal muscle, is an example of this class of master regulatory proteins. The introduction of DNA containing the *MyoD1* gene into cultured fibroblasts (cells that never express muscle-specific gene products), can convert the fibroblast to muscle cells, phenotypically as well as biochemically. The exact process by which MyoD1 is able to convert cells to a skeletal muscle phenotype is unknown; however, this protein binds to the regulatory region of several (but not all) skeletal muscle-specific genes. That this protein does not bind to the regulatory regions of all skeletal muscle genes implies that it must trigger other regulatory events required for the expression of the muscle phenotype. Furthermore, the regulation of specific gene sequences must be governed by a combination of regulatory-binding activity. Regardless of the mode of action, the transcriptional regulation leading to cellular specialization is subject to a coordinate activation of gene sequences, with the subsequent modulation by the binding of sequence-specific regulatory protein factors.

Primary Transcripts Are Modified to Form Mature Messenger RNAs

The primary transcripts of RNAs synthesized by RNA polymerase II (mRNAs) are modified in the nucleus by three distinct reactions: the addition of a 5′ cap; the addition of a polyA tail; and the excision of the noninformational intron segments. These modifications are required to form a mature RNA capable of supporting the translation of a protein, and the entire set of events is called *RNA processing*.

Figure 5–19. DNA segments that can modulate transcription by binding gene regulatory proteins. There are several DNA sequences that have the potential to bind gene regulatory proteins and are important in the modulation of transcription. **A:** These can be divided into two categories: core promoter elements that are conserved A–T-rich DNA sequences required for basal transcription activity, and enhancer DNA elements that may be placed 5′ (upstream) or 3′ (downstream) relative to the gene. **B:** Two general mechanisms can be envoked to explain the action of DNA-binding sites: the binding of a gene regulatory protein may change the surrounding DNA, such that it is more favorable for binding the transcriptional machinery (polymerases); or the bound regulatory protein may directly interact with the transcriptional machinery.

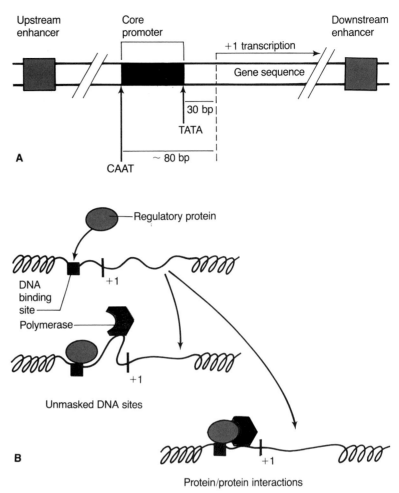

The 5′ end of the mRNA (the end that is synthesized first during transcription) is capped by the addition of a methylated guanosine nucleotide. The addition of the 5′ cap is the first modification of the mRNA primary transcript and occurs almost immediately with the onset of transcription (Fig. 5–22). The formation of the cap involves the condensation of the triphosphate moiety of a guanosine triphosphate (GTP) molecule with the diphosphate group of the nucleotide at the 5′ end of the initial transcript. The enzymes responsible for the capping reaction(s) are thought to reside within the subunit structure of RNA polymerase II. The addition of the 5′ cap structure is critical for a mRNA to be translated in the cytoplasm and appears to be needed to protect the growing RNA chain from degradation in the nucleus.

The second modification of a mRNA transcript occurs at its most 3′ end, the addition of a polyA tail. The 3′ end of most polymerase II transcripts is not defined by the termination of transcription, but by the specific cleavage of the RNA molecule and the addition of adenosine residues to the cleaved molecule by a separate polymerase, polyA polymerase. The signal for cleavage is the appearance of the sequence AAUAAA in the growing RNA chain, with the actual cleavage occur-

ring about 10–30 nucleotides away from this signal sequence. Immediately upon cleavage, polyA polymerase adds 100–500 residues of adenylic acid to the 3′ end of the cleaved RNA molecule. The RNA polymerase II appears to continue transcription well beyond the cleavage site, with the subsequent RNAs being rapidly degraded, presumably because they lack a 5′ cap structure. The exact functions of the polyA tail are not well defined; however, experimental evidence suggests that it plays an important role in the export of mature mRNAs from the nucleus to the cytoplasm. Additionally, it may serve a regulatory function, in that some genes contain multiple sites for polyA addition.

Following modification of the 5′ and 3′ ends of the primary transcript, the noninformational intron segments are removed, and the coding exon sequences are joined together by RNA splicing. The specificity of exon joining is conferred by the presence of signal sequences marking the beginning (called the 5′ donor site) and the end (called the 3′ acceptor site) of the intron segment (Fig. 5–23). These signal sequences are highly conserved (they are approximately the same in all known intron segments) and, as might be predicted, alterations in these sequences lead to aberrant mRNA molecules. For example, a group of genetic diseases,

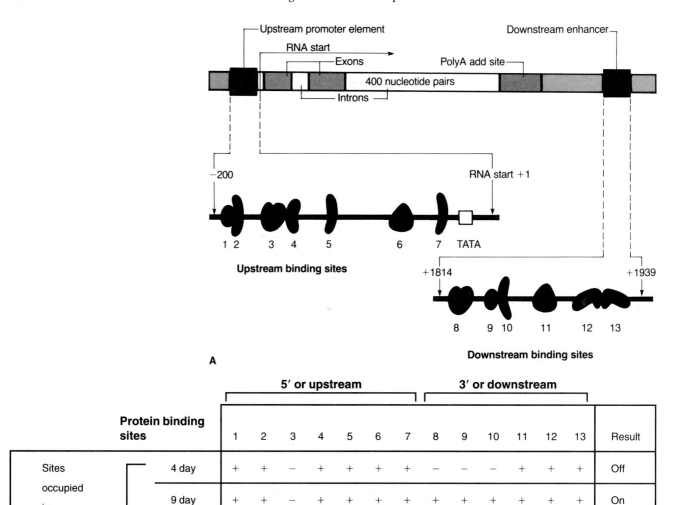

Figure 5–20. Site of gene regulatory protein binding that controls the expression of the chicken β-globin during development. **A:** A diagram of the chicken β-globin gene with the known binding sites for the 13 different gene regulatory proteins. Notice that some of the regulatory proteins (shown by the different shapes) have two binding sites—for example, sites 3 and 8—and that some binding sites are very close to each other—sites 1 and 2, 3 and 4, 9 and 10, 12 and 13—to promote protein–protein interactions between the two regulatory factors. **B:** The occupation of regulatory protein-binding sites during development is shown by the *plus* (+) occupied, and *minus* (–) unoccupied sites. As indicated by this chart, the difference in the β-globin gene being on (9 days) and off (4 days and adult) is due to a balance of the activities of multiple gene regulatory proteins that may exhibit binding to different DNA sequences near the gene. (Modified from Alberts B, Bray D, Lewis J, et al. *Molecular Biology of the Cell*, 2nd ed. New York: Garland Publishing, 1989.)

collectively called the β-*thalassemia syndromes* (characterized by the abnormally low expression of hemoglobin), are directly attributable to single-base changes in the genome at splice junctions of the β-globin gene that disrupt the appropriate joining of exon segments. Therefore, splicing reactions must occur with exquisite precision to ensure that a functional RNA molecule is formed.

The excision of an intron segment from an RNA is carried out by a ribonucleoprotein complex called the *spliceosome*. The spliceosome is formed from a set of undefined proteins complexed with a series of small RNA molecules referred to as *U1* through *U12*. The splicing reaction occurs in steps that include (a) recognition of consensus 5′ donor and 3′ acceptor sequences; (b) cleavage of the 5′ splice site and formation of a looped

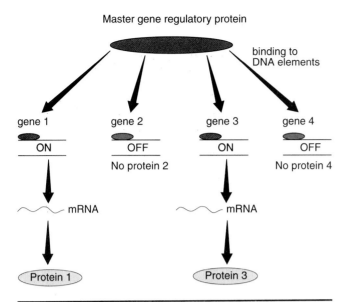

Master gene regulatory protein

binding to
DNA elements

gene 1   gene 2   gene 3   gene 4

ON   OFF   ON   OFF

No protein 2   No protein 4

mRNA   mRNA

Protein 1   Protein 3

Figure 5–21. Scheme of the activity of a master gene regulatory protein. The expression of a protein, such as MyoD1, that has the ability to regulate the expression of multiple genes—some positive, some negative—could lead to cellular specialization. The activities of proteins that are subject to regulation by the master gene protein can be a biochemical trait for a differentiated cell (actin and myosin in skeletal muscle) or can be a protein that is required in the nucleus to regulate specific genes of the differentiated cell.

RNA structure termed the *lariat*; and (c) the cleavage of the 3′ splice site and subsequent ligation of the RNA molecule. The exact role of individual components of a spliceosome is under study; however, it is known that the excision of introns requires the energy of ATP hydrolysis. The excised intron is degraded almost immediately after its release from the primary RNA transcript.

Although it would be logical that the splicing of an RNA proceeds by the removal of the most 5′ intron to the last or 3′ intron, experimental evidence has demonstrated that the removal of introns from any given transcript follows a preferred path, often beginning with introns internal to the transcript. This seems to be an inefficient mechanism and, initially, was viewed with skepticism; however, the recent discovery that a single gene may express multiple different proteins by the selected joining of exon sequences to form different mRNAs has shed some new light on the pathways of intron removal. For example, a single gene encoding the protein troponin T can produce at least ten distinct forms of the molecule by simply joining different combinations of encoded exon segments. This variability can be influenced by cell type or by factors extrinsic to a cell, enabling the expression of protein isoforms needed to compensate for alterations in cellular metabolism. The ability of certain genes to form multiple proteins by joining different exon segments in the primary transcript is called *alternate splicing* and has caused a reexamination of the concept of one gene, one protein.

RNA Transport to the Cytoplasm Occurs Via the Nuclear Pore Complex

Following the modifications of the primary transcript, the functional mRNA must transverse the nuclear envelope to the cytoplasm to direct the synthesis of a protein. Although this step is important, it is probably the least understood step of the gene expression pathway. One model for transport of mRNAs to the cytoplasm suggests that RNAs pass the nuclear envelope through the nuclear pores, presumably by an active transport mechanism that requires the recognition of the RNA, or a protein bound to it, by a receptor molecule either, within the nucleoplasm or directly associated with the pore complex (Fig. 5–24). Experiments have shown that primary transcripts that cannot be appropriately processed are retained in the nucleus and will be degraded; they are not allowed transport into the cytoplasm until all processing steps are completed. Therefore, there must be components within the nucleus that cause selective retention of RNA molecules. This selective retention might be operative on a broader scale, allowing the possibility of nuclear components serving as a filtering mechanism in the determination of which RNAs are transported. Currently, the mechanisms that allow the transport of RNAs are largely unclear although, without this transport, an RNA cannot complete the path of information transfer. Thus, it is a necessary step that has potential regulatory control of gene expression.

RNAs in the Cytoplasm Are Subject to Degradation

Once an RNA reaches the cell's cytoplasm, it is subject to degradation by nuclease components resident in the cytoplasm; that is, RNAs, as well as other cellular components, are continuously being replenished by a balance of their synthesis and degradation. In eukaryotic cells, mRNAs are degraded at different selective rates. A measure of degradative rate for a particular mRNA is called its *half-life* (the period of time it takes to degrade an RNA population to half of its initial concentration), and RNAs with longer half-life measurements are said to be more stable. For example, β-globin mRNA has a half-life longer than 10 hours, whereas RNAs encoding the growth factors called *fos* and *myc* have measured half-lives in the same cells of approximately 30 minutes. Therefore, the β-globin RNA is more stable than are *fos* and *myc*, and these experiments demonstrate the ability of selective mRNA degradation within the cell. Additionally, this example is indicative of how selective degradation of mRNAs might control expression. Simply stated, a β-globin mRNA, by virtue of its longer resident time in the cytoplasm, has the ability to direct the synthesis of more protein because it is in contact with the synthetic machinery (the ribosomes) for a longer period than are the *fos* and *myc* RNAs.

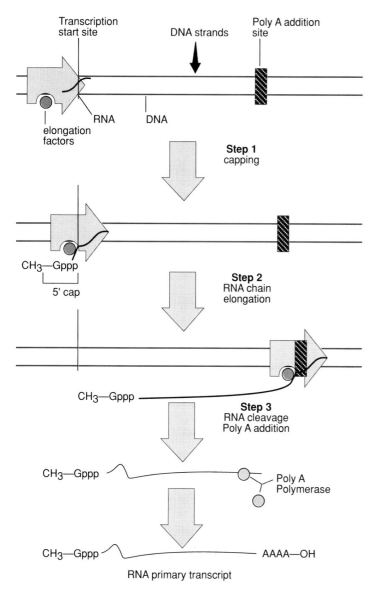

Figure 5–22. Synthesis of a primary RNA transcript involves two modifications of the RNA strand. Almost immediately after RNA synthesis is initiated, the 5′ end of the RNA is capped by a guanosine residue (*step 1*), which protects it from degradation during the elongation of the RNA chain (*step 2*). Upon reaching the signal sequence for the addition of the 3′ polyA tail, the RNA is cleaved (*step 3*), allowing polyA polymerase to add multiple adenosines to the 3′ end of the RNA. This RNA is the primary transcript and ready for splicing of intron segments to form the mRNA. Although the steps involved in transcriptional termination are not well defined, one model is that the polymerase is altered in its activity and continues to synthesize RNA, but this synthesis is not productive because the RNA is degraded.

The stability of a mRNA can be influenced by extracellular signals. The primary response of cells to steroid hormones is an increased transcription rate of selective genes. However, the hormone also can influence the expression of these gene products by increasing the stability of their mRNAs in the cytoplasm of the cell. Certain signals may cause the selective degradation of RNAs, leading to less protein being expressed. For example, the addition of iron to cells decreases the stability of the mRNA that encodes the ion-scavenging transferrin receptor (Fig. 5–25). The altered stability of this mRNA is mediated by a specific nucleotide sequence within the 3′ nontranslated region of the molecule. This region of the transferrin receptor mRNA is bound by an iron-sensitive receptor protein that, when resident, protects the RNA from degradation. In the presence of excess iron, the receptor is dislodged from the 3′ nontranslated binding site, and the RNA is rapidly degraded, preventing the synthesis of the transferrin receptor protein.

As suggested by experiments examining transferrin mRNA stability, the selective degradation of many mRNAs is controlled, at least in part, by specific nucleotide sequences within the 3′ nontranslated region of the RNA molecule. This concept was demonstrated by genetic engineering, mixing and matching specific regions from RNAs displaying different stabilities, such as the experiments shown in Fig. 5–26. When the 3′ noncoding segment of a stable mRNA, such as globin, is substituted for the analogous region of a nonstable growth factor mRNA (e.g., *fos*), the resultant growth factor mRNA displays a stability similar to that of the globin RNA. That is, the engineered growth factor RNA becomes more stable solely because of its new 3′ nontranslated segment. Similarly, when the 3′ terminus of histone RNA, an RNA that shows selective stabilization during the DNA synthetic phase (S phase) of the cell cycle, is placed onto a globin mRNA, the globin mRNA then acquires the cell cycle-dependent degrada-

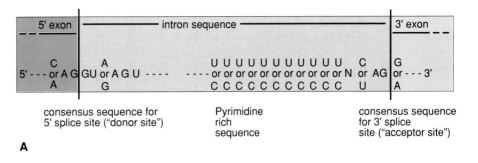

Figure 5–23. Mechanism of RNA splicing to form mature mRNA molecules. **A:** RNA splicing occurs at very discrete locations that are marked by conserved sequences. The consensus sequences for RNA splicing, listed here, have been determined by comparison of many eukaryotic polymerase II gene sequences. The most conserved nucleotides (*shaded*) mark the boundaries of the intron sequence. **B:** The mechanics of RNA splicing involve the recognition of signal sequences by the U1 (5′ donor) and U2 (polypyrimidine sequence), which leads to the formation of the spliceosome (a combination of many SnRNP molecules). Once the spliceosome is formed, the 5′ donor is cleaved by the formation of an RNA lariat, the 5′ donor is then ligated with the 3′ acceptor, and the spliced intron is degraded in the nucleus. (Modified from Alberts B, Bray D, Lewis J, et al. *Molecular Biology of the Cell*, 2nd ed, New York: Garland Publishing, 1989.)

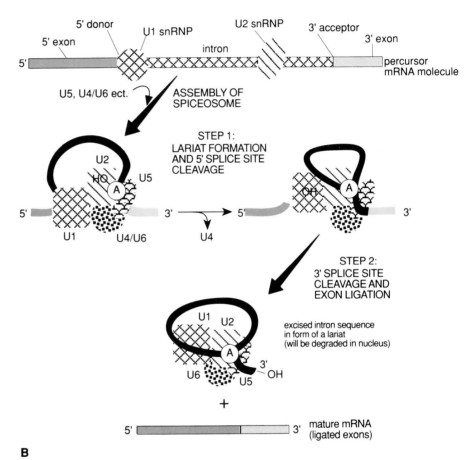

tion characteristics of the histone mRNA. The conclusion drawn from these mix-and-match experiments is that the specific stability of an RNA is governed, in part, by sequences resident within the 3′ noncoding portion of the RNA molecule. However, the cellular components responsible for this selective degradative process are not well established.

Gene Expression Can Be Controlled By Selective Translation of Messenger RNAs

The second basic step of gene expression is the translation of mRNAs into protein. In eukaryotes, the translation of proteins occurs in the cytoplasm; however, not

all mRNAs are translated upon arrival to the cytoplasm. The RNA molecules in the cytoplasm, as in the nucleus, are constantly associated with proteins, some of which may function to regulate translation. Most of the defined mechanisms that regulate translation operate to repress protein synthesis (negative translational controls), although there is some evidence (derived from studies examining viral RNAs) that positive or enhanced translation of certain RNAs may be operative. Translational controls are important in many fertilized eggs, which must rapidly switch from making the proteins required for maintenance of a quiescent oocyte to making proteins required for cell division and growth. These eggs have stored as RNA–protein com-

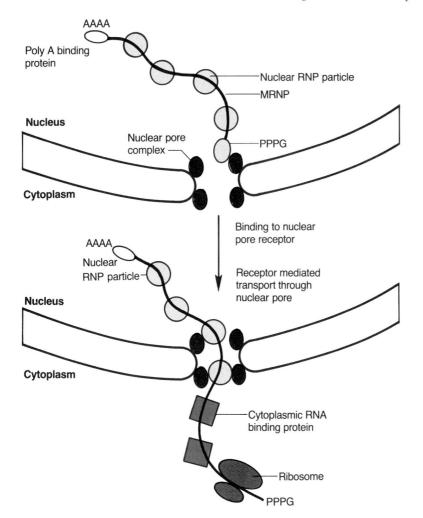

Figure 5–24. Potential mechanism of mRNA transport through the nuclear pore complex. A mRNA that is ready for transport to the cytoplasm is bound by a variety of proteins, including a polyA-binding protein, ribonuclear protein (RNP) particles, and perhaps others. These proteins protect the mRNA from degradation and perhaps bind with a receptor molecule on the nuclear pore complex. Once bound, the nuclear pore receptor may facilitate the transport of the mRNA to the cytoplasm, similar to the mechanism of protein import to the nuclear compartment.

plexes maternal mRNAs that are not translated until the egg is fertilized.

Another important negative translational control has been demonstrated for the expression of iron storage protein ferritin. Ferritin mRNA in the cytoplasm shifts from an inactive RNA–protein complex to a translationally active polyribosome on elevation of intracellular iron concentration. The block of ferritin mRNA translation is mediated by a 30- to 40-nucleotide segment of the RNA at the 5′ leader (5′ nontranslated region) segment of the molecule (see Fig. 5–25). This segment of the RNA binds a repressor protein that blocks the ability of ribosomes to form active complexes on the mRNA. This repressor, called the *iron response molecule*, is the same protein that is bound to the 3′ nontranslated segment of the transferrin receptor mRNA discussed in the previous section. Thus, this iron response protein allows exquisite control of intracellular iron metabolism by increasing the degradation of the mRNA encoding an iron-salvaging protein, the transferrin receptor, and simultaneously releasing the translational block of the iron-binding protein, ferritin. This provides rapid sensitive controls of gene expression without affecting the synthetic rates (transcription) of these mRNAs.

Modifications of Proteins Posttranslationally Can Affect the Expression of an Active, Functional Molecule

Once a protein has been synthesized by the ribosomal complexes in the cytoplasm, the functional capabilities of the protein often are not realized until the protein has been modified. Although these mechanisms, collectively referred to as posttranslational modifications, are not often thought of as gene expression controls, it is recognized that the manifestation of gene expression is not complete until a protein is carrying out its function within the cell. An example of posttranslational modifications occurs in the formation of a functional insulin protein. Insulin is a secreted polypeptide hormone that is synthesized on ribosomes associated with the endoplasmic reticulum. It is synthesized as a single polypeptide referred to as preproinsulin. The prefix pre refers to a signal peptide sequence that directs the translocation of the proinsulin molecule across the endoplasmic retic-

Figure 5–25. Iron metabolism in cells is regulated by modulation of ferritin translation and destabilization of transferrin mRNA. An iron-sensitive receptor protein is able to bind to specific sequences of the ferritin and transferrin mRNAs. A: The ferritin sequence is located in the 5′ nontranslated region of the mRNA, and the bound protein blocks the translation of this mRNA. B: The transferrin mRNA contains similar binding sequences, and the binding of the same iron-sensitive receptor protein to these sequences stabilizes this mRNA, allowing more protein to be translated. An increase in iron concentration in cells is sensed by the receptor and, on binding the excess iron, the conformation of the protein changes so that it no longer is bound to the mRNAs. The result is a release of the translational block of ferritin mRNA and a destabilization of the transferrin mRNA. Thus, the regulation of intracellular iron concentration is accomplished quickly by posttranscriptional regulation of gene expression.

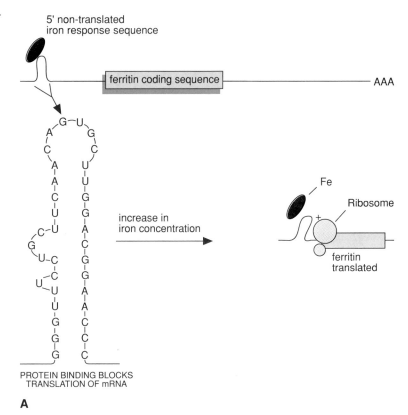

A

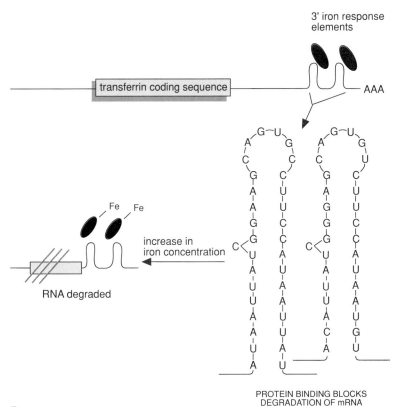

B

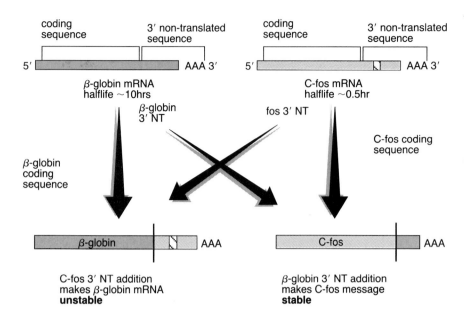

Figure 5–26. Genetic mix-and-match experiment examining the role of the 3′ nontranslated region in mRNA degradation. The experiment outlined shows that if the 3′ nontranslated region from a stable mRNA (such as β-globin) is substituted onto an mRNA that has a short half-life (such as *c-fos*), the resulting chimeric mRNA will display the characteristics of the β-globin mRNA, in that it becomes more stable when introduced into cells. The opposite effect is seen when the *c-fos* 3′ end is placed on the β-globin coding sequence. (Modified from Raghow R. *Trends Biochem Sci* 1987;12:358–360; and Shaw G, Kamen R. *Cell* 1986;46:659–667.)

ulum membrane. This pre sequence is immediately removed by protease cleavage, the first posttranslational modification. The proinsulin molecule then folds such that the NH2- and COOH-terminal ends are held in close proximity by sulfhydryl bonds. The proinsulin is inactive until a second cleavage event removes the connective, or C peptide, leaving two chains formed from the NH2-terminal domain and COOH-terminal domain of the proinsulin peptide. Thus, multiple posttranslational events are required to form an active insulin hormone. Moreover, the sequestering of the hormone into secretory vesicles can be viewed as a regulatory event, in that some proteins are required at specific cellular locations for them to exert the functional properties. There are any number of events or modifications that alter the activity of a protein (e.g., phosphorylation, methylation, glycosylation, etc.), or that may play a role in the selective degradation of proteins (the addition of a ubiquitin molecule) that can constitute posttranslational controls of gene expression.

Structural RNAs (tRNAs and rRNAs) Are Subjected to Regulatory Mechanisms

The preceding sections have limited discussions to RNA synthesized by polymerase II, the mRNAs. However, it is important to point out that the structural class of RNAs—rRNAs and tRNAs—is also subject to regulation. RNA polymerase III transcribes the tRNAs and a class of small RNAs, referred to as 5S RNA, each of which is made as precursor molecules that are subsequently modified to form functional molecules. Moreover, the transcription of these RNAs is facilitated by the specific binding of proteins to the promoters of these genes; one of which has been purified and is called transcription factor (TF) IIIC. Similarly, RNA polymerase I

transcribes ribosomal RNA, a process facilitated by the binding of factors. For example, there is a protein called TFID that regulates the transcription of the 45S rRNA precursor, which is modified in the nucleolus to form 18S (small) and 28S (large) rRNAs. Because each of the RNAs (mRNA, rRNA, tRNA) must work in concert to provide the cell with functional proteins, it is clear that regulatory events for each RNA class are important for the overall transfer of information from DNA to protein.

Genetic Therapy

There Are Many Obstacles to the Development of Effective Gene Therapies

With the development of recombinant DNA technology has come the promise, or potential, of it dramatically improving the practice of medicine. Indeed, as advances have been realized, with various aspects of DNA technologies, there are parallel applications of new methods in the clinical management of patients. Restriction fragment length polymorphism analysis is commonly used to aid in the diagnosis of specific inherited diseases. The sensitivity of the PCR technique makes it useful not only for the diagnosis of inherited diseases and for latent viral diseases (e.g., AIDS virus), but also for forensic medicine. An ability to understand the roles of a specific gene product in the pathogenesis of human disease has allowed precise and effective clinical intervention. Moreover, recombinant DNA technologies have led to the development of new therapeutic products made possible by the ability to engineer the overexpression of genetic material in bacteria and eukaryotic cells. Although recombinant DNA technologies have made significant inroads toward diagnosis and management of human disease, the treatment of human disease through transfer of genetic material to a patient (e.g., gene therapy) is not yet commonplace in medical practice.

In general, there is a variety of questions which must be addressed as a prelude to the implementation of an effective genetic based therapy. First, the gene that is the root cause of the genetic disease must be well studied. This would include its identification with cloning and sequencing, as well as an understanding of its regulation and the function of the expressed gene product within the cell. Second, the gene must be delivered to the appropriate cell(s) and maintained stably expressed at an appropriate level within the cell. For example, the targeting to and expression of a normal β-globin molecule in a neuron of the brain would not provide an effective therapy for sickle cell anemia. Finally, the expression of the gene product must be able to correct or reverse the disease process. This is important, particularly for somatic cell therapies (discussed later), when the expression of an appropriate gene product will be able to effect a curative response; that is to say, it will reverse the disease phenotype. Few inherited diseases are understood at the level of complexity outlined above. However, the daily contributions to our knowledge of the molecular bases underlying inherited diseases will soon bring genetic therapies to the forefront. Two considerations for successful genetic based therapies are a suitable model system for study and strategy for gene replacement.

Transgenic Mice Offer Unique Models of Genetic Diseases

Medical science has prospered dramatically when there is the ability to model a disease process in animals. The mouse has presented an excellent model for much of this work, in large part because of the tremendous knowledge base of mouse genetics. Many experiments mapping human genes and diseases have relied on somatic cell techniques using mouse cells. The finding of rodent genes that can complement the function of the human counterparts, and the high homologies observed for many genes at the primary structure (DNA sequences) and chromosome locals between mice and humans demonstrates the similarities between genetics, metabolism, and physiology, lending support for the use of mice as human disease models.

It has been demonstrated in a variety of eukaryotic systems that linear DNA fragments introduced to a cell will be ligated together to form tandem arrays (i.e., DNA molecules joined end to end) and can be integrated into the genome at random positions. This provides a technique to test the function of a gene within the confines of the recipient cell. If the gene-modified chromosome is allowed to enter the germ line cells, it can then be passed along to progeny. These animals, if maintained, contain permanently altered genomes and are said to be transgenic animals with the foreign DNA referred to as *transgenes*.

There have been many transgenic mice formed by the introduction of a human gene that houses altered information leading to disease, allowing the function of the human gene to be examined. Further, the recent ability to add a DNA that integrates at a specific location as a gene within the mouse genome (e.g., gene targeting) has allowed the ability to generate transgenic mice with altered expression of the mouse gene or to generate a mouse that expresses an altered gene product—altered at the information storage level or the genome. These types of experiments have provided powerful knowledge of human disease states, from the ability to model the disease, and have enhanced our understanding of genetics so that gene-based therapies for human diseases are now within reach.

Many Strategies Are Available for Gene-Based Therapies

The essence of a successful genetic therapy is the availability of a strategy for gene replacement. There are basically two types of gene replacement therapies: altering of germ line cells and altering of somatic cells. Whereas the fixing of germ line cells could potentially cure the disease, there are many ethical issues that must be addressed before this is a viable alternative. The best opportunity for gene therapy is the ability to alter somatic cell expression of the diseased gene. As stated above, the disease must be studied to the greatest of molecular detail before a strategy can be devised. Perhaps the best examples to date, and those for which strategies have been devised, are sickle cell anemia, adenosine deaminase deficiency, cystic fibrosis, Duchenne's muscular dystrophy, and familial hypercholesterolemia. In each case, the gene deficiency is known, its expression pattern studied, and the function of the expressed product defined.

To effect a successful gene-based therapy, one must be able to target the gene to the appropriate cell. This can be done in two ways, *ex vivo* (the cells are outside the organisms); or *in vivo* (the cells remain within the organism). Many protocols rely on *ex vivo* techniques because of the ability to control the cellular environment during the delivery of the genetic molecule. A potential *ex vivo* strategy for the genetic intervention of familial hypercholesterolanemia is outlined in Figure 5–27. As illustrated in Figure 5–27, the potential therapeutic DNA must be delivered to the cell in a way that it becomes part of the recipient cell's genome. There are essentially two methods for accomplishing this: physically adding the DNA and allowing it to integrate, or using a virus to facilitate DNA uptake. The viral approach is most utilized, due to the ability to have DNA integrate in higher numbers of cells; however, the virus (retrovirus or DNA virus are both used) can cause immunological problems later. In either

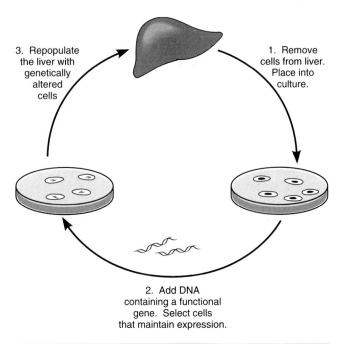

3. Repopulate
the liver with
genetically
altered
cells

1. Remove
cells from liver.
Place into
culture.

2. Add DNA
containing a functional
gene. Select cells
that maintain expression.

Figure 5–27. A potential *ex vivo* method for genetic-based therapy of a liver disease. In the example shown here, hepatocytes are first isolated from the liver and placed into culture dishes with an appropriate media, such that they remain as hepatocytes. The DNA carrying an altered gene is added to the cells—either by physical means or facilitated by virus vectors—and the cells that received the gene are selected for in the culture. The cultured cells containing the altered gene are then added back to the original liver tissue. It is necessary for the cells to have either a selected advantage or the ability to repopulate the liver tissue, so that the genetically altered cells will be effective in its therapeutic nature.

case, the cells containing the altered gene must be replaced into the organism.

There are as many strategies as there are genetic diseases, and no one approach will be applicable to all diseases. For example, it is possible to stably express protein for long periods in skeletal muscle cells. Thus, if the gene product is used outside the cell (for example α-1-antitrypsin is a secreted molecule made in the liver, but its absence leads to lung disease), it could be possible to use the skeletal muscle as a synthetic protein "factory." As the efficiency of genetic-based therapies becomes greater and our knowledge base expands, using better molecular techniques and more expansion model systems, there is great promise in the ability to treat human disease with a specifically designed DNA molecule.

Clinical Case Discussion

Adenosine deaminase deficiency is one of the first genetic diseases successfully treated by gene therapy. The syndrome of severe combined immunodeficiency (SCID)

encompasses a group of disorders, each due to different genetic defects that have in common a primary impairment of both humoral- and cell-mediated immunity. The majority of cases have X-linked inheritance and are due to mutations at the locus for the chain common to several receptors for interleukins. The remainder of cases show autosomal recessive inheritance, 50% of which are due to a genetic deficiency of the enzyme adenosine deaminase (ADA). Genetic deficiency of purine nucleoside phosphorylase (PNP), the next enzyme in the purine salvage pathway, is a second cause of immunodeficiency. The spectrum of clinical and immunologic abnormalities in patients with ADA-SCID include diarrhea, vomiting, and cough. The diarrhea causes failure to thrive. The cough is persistent and often due to *Pneumocystis carinii* infection. A Candida diaper rash is usually present. All patients with SCID have some degree of hypogammaglobulinemia. Fifty percent show an abnormality of the costochrondral junctions visible radiologically.

Adenosine deaminase is an enzyme of the purine salvage pathway that catalyzes the irreversible deamination of adenosine and deoxyadenosine. A diagnosis of ADA deficiency can be made by enzyme assay of erythrocytes, lymphocytes, or fibroblasts. Enzyme activity in patients with ADA-SCID is essentially undetectable with elevated concentrations of adenosine and deoxyadenosine (dATP), which inhibits lymphocyte function and results in chromosomal breakage. Deoxyadenosine is a known feedback inhibitor of ribonucleotide reductase, an enzyme activity required for normal DNA synthesis. Over 30 different mutations in the gene encoding for ADA have been identified. The treatment of choice for ADA-SCID is allogeneic bone marrow transplantation from an HLA-identical sibling with cure rates of 90–100%. If a donor is not available, PEG-ADA (polyethylene glycol) maintains high levels of ADA activity and uniformly corrects the intracellular metabolic abnormalities. More recently, several patients have received genetically repaired (with retrovirus vectors) T-cells and hematopoietic stem cells. The number of T-cells normalized in these patients, as did many cellular and humoral immune responses.

Suggested Readings

Gene Cloning and Sequencing

Gilbert W. DNA sequencing and gene structure. *Science* 1987; 214:1305.

Lewin B, ed. *Genes*, 4th ed. New York: Oxford University Press, 1990:1.

Nathans D, Smith HO. Restriction endonucleases in the analyses and restructuring of DNA molecules. *Annu Rev Biochem* 1975;44:687.

Gene Expression: The Transfer of Information from DNA to Protein

Leff S, Rosenfeld M, Evans R. Complex transcriptional units: diversity in gene expression by alternative RNA processing. *Annu Rev Biochem* 1986;55:1091.

Maniatis T, Goodbourn S, Fischer J. Regulation of inducible and tissue specific gene expression. *Science* 1987;236:1237.

Raghow R. Regulation of messenger RNA turnover in eukaryotes. *Trends Biochem Sci* 1987;12:358.

Watson J, Hopkins NH, Robert J, et al, eds. *Molecular Biology of the Gene*, 4th ed. Menlo Park: Benjamin-Cummings, 1987.

Gene Therapy

Kayes SG, Wolff JA. Genetic vaccination, gene transfer techniques. In: Meyers, RA, ed. *Encyclopedia of Molecular Biology and Molecular Medicine*. Weinheim: VCH Press, 1996:430.

Lee JH, Klein HG. Cellular gene therapy. *Transfus Med* 1995; 9:91.

Milligan RC. The basic science of gene therapy. *Science* 1993; 260:926

Review Questions

1. Adenosine deaminase deficiency leads to all of the following abnormalities in immune function **EXCEPT:**
   a. Hypergammaglobulinemia
   b. Inhibition of T-lymphocyte function
   c. Abnormal delayed type hypersensitivity
   d. Recurrent invasive bacterial infections
   e. Lymphopenia
2. Proteins directly involved in the modulation of gene transcription are called:
   a. Translational enhancers
   b. Histones
   c. Gene regulatory proteins
   d. Spliceosome proteins
   e. Promoter GTPases
3. You have engineered a vector molecule that allows high expression of proteins in skeletal muscle cells. Your main contribution was to design the vector so that it would remain within the cells for many years, and during this time frame it continuously expresses high levels of foreign proteins. You decide to clone the coding sequences for normal human α and β globins into your vector as a genetic-based cure for sickle cell disease. Other physicians have indicated to you that this would NOT be a good strategy for curing sickle cell disease because:
   a. The muscle cell would not place the heme-cofactor in the right orientation to carry oxygen
   b. The tetramers of α and β-globin would not form in the skeletal muscle cell
   c. If expressed in the skeletal muscle cell, the α and β globins could not help the patient's sickle cell disease
   d. Foreign proteins do not fold appropriately in skeletal muscle cells
   e. It is impossible to express both α and β globin proteins within a single skeletal muscle cell
4. The joining of two DNAs using DNA ligase forms a:
   a. Vector molecule
   b. Recombinant DNA molecule
   c. Restriction fragment
   d. cDNA molecule
   e. Genetic replicon molecule
5. The method of amplifying a DNA sequence using a thermostable DNA polymerase is called:
   a. Progressive charge reaction
   b. Polymerase chain reaction
   c. DNA sequencing reaction
   d. Transcriptional chain reaction
   e. DNA duplex formation
6. Which of the following statements concerning primary transcripts is false?
   a. DNA acts as a template for RNA synthesis
   b. Synthesis of mRNA, tRNA, and rRNA is catalyzed by three separate enzymes
   c. Synthesis of rRNA occurs in the nucleolus
   d. The intron sequences will be joined to form the mature mRNA
   e. Proteins are associated with mRNAs as they are synthesized
7. Which of the following concerning mRNA degradation (mRNA stability) is true?
   a. The polyA tail becomes longer to destabilize the mRNA
   b. The rate of splicing is the major factor in RNA stability
   c. The 5' nontranslated sequence is the only factor that influences mRNA degradation
   d. The rate of mRNA degradation is influenced mainly by the 3' nontranslated sequences
   e. Cytoplasmic spliceosomes are responsible for the degradation of the mRNAs

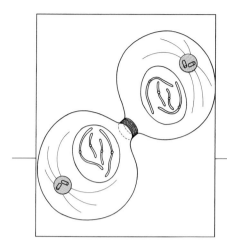

Chapter 6

Cell Adhesion and the Extracellular Matrix

Clinical Case

Ronald McDougal is a 13-year-old adolescent diagnosed with scoliosis on a routine screening examination by the school nurse. He was referred to a pediatrician for further evaluation. He has been in relatively good health except for several episodes of dislocation of his knees and ankles. He states this seems to be a problem that "runs in the family." Ronald has worn glasses for about 3 years for strabismus. His physical examination was remarkable for an extremely tall (>95 percentile), thin young man with unusually long fingers and toes (arachnodactyly), flat feet, hypermobility of all joints, a high arched palate and pigeon chest (pectus carinatum). In addition, an abnormality of the lens was seen in his right orbit. A 2/6 systolic ejection murmur at the left lower sternal border radiating to the back was heard. Dilatation of the aortic root and mitral valve prolapse was diagnosed by echocardiography. Referral to an ophthalmologist for slit-lamp examination confirmed bilateral subluxation of the lenses.

Vertebrates respond to their environment in a complex manner unrealized by unicellular organisms. Individual cell types are specialized to perform different functions. However, certain problems must be overcome for this division of labor to work efficiently. Tissues must form in a manner that enables component cells to adhere to one another. Mechanisms of cellular adhesion have evolved, allowing vertebrates to address this task and are the focus of this chapter. In later sections, we will see how some specific examples of intercellular adhesion apply to the construction and maintenance of muscle and neural tissue.

Cell Junctions

Extracellular Calcium Is an Important Mediator of Cell–Cell Adhesion

Several mechanisms are involved in the adhesion of cells into tissues, and these adhesive mechanisms are mediated by numerous specific adhesive molecules. Adhesive molecules can be randomly scattered within the plasma membrane or they can be organized into morphologically discrete and specialized attachment structures that provide tissues with considerable strength. Proteins that are involved in tissue formation are classified as being either cell adhesion molecules, substratum adhesion molecules, or cell–cell junctional molecules, although many of the cell adhesion molecules also are involved in the formation of intercellular junctions. The adhesion molecules include, but probably are not limited to, the glycoproteins that form cell junctions, nonjunctional adhesive proteins, and integral membrane proteoglycans.

Cell–cell adhesion proteins can be either Ca^{2+}-dependent or Ca^{2+}-independent. Many of the Ca^{2+}-dependent adhesion molecules belong to a family of proteins known as the *cadherins*. Several different cadherins have been identified, and it has been demonstrated that the type of cadherin varies from tissue to tissue. Although cadherins are often distributed randomly within the plasma membrane, at times they can be localized into junctional complexes, as when a subclass of cadherins is organized into adhesion belts. During adhesion processes, a cadherin on the surface of one cell generally binds to an identical cadherin being expressed on the surface of an adjacent cell. This type of binding, in which like molecules adhere, is referred to as a *homophilic binding reaction*. Cadherins are single-pass transmembrane proteins that assume a stable conformation in the presence of extracellular Ca^{2+}. When Ca^{2+} is removed from the extracellular environment, cadherins undergo a conformational change and are rapidly degraded. This proteolytic event can lead to tissue dissociation. Experimental investigations have shown that many of the other Ca^{2+}-dependent adhesive molecules, such as those forming desmosomes, also are proteolyzed in the absence of Ca^{2+}. Furthermore, cDNA sequence analysis has determined that Ca^{2+}-dependent proteins involved in the formation of desmosomes show a high degree of sequence homology to the cadherins. These results suggest that the cadherins may constitute a large superfamily of cell–cell adhesive molecules.

A second family of Ca^{2+}-dependent adhesive proteins are the members of the selectin family of adhesive molecules. The selectin molecules are important in the movement of white blood cells out of the circulation and either into the connective tissue at sites of infection or into the organs of the immune system. The selectins are transmembrane proteins that, unlike cadherins, do not interact with target proteins on adjacent cells. Instead, selectins bind in a Ca^{2+}-dependent fashion to specific sugar residues on the surfaces of target cells. An example of selected-mediated adhesion is the binding of a neutrophil to the endothelial surface during an infection. At the site of inflammation, an endothelial cell will be triggered to begin the production of a particular selectin. The cell surface selectin will bind to sugar residues on the surface of neutrophils, thereby localizing the neutrophil to the site of inflammation. The neutrophil can then exit the bloodstream and begin to carry out its role in the immune response. This type of adhesion process, in which one type of molecule on the surface of a cell binds to a different type of molecule on an adjacent cell, is an example of a heterophilic binding reaction.

Other surface proteins interact through Ca^{2+}-independent mechanisms. Analysis of these adhesion molecules has demonstrated that they are members of the immunoglobulin superfamily. One of the most intensively studied Ca^{2+}-independent adhesive molecules is a protein called *neural cell adhesion molecule* (N-CAM). N-CAM is expressed on the surfaces of both neurons and glial cells and causes the adherence of these two cell types. Moreover, the expression of N-CAM varies during developmental stages, suggesting that this protein is important in the formation of the nervous system by promoting cell adhesion at appropriate times during embryogenesis. The activity of N-CAM will be discussed in more detail later in this chapter.

Cell Junctions Can Be Classified According to Function

Epithelial tissues are mostly cellular, having little extracellular matrix material. Instead, epithelia are held together principally by cell–cell interactions at specialized regions, termed *cell junctions*. Cell junctions are classified into three functionally distinct groups: occluding, anchoring, and communicating. *Occluding junctions*, also called *tight junctions*, form a physical barrier that prevents the leakage of even small molecules between cells. *Anchoring junctions*, which include the desmosomes, hemidesmosomes, focal contacts, and adhesion belts, couple neighboring cells either to each other or to the extracellular matrix through interactions that are mediated by the cytoskeleton. *Communicating junctions*, such as the gap junctions, attach cells together in a manner that allows the transfer of chemical or electrical signals between neighboring cells.

Junctional complexes can be distinguished by several other criteria. Each of the different classes of cell junctions has a characteristic ultrastructure, and biochemical analysis of the different types of junctions has identified proteins that are specific for the individual junctional complexes. In some instances, location can be used as an indicator of junction type. In the following sections, the biochemical composition and molecular organization of each class of junctional complex will be detailed.

Tight Junctions Form a Physical Barrier That Inhibits the Leakage of Molecules from One Side of an Epithelial Sheet to the Other Side

In many instances, the indiscriminate diffusion of molecules and ions across the surface of an epithelium would be detrimental to maintaining homeostasis. For example, considerable energy is expended during the filtration of wastes and toxins from the blood in the kidney.

If it were not for tight junctions between the epithelial cells of the urinary bladder, these highly concentrated toxins would seep into the extracellular spaces and severely damage surrounding tissue. Elsewhere, tight junctions serve as a barrier to diffusion in the intestinal lining. During transport through the gastrointestinal system, food is broken down into nutrients for use in metabolic pathways throughout the body. To access cells where these reactions occur, nutrients must first enter the circulatory system. Tight junctions in the intestinal epithelium serve as physical barriers that restrict foodstuffs to the lumen and allow amino acids, sugars, and other molecules to be directly routed to the circulatory system. Transport proteins and pumps in the membranes of intestinal epithelial cells facilitate the movement of specific nutrient molecules, such as glucose, from the lumen into the epithelial cells, then directly into the bloodstream.

Tight junctions are observed as regions where plasma membranes of adjacent cells come into direct contact with one another (Fig. 6–1). Freeze-fracture electron microscopy has demonstrated that tight junctions are composed of a complex, interwoven network of membrane proteins that encircles epithelial cells along their apical surfaces (see Fig. 6–1D). Several of the proteins that comprise tight junctions, including zo-1, zo-2, and occludin, have been identified, although the exact molecular organization of tight junctions is unknown. The integrity of occluding junctions is dependent on the presence of Ca^{2+} ions, suggesting that cadherin-like molecules may be involved in their construction. Furthermore, the permeability of epithelia to small molecules varies from organ to organ. For example, the epithelial lining of the urinary bladder is much more resistant to the passage of ions than is the epithelium in the intestine. It is thought that these differences in the permeability of epithelia to small solutes is due to the number of anastomosing strands of membrane proteins that compose the tight junctions between neighboring plasma membranes. As a rule, membranes with more rows of tight junction proteins are less permeable to small molecules than are membranes with fewer rows of anastomosing protein strands.

Tight Junctions Form the Blood–Brain and Blood–Cerebrospinal Fluid Barriers

Unlike in nonneural tissues, microvascular endothelial cells in most regions of the brain and spinal cord are joined by continuous tight junctions. Because of the

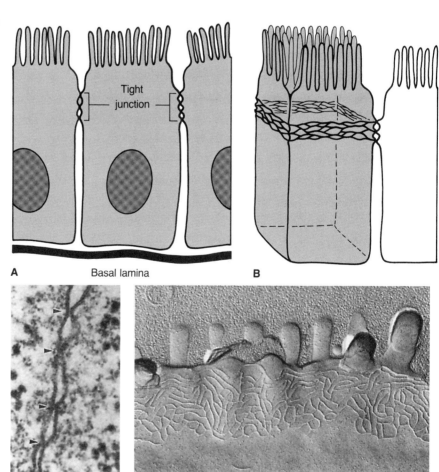

Figure 6–1. A schematic drawing and electron micrographs showing the location and appearance of tight junctions. **A:** A drawing showing a cross-sectional layer of epithelial cells. Tight junctions are located between neighboring cells at the most apical region of the cells and appear as points where the cell membranes come into direct contact with one another. **B:** Tight junctions actually are composed of anastomosing strands of transmembrane proteins that completely encircle the cells. **C:** A TEM image of a tight junction between epithelial cells. Several regions where the membranes of the adjacent cells appear to come into direct contact with one another are shown. These regions of focal membranous connection are sites at which glycoproteins from the neighboring cells interact. **D:** A freeze-fracture electron micrograph showing a tight junction. As shown, the tight junction is composed of a series of strands of integral membrane proteins. The projections on the apical surface of the cell are microvilli. (C, Gilula NB. In: Cox RP, ed., *Cell Communication.* New York: John Wiley & Sons, 1974:1–29. D, Hull BE, Staehelin LA, *J Cell Biol* 1976;68:688–704, with permission.)

absence of fenestrations and transcellular channels in the capillaries, transcapillary movement of molecules, ranging in size from ions to proteins, is consequently prevented. Thus, a physical barrier, termed the *blood–brain barrier*, separates the blood from the interstitial fluid.

In addition, a biochemical barrier formed by various transport carriers and active transport systems allows the brain to concentrate metabolites such as glucose and neurotransmitter precursors (e.g., the drug, L-dopa), while ridding itself of other substances. Surrounding the external surface of these capillaries is a layer of extracellular matrix sandwiched between the apposed foot processes of astrocytes (Fig. 6–2). Experimental evidence suggests that astrocytes induce endothelial cells to form tight junctions and to adopt many of their specialized biochemical properties in brain. Consistent with this hypothesis are observations that the blood–brain barrier is absent where this astrocyte–endothelial cell relationship is disrupted pathologically (in tumors) or normally (e.g., the pituitary gland).

Another area at which brain capillaries are not bound by tight junctions is in the choroid plexus, which secretes cerebrospinal fluid (CSF) into the ventricles. Although molecules can freely move out of these capillaries through fenestrations, tight junctions between the surrounding epithelial cells form a blood–CSF barrier (see Fig. 6–2B). Plasma that leaks through the capillaries is filtered free of blood proteins, which cannot diffuse across the blood–CSF barrier. These epithelial cells, in turn, secrete CSF, which they concentrate with ascorbic acid, glucose, and other nutrients by epithelial carrier and active transport systems. Active clearance systems in the choroid epithelium cells selectively transport K^+, weak organic acids, and other potential neurotoxic substances from the CSF to the blood.

Anchoring Junctions Have A Common Organization

Anchoring junctions provide great strength and are most numerous in tissues that are subjected to extreme shear or abrasive stress, such as the epithelium of skin. All anchoring junctions exhibit the same basic architectural design. Cytoskeletal filaments are attached to the

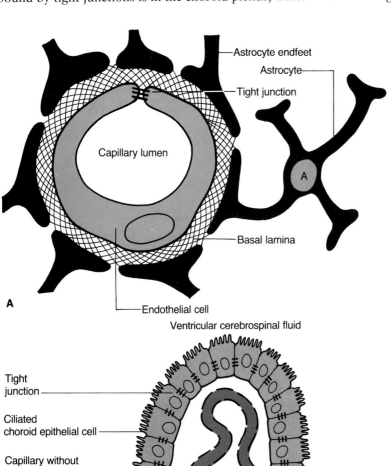

Astrocyte endfeet
Astrocyte
Tight junction
Capillary lumen
A
Basal lamina
A
Endothelial cell
Ventricular cerebrospinal fluid
Tight junction
Ciliated choroid epithelial cell
Capillary without tight junctions
B

Figure 6–2. Schematic diagrams of the blood–brain and blood–CSF barriers. **A:** Tight junctions joining capillary endothelial cells in most regions of the central nervous system limit transcapillary diffusion to water. Surrounding the capillary in the interstitial fluid is a layer of extracellular matrix bounded by the end feet of astrocytic (**A**) processes. These three components contribute physically and biochemically to form the blood–brain barrier. **B:** In the choroid plexus, a single layer of secretory epithelial cells envelopes leaky capillaries. The blood–CSF barrier is formed by tight junctions between the choroid epithelial cells that secrete CSF. As with endothelial cells of the blood–brain barrier, choroid epithelial cells contain carrier-mediated and active systems to transport additional nutrients and metabolites from one side of the barrier to the other. (Modified from Siegel G, Agranoff B, Albers RW, Molinoff P, eds. *Basic Neurochemistry*, 4th ed. New York: Raven Press, 1989.)

cytoplasmic surface of the plasma membrane by a family of attachment proteins that, in turn, are bound to a transmembrane glycoprotein. The transmembrane glycoprotein either directly interacts with transmembrane glycoproteins on the surface of a neighboring cell or binds to filamentous proteins of the extracellular matrix. The net effect is that the cytoskeleton of a cell is attached, through the plasma membrane via a network of proteins, either to the cytoskeleton of a neighboring cell or with the filaments of the extracellular matrix, resulting in enhanced tissue strength.

Anchoring junctions can be subclassified according to the type of cytoskeletal element that is involved.

Those formed with intermediate filaments linked to cell surface attachment proteins are called either *desmosomes* or *hemidesmosomes*. In desmosomes, the transmembrane linker glycoproteins interact with linker proteins that extend from the surfaces of neighboring cells, whereas transmembrane glycoproteins of hemidesmosomes bind to components of the extracellular matrix. Anchoring junctions also can be formed due to an association of actin microfilaments with cell surface proteins, and these types of junctional complexes are called *adherens junctions*. Similar to desmosomes and hemidesmosomes, adherens junctions can be of both the cell–cell and cell–matrix varieties.

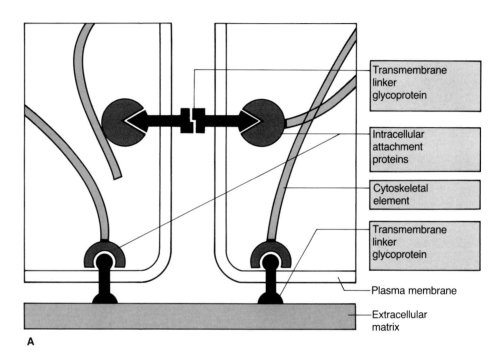

Junctional complex	Type	Cytoskeletal element	Transmembrane linker glycoprotein	Intracellular attachment protein
Adhesion belt/ Zonula adherens	Cell–cell	Microfilament	Cadherins	Catenins
Focal contacts/ adhesion plaques	Cell–matrix	Microfilament	Fibronectin receptor	Vinculin, Talin, and α-actinin
Desmosome/ Macula adherens	Cell–cell	Intermediate filament	Desmocollins and Desmogleins	Plakoglobin and Desmoplakins
Hemidesmosome	Cell–matrix	Intermediate filament	Laminin receptor	?

B

Figure 6–3. A schematic summarizing the general organization of the adherens junctions. **A:** Each different type of adherens junction is composed of transmembrane linker glycoproteins that interact with a cytoskeletal component through an intracellular attachment protein. **B:** A table listing the appropriate cytoskeletal element, transmembrane linker glycoprotein, and intracellular attachment protein for each type of adherens junction. (**A,** modified from Alberts B, Bray D, Lewis J, et al. *Molecular Biology of the Cell*, 2nd ed. New York; Garland Publishing, 1989.)

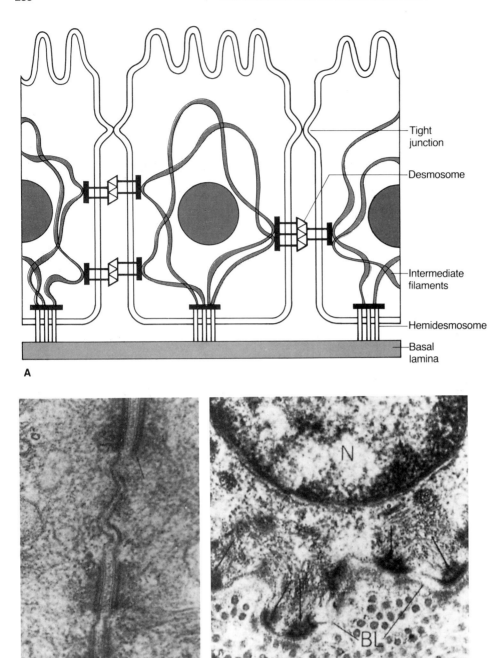

Figure 6–4. A schematic drawing and morphology of desmosomes and hemidesmosomes. **A:** In desmosomes, the transmembrane linker glycoproteins of one cell interact with identical transmembrane linker glycoproteins that extend from neighboring cells, whereas in hemidesmosomes, the transmembrane linker proteins interact with protein components of the basal lamina. **B:** A transmission electron micrograph through a series of desmosomes. Intermediate filaments make contact with a series of proteins that form a plaque structure on the cytoplasmic surface of the plasma membrane. **C:** Hemidesmosomes on the basal surface of epithelial cells. This TEM image shows intermediate filaments converging on dense cytoplasmic attachment plaques. Collagen fibrils can be seen in the extracellular space: *N*, nucleus; *BL*, basal lamina. (B, Gilula NB. In: Cox RP, ed. *Cell Communication*. New York: John Wiley & Sons, 1974;1–29. C, Ross MH, Romrell LJ. *Histology: A Text and Atlas*, 2nd ed. Baltimore: Williams & Wilkins, 1989; with permission.)

Cell–cell adherens junctions are called *zonula adherens* (or *adhesion belts*), whereas cell–matrix adherens junctions are termed either *focal contacts* or *adhesion plaques*. Figure 6–3 shows the general organization of anchoring junctions.

Desmosomes Are "Spot-Welds" That Hold Cells Together

Electron-microscopic analysis has shown desmosomes to be fine focal points of intercellular contact at which neighboring cells are tightly bound to one another (Fig. 6–4). Extending from the intracellular surface of a desmosome are numerous intermediate filaments that are usually composed of keratin subunits. These filaments terminate in electron-dense plaques composed of intracellular attachment proteins. The transmembrane linker glycoproteins bind neighboring cells together through a Ca^{2+}-dependent homophilic mechanism.

Hemidesmosomes, as their name indicates, appear as half-desmosomes under the electron microscope (see Fig. 6–4C). Like desmosomes, intermediate filaments of hemidesmosomes can be observed terminating in electron-dense foci of intracellular attachment proteins that dot the internal surface of the plasma membrane. However, unlike desmosomes, the binding domains of hemidesmosomal transmembrane linker glycoproteins do not attach to other cells. Instead, the linkers bind to components of the specialized extracellular matrix of the basal lamina. By doing so, the basal surface of epithelial cells is attached to the basement membrane. Although hemidesmosomes and desmosomes are similar ultrastructurally, recent biochemical evidence suggests that the two types of junctions are constructed from different types of intracellular attachment proteins and transmembrane linker proteins.

The importance of desmosomes in the maintenance of epithelial integrity is underscored in human patients with the autoimmune disease pemphigus. These patients produce autoantibodies that react specifically with the desmosomal transmembrane linker glycoproteins. This disrupts desmosomes in the skin, resulting in severe blistering and body fluid loss that can be fatal.

Adherens Junctions Are Important for Both Cell–Cell Adhesion and Cellular Locomotion

Adherens junctions that form between cells and the extracellular matrix are called *focal contacts* or *adhesion plaques*. In this type of junction, actin bundles are bound through a series of cortical membrane-associated polypeptides to a transmembrane glycoprotein, the fibronectin receptor. The extracellular region of the fibronectin receptor binds to fibronectin, a multifunctional extracellular protein that contains both a receptor-binding domain and one that interacts with structural proteins in the extracellular matrix, such as collagen.

Focal contacts are important for allowing cells to remain intimately associated with the filaments of the extracellular matrix. In addition, they facilitate certain types of cellular movement. Migratory cells, such as white blood cells that move toward chemotactic agents during immune responses, send out pseudopods in the direction of migration. If these pseudopods do not make contact with collagen fibers to form adhesion plaques, the pseudopod is rapidly resorbed. However, if the pseudopods attach to collagen fibers to form an adhesion plaque, the cell is able to migrate along the surface of the collagen fiber. A more detailed description of chemotaxis and cell migration is presented in Chapter 10.

Cell–cell adherens junctions, or zonula adherens, are found between the neighboring cells in epithelial sheets. They are located in the apical regions of the cells, just below the tight junctions. Because zonula adherens form a continuous ring around the circumference of each cell, they are sometimes called *adhesion belts* (Fig. 6–5). The adhesion belts in neighboring cells are directly apposed to one another, and the extracellular domains of the transmembrane linker glycoproteins bind to an adjacent cell's linker glycoproteins through a Ca^{2+}-mediated interaction. The actin microfilament network that forms a portion of the adhesion belt is contractile, and it is believed that the contraction of the zonula adherens is responsible for many of the morphologic changes that occur during development. For example, the contraction of the microfilament bundles underlying the adhesion belt region would cause a constriction of the apical region of a cell. If all of the cells in a epithelial sheet contracted synchronously, thereby constricting all of their individual apical regions, the cell sheet would convert into a tube. Such a mechanism could explain the morphogenesis of the neural tube during the embryonic development of vertebrates.

Gap Junctions Are Involved In Adhesion and Cellular Communication

Gap junctions allow the electrical and metabolic coupling of neighboring cells of a single type. They are formed by a grouping of transmembrane proteins into a structure called a *connexon* (Fig. 6–6). Connexons of neighboring cells adhere to one another through extracellular domains that align the proteins to form an aqueous pore or channel through the membranes of the two cells. The result is that the cytoplasm of one cell is connected directly to that of its neighbor by aligned hemichannels contributed by each cell. Small molecules of relative molecular mass (M_r below 1,500 Da) can readily move between cells through gap junctional complexes. The result is that cells can be metabolically and electrically coupled. Like other ion channels, gap junction permeability can be regulated (i.e., gap junctions behave like gated ion channels).

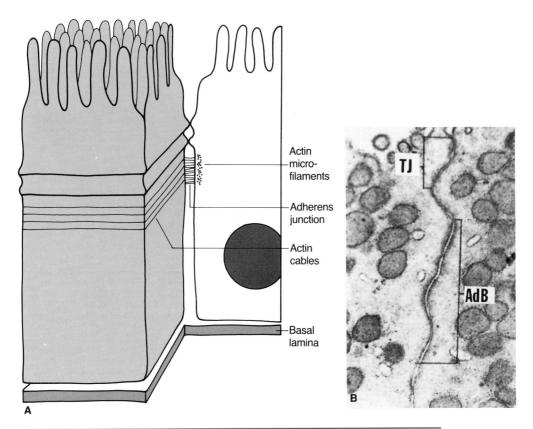

Figure 6–5. The organization of an adhesion belt. **A:** A schematic demonstrating that cables of actin microfilaments completely encircle the inner surface of the cell and are connected to transmembrane linker glycoproteins (cadherins) by a group of actin-binding proteins. The cadherins extending from one cell interact with cadherins extending from neighboring cells. Adhesion belts generally are located in the apical region of a cell, directly below the tight junctions. **B:** A transmission electron micrograph showing the region of an adhesion belt. The plasma membranes of neighboring cells are separated by a uniform space. Underlying the plasma membrane in the region of the adhesion belt are electron-dense microfilaments cut in cross section. *AdB*, adhesion belt; *TJ*, tight junction. (**B**, Ross MH, Romrell LJ. *Histology: A Text and Atlas*, 2nd ed. Baltimore: Williams & Wilkins, 1989, with permission.)

Ultrastructurally, gap junctions are observed at regions where plasma membranes of coupled cells are closely juxtaposed (see Fig. 6–6). However, unlike tight junctions for which the membranes appear to touch, membranes between adjacent cells are separated in regions by slight gaps; hence, the name *gap junction*. By freeze-fracture electron microscopy, gap junctions appear as a patch of intramembranous protein particles representing up to several hundred individual channel-forming connexons. Biochemical analysis has demonstrated that each connexon is comprised of six identical multipass transmembrane proteins, termed *connexins*, that are arranged in a hexagonal pattern surrounding a channel pore. Gap junctions between cells in different tissues, such as the lens, heart, and liver, are constructed from different connexins derived from one large gene family. Although exhibiting highly homologous regions, different members of the connexin family vary in mole-cular mass and exhibit differences in their cytoplasmic regions. Differences in the biochemical properties of various connexins may account for why the conductance of substances through gap junctions is modulated by the cytoplasmic Ca^{2+} concentration in some tissues, whereas others respond preferentially to a change in the intracellular pH or neurotransmitter-activated second-messenger systems.

The Extracellular Matrix and Its Role in Tissue Formation

Tissues contain varying amounts of material deposited outside cells, called the *extracellular matrix*. Some tissues, such as epithelia, are principally cellular in composition, although a small amount of matrix material is present. Termed the *basal lamina*, this substance pro-

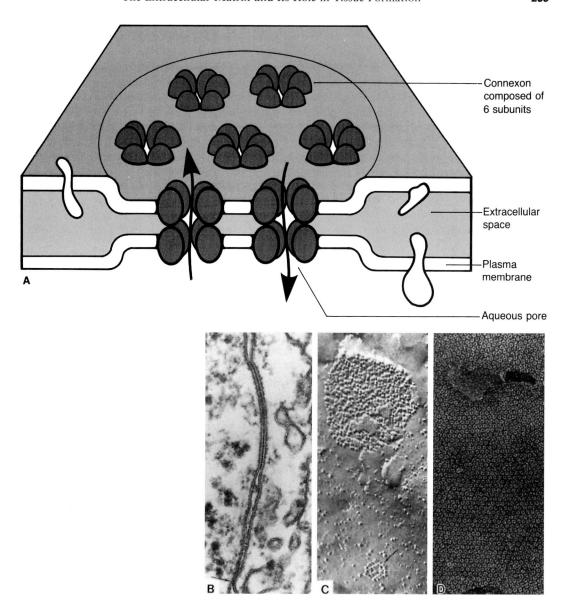

Figure 6–6. A schematic and various electron microscopic representations of gap junction complexes. **A:** Gap junctions are formed by connexons, and each connexon is composed of six transmembrane protein subunits. A connexon from one cell aligns with a connexon from a neighboring cell, and an aqueous pore is formed that connects the cytoplasms of the two cells. Unlike tight junctions, in which the membranes of adjacent cells appear to be touching, the plasma membranes of adjacent cells are separated by a slight space in gap junctions. **B:** A transmission electron micrograph through regions of gap junctions. **C:** A freeze-fracture electron micrograph through gap junctions demonstrates that gap junctions are composed of numerous clusters of transmembrane proteins. **D:** A negatively stained electron micrograph showing isolated gap junctions. The protein components of the connexons form porous channels between neighboring cells. (**A,** modified from Darnell J, Lodish H, Baltimore D. *Molecular cell biology,* 2nd ed. New York: Scientific American Books, 1990; **B, C,** Gilula NB. In: Cox RP, ed. *Cell Communication.* New York: John Wiley & Sons, 1974;1–29; **D,** Gilula NB. In: Feldman J, Gilula NB, Pitts JD, eds. *Intercellular Junctions and Synapses. Receptors and Recognition Series B,* vol 2. London: Chapman & Hall, 1978;3–22, with permission.)

vides a substrate for cellular attachment to underlying connective tissues. In other tissues, such as connective tissue, cells make up only a small percentage of the tissue mass, with the bulk of the tissue material being composed of extracellular matrix. The physical nature of the extracellular matrix also varies from tissue to tissue. It can be very hard in bone and teeth, owing to calcification, or quite spongy in cartilage and in the loose connective tissue of the dermis. In tendons, the extracellular matrix material is ropelike, whereas in blood it is fluid and composed of plasma.

Producers and Components of the Extracellular Matrix

The extracellular matrix is produced by the cells that inhabit it. In connective tissues, the principal extracellular matrix-producing cell is the fibroblast. Fibroblasts produce and secrete the macromolecules that compose the matrix and are thought to then pull and tug on these components so that structural fibers in the matrix are organized in a manner that gives strength and resilience to tissues. In specialized connective tissues, macromolecules that compose the extracellular matrix are secreted by other members of the fibroblast family. Specific examples include the chondroblasts in cartilage, osteoblasts in bone, and odontoblasts and ameloblasts in the developing tooth that produce dentin and enamel, respectively.

Three major kinds of extracellular macromolecules compose the extracellular matrix. The first type, the glycosaminoglycans (GAGs) and proteoglycans, give connective tissues a gel-like nature. The second type of extracellular macromolecule is the fibrous protein. Principal fibrous proteins are collagen, which gives strength to extracellular matrices, and elastin, which provides resiliency. The final class of matrix macromolecules are the adhesive proteins. These molecules form a direct bridge between cells and the fibrous proteins of connective tissues. The major adhesive molecules are fibronectin, which was described briefly earlier in this section, and laminin, which is found in basal laminae. By varying the relative amounts and organization of these macromolecular components, the characteristics of the extracellular matrix can be modified so that the matrix can fulfill numerous roles in the body.

Glycosaminoglycans and Proteoglycans Form a Hydrated Gel That Is the Ground Substance of the Extracellular Matrix

Glycosaminoglycans (GAGs) are so named because they are composed of repeating disaccharide chains in which one of the two sugars is an amino sugar (either *N*-acetylglucosamine or *N*-acetylgalactosamine). These

disaccharide units are organized into long unbranched chains, and the amino sugar is usually sulfated. Sulfates and carboxyl groups in the sugar residues make GAGs highly negatively charged. As a result, they attract osmotically active cations such as Na^+. This causes water to be drawn into the matrix, resulting in swelling, or turgor. Pressure from turgor helps the extracellular matrix resist opposing forces of tissue compression. The importance of this is underscored by the resiliency of cartilage that lines the surfaces of the long leg bones meeting at the knee joint. During ambulation, considerable force is absorbed at this joint by the tough, spongy cushion of cartilage that separates the two bones. If not for the presence of cartilage, these bones would abrade with contact and shatter during running.

Four main groups of GAGs can be identified on the basis of their sugar residues and other characteristics, such as the number of sulfate groups per disaccharide unit (Table 6–1). Of these, hyaluronic acid represents the largest type of GAG, consisting of repeating disaccharide units that are several-thousand sugar residues in length. In addition, hyaluronic acid is unique because it is neither sulfated nor covalently linked to protein. A component of virtually all connective tissues, hyaluronic acid is especially prominent in joint fluids and in early embryonic connective tissues.

Chondroitin sulfate and dermatan sulfate, heparan sulfate and heparin, and keratan sulfate constitute the other three classes of GAGs. Compared with hyaluronic acid, which may have an M_r of several million daltons, GAGs in the other classes are much smaller, with M_r's of 5,000–50,000 Da. In addition, these GAGs all are highly sulfated and are distributed throughout the body, where they form a hydrated gel that provides structural support for tissues while allowing the rapid diffusion of nutrients, wastes, and signaling molecules.

Glycosaminoglycans, with the exception of hyaluronic acid, are covalently bound to proteins to form proteoglycans. These macromolecules consist of a core protein to which numerous unbranched GAG side chains are covalently attached (Fig. 6–7). Core protein synthesis occurs in the rough endoplasmic reticulum. Polysaccharide chains of GAGs are added in the Golgi apparatus, producing extremely large molecules that have the appearance of a bottle-washing brush. By analogy, the bristles of the brush correspond to the glycosaminoglycan side chains, whereas the central shaft corresponds to the core protein.

In addition to forming the hydrated ground substance of the extracellular matrix, proteoglycans are thought to have additional functions. For example, proteoglycans can bind signaling molecules, such as growth factors. This characteristic probably allows signaling molecules to be retained in a specific area, thereby localizing the action of the molecule or colocalizing it with a

Table 6–1
Glycosaminoglycans

GAG	Approximate M_r	Repeating Disaccharide Unit	Protein Linkage	Sulfated	Tissue Distribution
Hyaluronic acid	5×10^3–1×10^7	Glucuronic acid and N-acetylglucosamine	–	–	Skin, numerous connective tissues, vitreous body, and synovial fluid
Chondroitin sulfate	5×10^3–5×10^4	Glucuronic acid and N-acetylgalactosamine	+	+	Cartilage, cornea, arteries, skin, and bones
Dermatan sulfate	1×10^4–6×10^4	Either glucuronic or iduronic acid and N-acetylgalactosamine	+	+	Skin, heart, heart valves, and blood vessels
Keratan sulfate	5×10^3–2×10^4	Galactose and N-acetylglucosamine	+	+	Cartilage, vertebral disks, and cornea
Heparin	5×10^3–3×10^4	Either glucuronic or iduronic acid and N-acetylglucosamine	+	++	Lung, liver, and skin (also highly concentrated in mast cell granules)
Heparin sulfate	5×10^3–1×10^4	Either glucuronic or iduronic acid and N-acetylglucosamine	+	+	Lungs, arteries, basement membranes, and cell surfaces

GAG, glycosaminoglycan.

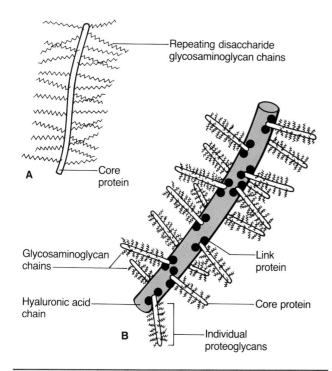

Figure 6–7. Schematic representations of proteoglycans. **A:** Proteoglycans are formed by the addition of numerous repeating disaccharide chains to a core protein. **B:** Proteoglycans can be further organized by the noncovalent attachment of several proteoglycan monomers to a large chain of hyaluronic acid molecules. The association of the proteoglycan molecules with hyaluronic acid is mediated by link proteins. (**B**, modified from Alberts B, Bray D, Lewis J, et al. *Molecular Biology of the cell*, 2nd ed. New York: Garland Publishing, 1989.)

cell surface receptor (see Chapter 8). In addition, it is thought that proteoglycans are important for forming the filtering apparatus in the kidney, where they act as a molecular sieve in the basement membrane of the glomerulus.

Attempts to visualize proteoglycans by electron microscopy have been only marginally successful. Proteoglycans are water-soluble and, as a result, are extracted during the routine processing of tissue sections for electron microscopy. Biochemical studies have demonstrated that proteoglycans can bind to other proteoglycans and to structural proteins in the extracellular matrix in relatively specific fashions. Presumably, these molecular interactions are important in organizing the extracellular matrix. Not all proteoglycans are secreted by cells. Some are retained as integral membrane proteins, for which they are thought to be involved in the attachment of cells to the extracellular matrix and in the immobilization of other molecules.

Collagen Provides Tensile Strength to the Extracellular Matrix

Whereas the function of the glycosaminoglycans and proteoglycans is to form a hydrated gel that allows the extracellular matrix to resist compressive forces, collagen forms tough, protein fibers that are resistant to shearing forces. The organization of collagen molecules into fibers is probably determined by the fibroblasts that secrete them. Presumably, pulling and tugging forces exerted by the fibroblasts orient secreted collagen into different arrangements. For example, collagen fibers are arranged in layers in bone, with the successive layers of collagen being oriented at right angles to each other in a fashion that reminds one of the organization of alternating wood layers in plywood. This molecular arrangement allows bone to resist forces exerted from any number of directions and gives it great strength.

The collagens are a family of proteins that are perhaps the most abundant in the body. Because of their molecular organization, collagen molecules exhibit a characteristic, repeating structure that can be identified readily in electron micrographs (Fig. 6–8). At least 12 different types of collagen have now been identified, although the functions of only a few are understood. Collagen types I, II, and III are the fibrillar collagens. Following secretion into the extracellular space, fibrillar collagens are organized into fibrils, which resemble thin cables in electron micrographs. Type I collagen fibrils are often aggregated into larger bundles, called collagen fibers, and these collagen fibers can be observed with the light microscope. Collagen type I is found in the connective tissue of skin, bone, tendon, ligaments, the connective tissue layers of muscle and nerves; and in the capsules that surround various internal organs. Type II collagen is found exclusively in cartilage, and type III collagen assembles into fine spicules called *reticular fibers*. Reticular fibers form the skeletal framework for many blood vessels and internal organs. Collagen type IV is found exclusively in basal laminae. Rather than form fibrils, this latter type of collagen is organized into sheets.

Collagen molecules are made up of three helical polypeptide chains, called α-chains, that wind around one another to form a superhelical triple-helix. Every third amino acid in an α-chain is a glycine, which allows the tight packaging of the three α-chains. In addition, collagen α-chains contain numerous hydroxyproline and hydroxylysine residues. Hydroxyproline residues are thought to be important for hydrogen bonding between α-chains, and hydroxylysine residues are essential for an unusual lysine-linked glycosylation that is a characteristic of collagen. Hydroxylysine residues also are necessary for the binding of collagen molecules into fibrils. These assembly events are summarized in Fig. 6–9.

α-Chains are synthesized as propeptides that contain extra amino acids. Three pro-α-chains assemble into a procollagen molecule that is then secreted. Outside the cell, the propeptides are enzymatically cleaved to form tropocollagen molecules, which then assemble into collagen fibrils. The packaging of the collagen molecules within the fibrils produces a characteristic, striated-banding pattern with a 64-nm repeat when observed in the electron microscope.

Abnormalities in collagen synthesis underlie several human diseases. For example, deficiencies in vitamin C intake result in decreased proline hydroxylation and aberrant collagen production. Patients with a vitamin C deficiency often exhibit scurvy, a disease in which blood vessels become fragile, teeth fall out, and wounds fail to heal properly. In addition, several human diseases, such as Ehlers Danlos syndrome, osteogenesis imperfecta, and systemic sclerosis, are characterized by abnormal collagen production.

Elastin Provides Tissues with Elasticity

The other major fibrous protein in the extracellular matrix is elastin. It is present in varying amounts in tissues and organs, such as blood vessels, skin, and the lungs, which must be able to stretch and recoil. Elasticity in tissues is provided by a network of elastic fibers in the extracellular matrix, the principal protein component of which is elastin.

Elastin is produced and secreted principally by fibroblasts. Once in the extracellular space, elastin molecules are cross-linked to one another to form an extensive network of sheets and filaments. Unlike most proteins, the elastin molecule does not assume any one particular shape. Instead, it can assume any one of several relaxed, random-coiled orientations. As a tissue is stretched, the elastin molecule is elongated into a more linear, unstable conformation (Fig. 6–10). When the stretching force is released, the elastin molecule returns to any one of its random-coil structures. Because the elastin proteins are all covalently cross-linked to one another to form fibers, the entire elastin network can be stretched and then recoil like a rubber band. Several other proteins have been identified in elastic fibers, including fibrillin, a protein that is thought to be defective in Marfan's syndrome (MFS). The function of fibrillin is not known at this time.

Adhesive Glycoproteins Are the Glue That Holds Connective Tissues Together

Cells produce and secrete several adhesive glycoproteins into the extracellular space. Adhesive proteins are multifunctional molecules that contain several specialized domains. Molecular analysis has demonstrated that all adhesive glycoproteins contain a domain that binds to cell surfaces, another domain that interacts with collagen, and, finally, a third subregion that binds to proteoglycans. The multifunctional nature of adhesive glycoproteins allows them to act as the extracellular glue that holds many tissues together (Fig. 6–11).

Recombinant DNA techniques and biochemical studies have identified a specific sequence motif that serves as the cell-binding motif of many adhesive glycoproteins, including fibronectin (see below). This tripeptide sequence of arginine, glycine, and aspartic acid (RGD sequence) is essential for cellular adhesion to the

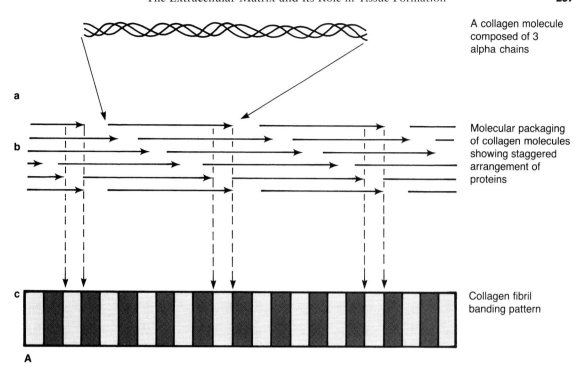

A collagen molecule composed of 3 alpha chains

Molecular packaging of collagen molecules showing staggered arrangement of proteins

Collagen fibril banding pattern

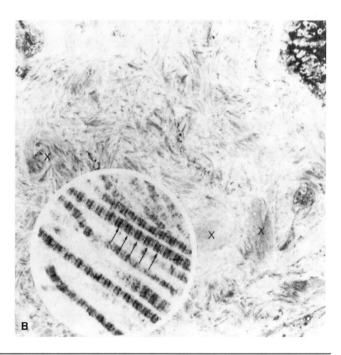

Figure 6–8. The packaging of collagen molecules into fibrils gives rise to the striated banding pattern that is characteristic of collagen seen in electron micrographs. **A:** The collagen molecule is composed of three intertwined polypeptide chains (*a*). The collagen molecules are organized in a manner that gives rise to regions of total overlap and regions in which the packaged collagen molecules have empty spaces (*b*). This results in a striated pattern in electron micrographs (*c*). **B:** A transmission electron micrograph of a connective tissue capsule. Numerous threadlike collagen fibrils (*x*) can be seen in the extracellular space. The inset shows a higher-power EM image of some of the longitudinally oriented collagen fibrils showing the repetitive-banding pattern (*arrows*) of collagen. (Ross MH, Romrell LJ. *Histology: A Text and Atlas*, 2nd ed. Baltimore: Williams & Wilkins, 1989, with permission.)

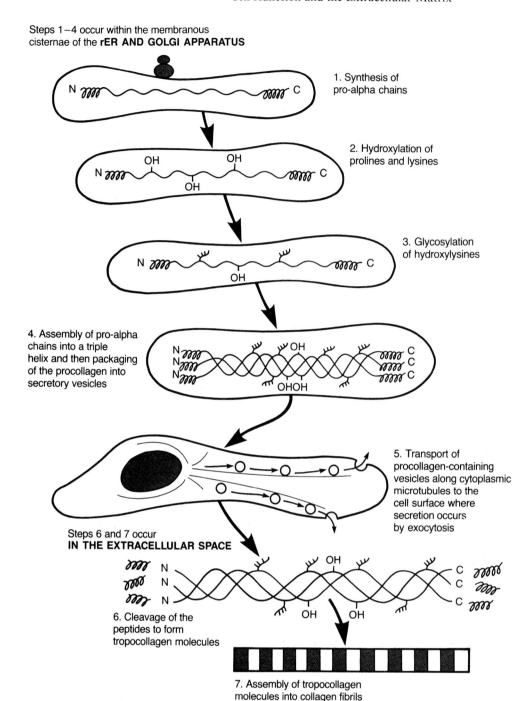

Steps 1–4 occur within the membranous
cisternae of the **rER AND GOLGI APPARATUS**

1. Synthesis of
pro-alpha chains

2. Hydroxylation of
prolines and lysines

3. Glycosylation
of hydroxylysines

4. Assembly of pro-alpha
chains into a triple
helix and then packaging
of the procollagen into
secretory vesicles

5. Transport of
procollagen-containing
vesicles along cytoplasmic
microtubules to the
cell surface where
secretion occurs
by exocytosis

Steps 6 and 7 occur
IN THE EXTRACELLULAR SPACE

6. Cleavage of the
peptides to form
tropocollagen molecules

7. Assembly of tropocollagen
molecules into collagen fibrils

Figure 6–9. **A summary of the events that result in collagen fibril formation.** Events
1–4 occur in intracellular compartments (RER, Golgi, other), and steps 6 and 7
occur extracellularly. (*1*) Pro-α-chains are synthesized on the RER and then (*2* and
3) selected prolines and lysines are modified. (*4*) Three pro-α-chains are then
assembled into a triple-helical procollagen molecule, and the procollagen is packaged into secretory granules. (*5*) The secretory granules are transported to the cell
surface along cytoplasmic microtubules, and then the procollagen is secreted. (*6*)
The propeptides are cleaved from the procollagen to form tropocollagen, and (*7*)
the tropocollagen molecules are assembled into collagen fibrils.

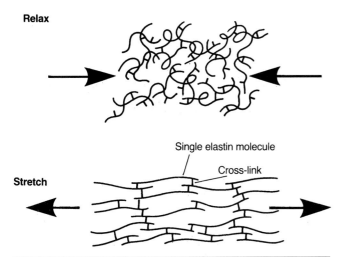

Figure 6–10. A schematic showing the organization and behavior of elastin molecules. Individual elastin molecules exist in various random-coiled conformations, and the elastin proteins are cross-linked together to form a covalently bonded elastin network. When a force is applied, the entire network stretches, and when the force is removed, the elastin network recoils into a relaxed state in which all of the elastin molecules assume random conformations. (Alberts B, Bray D, Lewis J, et al. *Molecular Biology of the Cell*, 2nd ed. New York: Garland Publishing, 1989, with permission.)

extracellular matrix. If synthetic peptides that contain the RGD sequence are produced and added to cultured cells, the synthetic peptides will compete with the adhesive glycoproteins for the binding sites on cell surfaces, thereby disrupting cell attachment. Other types of adhesive molecules do not rely on the RGD sequence and, instead, use other amino acid sequences to govern their cell-binding behavior.

The principal adhesive glycoprotein in connective tissues is fibronectin, a large molecule composed of two polypeptide chains. The fibronectin subunits are disulfide-bonded, then the molecule is folded into several globular domains. It is bound to cells by a large transmembrane glycoprotein called the *fibronectin receptor*. This adhesion is mediated by the RGD amino acid sequence in one of the fibronectin domains. The fibronectin receptor, as well as the laminin receptor (described below), is a member of a family of transmembrane proteins called *integrins*. The cytoplasmic portion of an integrin interacts with the cytoskeleton to form a transmembrane connection between the cytoskeleton of a cell and filaments of the extracellular matrix. Molecular analyses have demonstrated that other domains of fibronectin bind to heparin and collagen. In addition to its role in cell adhesion, fibronectin is also essential for cell migration.

Laminin is the adhesive molecule that is specific for basal lamina. Like fibronectin, laminin is a large molecule that contains several functional domains (see Fig.

6–10). One domain binds to laminin receptors on the cell surface. Another domain of the molecule binds to heparan sulfate, and yet another interacts with type IV collagen. Laminin is produced by the epithelial cells that rest on the basal lamina.

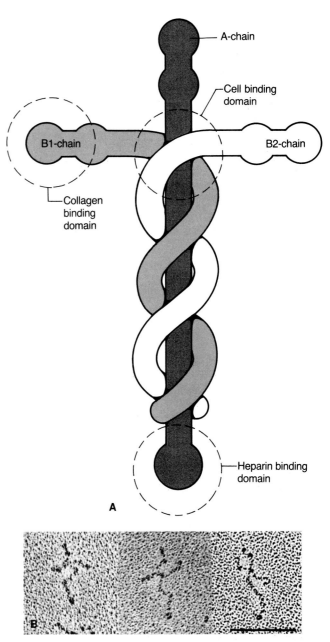

Figure 6–11. The structure of laminin. **A:** Laminin is composed of three polypeptide chains—A, B1, and B2—that are disulfide-bonded to one another to form a crucifix-shaped molecule. Several functional domains have been identified on the laminin molecule, including a cell-binding region, a heparin-binding domain, and a collagen-binding site. **B:** Negatively stained electron micrographs of isolated laminin molecules. (**A**, modified from Hogan BL, et al. In: Shibata S, ed. Basement membranes. Amsterdam: Elsevier Press, 1985; **B**, courtesy of Dr. D. R. Abrahamson.)

Basal Laminae Are Regions of Specialized Extracellular Matrix That Underlie Epithelia

Basal laminae are thin sheets of extracellular matrix that underlie all epithelia. In addition, they surround all muscle cells, adipose cells, and many of the supporting cells of the nervous system. The basal lamina is composed principally of type IV collagen, proteoglycans, and laminin, and these proteinaceous components of the basal lamina are produced by the cells that rest on this specialized form of extracellular matrix material. In addition to their role as a site of adhesion for epithelial cells, basal laminae perform several other functions. For example, the basal lamina of the kidney serves as a molecular sieve that filters molecules as urine is formed. The basal lamina also forms a scaffolding that plays an important role in wound healing and regeneration.

Electron microscopy has demonstrated that the basal lamina is composed of two distinct layers (Fig. 6–12). The layer directly adjacent to epithelial cells is electron-lucent and is called the *lamina lucida* (or *lamina rara*). Immunocytochemical studies have shown that the lamina lucida is composed principally of laminin and proteoglycans. Deep to the lamina lucida is an electron-dense layer, the *lamina densa*, composed of type IV collagen and proteoglycans. In some preparations, a third layer is observed adjacent to and deeper than the lamina densa. This layer is called the *lamina reticularis* because it is composed of reticular fibers, which are very thin extracellular fibrils composed of collagen. The term *basement membrane* is sometimes used to distinguish the thin basal lamina from the much thicker extracellular mats that are composed of both basal lamina and lamina reticularis.

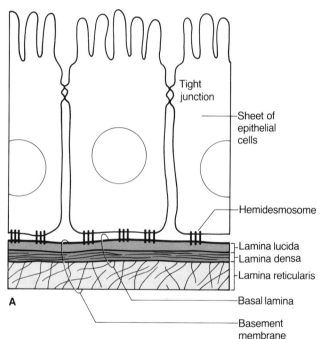

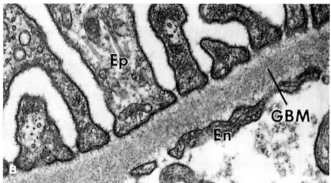

Figure 6–12. Structural organization of basal laminae and basement membranes. **A:** All epithelial sheets are supported by a basal lamina. The basal lamina is composed of two layers: the lamina lucida (also interna or rara), which abuts against the epithelial sheet, and the lamina densa, which is located below the lamina lucida. In some tissues and organs, a third layer can be identified. This third layer, the lamina reticularis (or externa), is composed of thin collagen fibrils called *reticular fibers*. The term *basement membrane* is used when a sheet of epithelial cells is underlined by all three laminae. **B:** A thin-section electron micrograph through the basement membrane in a kidney glomerulus. The epithelium (*Ep*), endothelium (*En*), and glomerular basement membrane (*GBM*) are shown. The basement membrane separates an endothelial cell (*En*) from the foot processes of a podocyte. (**A,** modified from Alberts B, Bray D, Lewis J, et al. *Molecular Biology of the Cell*, 2nd ed. New York: Garland Publishing, 1989; **B,** Abrahamson, DR. *J Pathol* 1986;149:257, with permission.)

Cell Adhesion in the Operation of Muscle

The Intercalated Disk Is the Site of Cell–Cell Adhesion in Cardiac Muscle

One of the distinguishing features of cardiac muscle is the intercalated disk (see Chapter 3). This structure forms at the site where two cardiac muscle cells adhere in end-to-end apposition. Junctional complexes of the intercalated disk are important for maintaining the integrity of the heart muscle.

Two junctional types exist between cardiac cells, adherens (anchoring) junctions and gap junctions. The adherens junctions within the intercalated disk are divided into two subtypes—the macula adherens and the fascia adherens, based on their structure and function. In electron micrographs, the macula adherens displays the structural features of desmosomes found in many tissues. Typically, it appears as a thickened region of the plasma membrane bordered by an adjacent region of electron-dense material where numerous intermediate filaments terminate, suggesting that additional linker proteins are present. Contact between adjacent myocytes is mediated by transmembrane glycoproteins of the macula complex that interact in the extracellular space through a Ca^{2+}-dependent mechanism. Macula adherens junctions are thought to constitute the major adhesion sites of cardiac muscle.

Similar to the macula adherens, the fascia adherens of the intercalated disk are also mediated by transmembrane glycoproteins. However, they serve as contact points for actin thin filaments, rather than intermediate filaments found in the macula junctional complexes. The actin thin filaments are anchored at their plus ends to the inside surface of the plasma membrane by a class of attachment proteins, principally α-actinin. This allows the coupling of the actin cytoskeleton between two adjacent cardiac muscle cells. Most importantly, the fascia adherens junctions maintain the correct polarity and register of actin filaments, relative to the myosin filaments that they interact with, to form the contractile apparatus of the cardiac myocyte. Thus, the fascia adherens junctions function as a specialized Z-disklike structure that allows the contractile apparatus of the coupled myocytes to work together.

The second major junctional complex within the intercalated disk of cardiac muscle is the gap junction. Gap junctions between cardiac cells allow individual cardiac cells to be metabolically and electrically coupled. This allows the synchronous contraction of the heart muscle that is necessary for it to effectively pump blood. In the heart, the gap junction channel is formed from connexin 43 (43-kDa M_r). The extracellular domain of connexin 43 is involved in the recognition and binding of cognate molecules on adjacent cells. The result of this binding is the formation of an aqueous channel allowing communication between the cells.

Adhesion in Smooth Muscle

Smooth muscle is constructed from the association of individual cells into "sheets," or layers characteristic of the smooth muscle tissue. Because of this arrangement, intercellular communication among myocytes is necessary to effectively coordinate their contractile activity (see Chapter 3). This is accomplished by the numerous gap junctional contacts, which are the primary means of cell–cell adhesion in smooth muscle. As discussed in previous sections, gap junctions are formed from specific associations of transmembrane connexins, which results in the creation of an aqueous channel of communication between the cytoplasms of adjacent cells.

Most smooth-muscle tissue is surrounded by extracellular matrix that it contacts primarily at focal adhesion sites. There, a transmembrane receptor of the integrin family binds to glycoproteins, such as fibronectin and laminin, in the extracellular matrix and is attached internally to the cytoskeleton by intracellular binding proteins (vinculin, talin, and α-actinin). Focal adhesion contacts with the extracellular matrix are important for smooth muscle function. Smooth muscle contraction results in the inward pulling of the cell's membrane and a change in cell shape (see Chapter 3). Adhesive contacts with the extracellular matrix ensure that all contraction is matched by a corresponding change in shape of the surrounding matrix. In an organ such as the intestine, smooth muscle contraction is translated to alterations in the diameter of the lumen through a molding of the surrounding extracellular matrix.

In addition to contraction, focal adhesion contacts with the extracellular matrix are apparently required to maintain the smooth muscle phenotype. When smooth muscle cells are placed in culture on a surface enriched in fibronectin, they change from a differentiated phenotype, capable of contraction, to a proliferative cell that no longer contracts. In contrast, similar cells placed on a surface of collagen fibers retain their differentiated phenotype. Although this is an artificially induced situation (i.e., the smooth cells are grown *in vitro*), the pathophysiology of conditions like atherosclerosis is believed to involve a change in smooth muscle cell phenotype related to components of the extracellular matrix affected by atherogenic lesions.

Cell Adhesion Mechanisms in Nervous System Development

A functional mammalian nervous system is built with circuits of neurons and other innervated cells, the connections of which are specified and constructed on a developmentally precise schedule. One of the central problems in modern biology is to understand what mechanisms determine the final positioning of nerve cells and guide their cytoplasmic processes to distal targets. The multitude of intricate contacts formed by a single neuron makes it unique among other cell types. When the problem encompasses hundreds of millions of nerve cells in the human brain, the task at hand seems overwhelming.

Much of the impetus to search for the molecular determinants that specify neuronal contacts is based on work, in the 1950s and 1960s, of the Nobel laureate, Roger Sperry. His experiments demonstrated that axons from the retina can form proper connections in the brain, regardless of whether they are faced with altered routes or optional targets. From these results, Sperry derived his *Chemoaffinity Hypothesis*, in which a particular axon is proposed to grow toward its target cell by following a diffusible gradient of target-derived molecules. Many of the contributory molecules responsible for this chemoattraction have been identified (Chapter 7). Adhesion molecules also contribute to axon guidance. Some are components of the axon surface, whereas others reside in the extracellular matrix or surface membrane of cells along which axons grow. Other adhesion molecules mediate interactions between glial cells and neurons during neuronal migration and the myelination of selected axons.

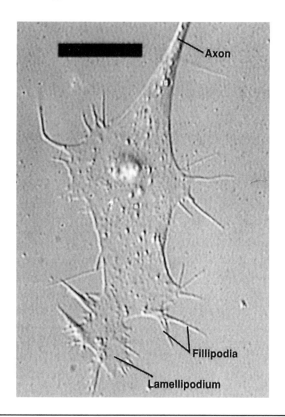

Figure 6–13. Growth cone viewed by high-resolution differential interference contrast microscopy. Long, fine-diameter filopodia extend from the body of the neuronal growth cone. The example shown is that of a neuron-derived tumor (neuroblastoma) cell, growing against the flattened surface of a tissue culture dish. (From R. S. Bedlack, Jr. and L. M. Loew.)

Growth Cones Are Motorized Extensions at the Leading Edge of Outgrowing Neurites

The growth cone is a distended structure located at the growing ends of a formative axon or dendrite. It functions to propel and guide these outgrowing processes through surrounding tissue along an appropriate path. Although growth cones are detectable in the developing nervous system, most of their properties have been discovered in simplified tissue culture in which embryonic neurons can thrive and be observed individually. Under these conditions, neurons send out *neurites*, a generic term for all outgrowing processes that have yet to differentiate into identifiable axons or dendrites. At the leading edge of each neurite is a growth cone resembling a hand flattened against the planar surface of the culture dish (Fig. 6–13). Radiating from the central portion, or palm of the growth cone, are several fingerlike filopodia. Being highly motile, filopodia continuously extend and retract to appear as though they are "feeling out" the surface to which the growth cone is attached. Between filopodia are often spread lamellipodia, weblike membranes that move in a ruffling manner.

Ultrastructural examination of the growth cone by electron microscopy reveals numerous mitochondria and microtubules in the body of the structure. In the periphery, bundles of actin microfilaments project into the filopodia. Good evidence exists that growth cone motility is generated by the continuous polymerization and depolymerization of actin filaments. Treatment with the drug cytochalasin, an inhibitor of actin polymerization (see Chapter 3), will arrest growth cone movement, although neurites continue to advance because microtubule extension is not interrupted. Without a functional growth cone, however, neurites extend randomly in the absence of a guidance mechanism to direct them to specific targets. How growth cone movement is coordinated with neurite extension remains poorly understood. Evidence, borne out of observations in tissue culture, suggests that the neurons attach more strongly to an adhesive surface at their growth cones than at the neurite shaft or cell body. As it advances, the growth cone tends to pull on the neurite and cell body because the rate of neurite elongation is slower. Tension created at the neurite tip, in turn, may control the direction and rate of microtubule assembly.

Growth Cones Are Pathfinders Allowing Axons To Project To Their Appropriate Target Areas

In addition to providing a locomotive function, growth cones also guide growing neurites toward their destinations by mediating two types of interactions. First, they attach to the most favorable adhesive substrate on which to bind and direct neurite extension. Second, surface receptors on the growth cone bind diffusible molecules that signal neuronal growth and act as chemoattractants. We will focus here on close-range cell contact-mediated interactions. The role of neuronal growth factors in this regard is presented in Chapter 7.

The course taken by a single axon in the mammalian brain is extremely difficult to follow. Most of what we know about basic mechanisms that establish specific neuronal connections is derived from model studies of simpler invertebrates, the neurons of which are few and readily mapped as they develop during embryogenesis. In general, axons in the central nervous system grow in bundles, or fascicles. Most axons that are recruited into a particular fascicle use as their environmental guide other axons that have already preceded them to a target. Other cues for directional growth can be provided by adjacent nonneuronal cells and the extracellular matrix.

Some of the most elegant experiments demonstrating how axons find their targets have come from the work of Goodman and colleagues, who have used the limb of the embryonic grasshopper as a model for studying patterns of sensory neuron innervation. "Pioneer axons," the first axons of a fascicle to emerge and grow toward their target, are thought to follow a gradient of an adhesive substrate formed by a scaffolding of glial neuroepithelial cells. As the pioneer axon grows toward the gradient, it changes course abruptly (Fig. 6–14) in the vicinity of the first of several specialized "guidepost" cells, which are frequently immature neurons. During this time, filopodia of the growth cone increase their motility and length. Once contact is made with the guidepost cell, axonal growth proceeds toward and beyond it until the next guidepost cell is contacted and serves to reorient further growth. When guidepost cells are specifically killed with a laser, a growth cone arriving at its former site will veer off its predicted course and proceed in the wrong direction. Once the initial axon(s) has reached its target by way of these stepping stones, subsequent axons are believed to trail along adhesive cues provided by the pioneers, rather than the epithelial guides.

The concept of axonal path-finding cued by guidepost cells extends to the developing vertebrate nervous system. During embryonic development, approximately one-half of growing retinal ganglion cell axons cross the midline of the mammalian optic chiasm to the contralateral side, whereas the other half continues to project ipsilaterally (Fig. 6–15). Studies by Reichardt and colleagues have demonstrated that the decision of path-finding axons in the optic nerve to cross the midline is governed by local guidance cues from a population of neurons residing within the formative chiasm. Experimental occlusion of specific cell adhesion molecules on their surface with antibodies will block further growth and routing of retinal ganglion cell axons penetrating the chiasm.

Key work in the chick embryo demonstrates that axons of developing motor neurons destined to become spinal nerves must also be directed to the appropriate skeletal muscle of innervation. Cell bodies of neurons that supply a specific muscle are typically pooled together in the spinal cord. However, when the growing axons first leave the cord toward the periphery, those destined for many different muscles migrate together in one bundle. Only when nearing the individual muscles at the base of the developing limb does the axon bundle segregate into new branches, the axons of which supply individual muscles. This pattern of outgrowth suggests that important environment cues reside in a more distal region, which ensures that motor axons follow the correct course to their targets. Even when misplaced experimentally, embryonic chick motor neurons can still send axons to the correct muscle along a new initial route (Fig. 6–16). These findings underscore the existence not only of a high degree of specificity in nerve–target cell interactions, but also mechanisms other than initial pathway guidance that axons use to navigate.

Specific Adhesion Molecules Promote Neurite Outgrowth and Fasciculation

Three major families of nondiffusible cell surface molecules mediating close-range adhesion of nonneural cell types also function in a similar capacity in the nervous system. Included are (a) cell adhesion molecules of the immunoglobulin superfamily, (b) cadherins, and (c) glycoproteins of the extracellular matrix and their receptors. Most of what we know about their function indicates that many serve as general promoters of growth cone attachment and neurite outgrowth, whereas others may be more selective, paving the way for specific subsets of axons while repelling others.

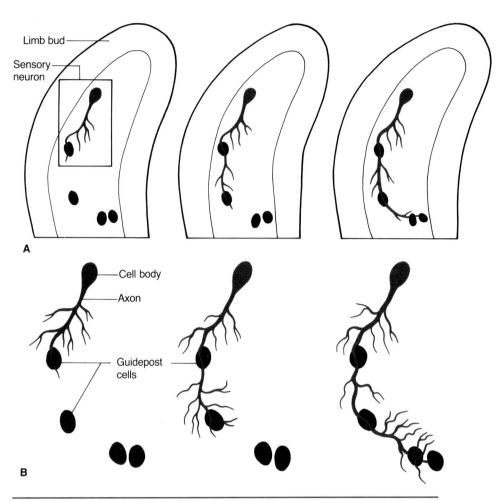

Figure 6–14. Outgrowth of developing sensory neurons in the grasshopper limb cued by "guidepost" cells. **A:** During successive developmental stages, the axon of a sensory neuron (*red*) in the limb follows a highly reproducible path as it contacts several specialized cells on its route toward the CNS of the grasshopper. **B:** Drawn at higher magnification, some of the filopodia emanating from the growth cone of the sensory axon preferentially adhere to another immature neuron and direct the axon to grow toward it. In this way, the vectorial growth of the axon proceeds in a step-wise fashion from one stepping stone to the next. (Modified from Taghert PH, Bastiani MJ, Ho RK, Goodman SC. *Dev Biol* 1982;94:139.)

Neural Cell Adhesion Molecules (N-CAMs) of the Immunoglobulin Superfamily Mediate Calcium-Independent Cell–Cell Adhesion

Of the many representatives of the immunoglobulin superfamily expressed in neural tissue (Fig. 6–17), the N-CAM is the most abundant. Discovered by Edelman in the 1970s, N-CAM is made by nearly all neural cells (nerve cells and glial cells) and probably promotes nonspecific adhesion involved in tissue assembly that begins with some of the earliest events in neural tube formation.

Multiple isoforms of N-CAM are encoded by distinct mRNAs, each derived from the same gene by alternative splicing events. Posttranslational modification causes extensive glycosylation of the extracellular polypeptide backbone, which is highly conserved among different N-CAMs. Of the three major membrane-bound N-CAMs that have been identified, two are transmembrane proteins, and the other is linked to the membrane by a covalent glycosylphosphatidylinositol linkage (Fig. 6–18).

The N-CAMs were originally discovered with antibodies to them that inhibited the self-aggregation of suspended embryonic retinal cells. Subsequent studies showed that when a single N-CAM species was purified and inserted into artificial lipid vesicles, the vesicles were able to bind to each other in a homophilic mechanism that is also inhibited with N-CAM-specific antibodies. These experiments indicate that surface domains of N-CAMs mediate adhesion between two

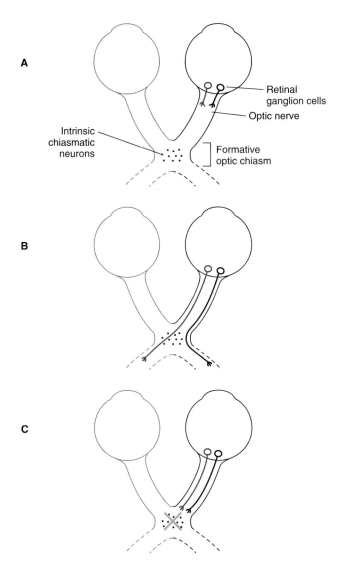

Figure 6—15. Axonal path-finding in the embryonic mouse visual system is interrupted by lesioning guidepost neurons positioned in the developing optic chiasm. **A:** At embryonic days 11–13, retinal ganglion cell axons begin to penetrate the future optic nerve, and a group of early generated neurons is present at the formative chiasm. **B:** By days 13–16, axons representing the nasal retinal field (*red*) from each eye (only one eye is emphasized) penetrate the chiasm and cross the midline. Axons representing the temporal retinal field turn away from the chiasm and project ipsilaterally. **C:** Selective lesioning of the chiasmatic neuronal population arrests retinal ganglion cell growth at the chiasm.

cells in a homophilic-binding mechanism; that is, N-CAMs adhere to each other (Fig. 6–19). The five immunoglobulin domains at the NH_2-terminus of N-CAM are believed to mediate self-association. Another modulator of adhesion is the number of negatively charged polysialic acid residues attached to the carbohydrate portion of the macromolecule. In the embryo, sialic acid constitutes nearly 30% of the mass of N-CAM protein and decreases as development progresses. A high degree of sialation causes N-CAMs to bind to each other with lower affinity.

Several other cell surface glycoproteins of the immunoglobulin superfamily are believed to engage in homophilic binding in neuronal development. These molecules are more restricted in their distribution among axonal tracts than N-CAM. Some, in addition to N-CAM and N-cadherin, may play specific roles in mediating neurite fasciculation by binding axons together (Fig. 6–20).

N-Cadherin Is the Principal Calcium-Dependent Adhesion Molecule in the Nervous System

The cadherins are another important family of neural cell adhesion molecules. The predominant one expressed neurally is termed *N-cadherin*. Like some N-CAMs, N-cadherin is a transmembrane glycoprotein that promotes cell adhesion by homophilic binding. Unlike N-CAM, however, N-cadherin-mediated cell adhesion depends on the presence of Ca^{2+} to bind and stabilize its external domain in a conformation that is otherwise susceptible to rapid proteolysis. N-cadherin is believed to promote the outgrowth of at least some axons along astrocytes (Fig. 6–21), which also express the N-cadherin. Elegant experiments have shown that fibroblasts, which normally do not make N-cadherin, can promote neurite outgrowth when transfected with an N-cadherin cDNA.

Extracellular Matrix Protein Interactions with Integrins Promote Neurite Outgrowth and Synapse Formation

The integrins, a third class of neurally expressed adhesion molecule, anchor cells to their substrates through a heterophilic-binding mechanism. Integrins protruding from the neuronal surface membrane function as receptors for laminin and fibronectin, among other components of the extracellular matrix, in neural as well as nonneural tissues (Fig. 6–22). Different subsets of neurons express the two polypeptide subunits (α and β) of the integrin molecule in different forms and combinations. This, in turn, changes the receptor specificity of integrins for various ligands in the extracellular matrix and thereby determines the preference of a neuron for a particular adhesive substrate. Some neurons whose axons migrate along a laminin-enriched pathway stop making integrin receptors after reaching their target. Laminin, in particular, promotes neurite outgrowth for a variety of neurons placed in culture (Fig. 6–23). Schwann cells in the peripheral nervous system (PNS) deposit laminin in a surrounding basal lamina. The interaction of neuronal integrin receptors with laminin

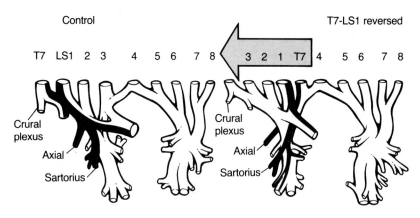

Figure 6–16. **Axonal path-finding of spinal motor neurons.** Routes taken by axons of chick motor neurons projecting from the spinal cord at the seventh thoracic (T-7) segment and first lumbosacral segment (LS-1) were traced by injecting into the spinal segments the dye horseradish peroxidase, which flows anterogradely (*red*) through the corresponding axons. On the left is the projection pattern 6 days after injection, corresponding in normal chick to stage 28 of embryonic development. On the right is the pattern taken by axons from the same segments that were surgically reversed in embryos a few days before dye injection. By stage 28, axons of the displaced motor neurons have taken a different course through the nerve plexus but eventually find the correct muscle branches. (Modified from Lance-Jones C, Landmesser L. *Proc R Soc Lond [Biol]* 1981;214:1.)

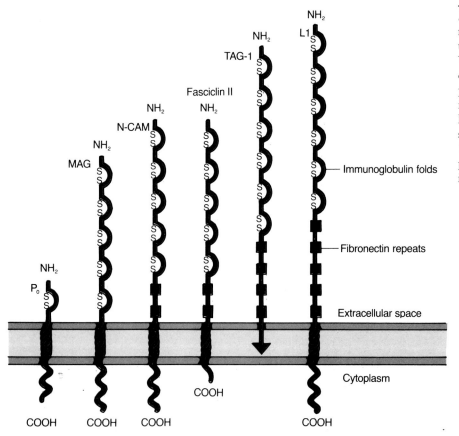

Figure 6–17. **Structures of some neural adhesion molecules belonging to the immunoglobulin superfamily.** Variable numbers of two different domains characterize the external portion of these proteins: (*1*) immunoglobulin-like repeats (*loops*) linked by disulfide bridges; and (*2*) sequence repeats of approximately 100 amino acids with homology to a portion of the fibronectin molecule in the extracellular matrix.

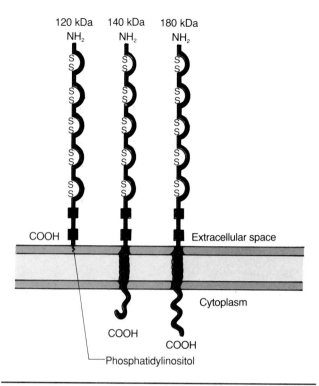

Figure 6–18. Membrane-bound isoforms of N-CAM. With highly conserved extracellular protein backbones, the three major isoforms of N-CAM are alternatively spliced from a single gene and vary in their cytoplasmic domains, association with the plasma membrane, and interaction with the membrane cytoskeleton.

near the Schwann cell membrane is an important stimulant of axonal growth in development and promotes neurite regeneration in response to peripheral nerve injury (Fig. 6–24).

The regulation of neuronal development by extracellular matrix molecules extends to synapse formation. Within the synaptic cleft of the neuromuscular junction, the junctional basal lamina induces formation of synaptic structures at the presynaptic and postsynaptic membrane. When muscle is denervated, remaining components of junctional lamina attract the regenerating nerve terminal to the location of the original motor endplate and promote the differentiation of presynaptic structures. The junctional lamina also causes acetylcholine receptors of the myocyte plasma membrane to concentrate at the synapse. Agrin, a glycoprotein deposited by motor nerve terminals into the junctional lamina, specifically induces aggregation of acetylcholine receptors in cultured myocytes. Many different types of neurons express agrin, suggesting that it also promotes synaptogenesis within the CNS.

Adhesive and Repulsive Forces Contribute to Growth Cone Guidance

Functioning in concert with adhesive substrates and those that, in addition, are permissive for axon growth, are inhibitory guidance cues. For example, members of the tenascin family of extracellular matrix glycoproteins have been implicated as contact-repulsive molecules whose expression *in vivo* temporally coincides with tract formation and cellular migration. A current model for the adhesive forces that drive axonal path-finding predicts that axons are restricted to a track of permissive substrate bounded by repulsive molecules that suppress aberrant growth of projections by causing growth cone deflection or collapse (Fig. 6–25).

Cell Adhesion Molecules Facilitate Myelination

One of the most remarkable cellular interactions in mammals is that which enables oligodendrocytes and Schwann cells to contact and encircle an axon with myelin membrane. Not only must the appropriate axons be identified (most axons in the CNS remain

Figure 6–19. Neuron neuron adhesion and axonal growth mediated by N-CAMs. N-CAMs are expressed by neurons and neuroepithelial cells such that (**A**) interneuronal adhesion or (**B**) the attachment of an axon to a layer of neuroepithelial cells (e.g., astrocytes) can be mediated by homophilic binding.

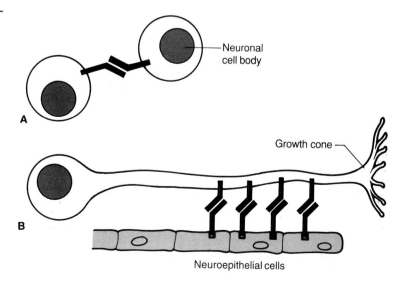

Figure 6–20. **Fasciculation along a pioneer axon.**
The attachment of a pioneer axon to an adhesive
substratum probably involves homophilic N-
CAM-mediated interactions occurring between
the axolemma and surface membrane of glial
neuroepithelial cells or astrocytes that function
as a scaffolding. Subsequent outgrowth of axons
that preferentially track along the pioneer axon
is thought to rely on other adhesion molecules
(e.g., TAG-1, fasciculin, L1) of the immunoglobulin superfamily that also self-associate interax-
onally in a homophilic or heterophilic manner.

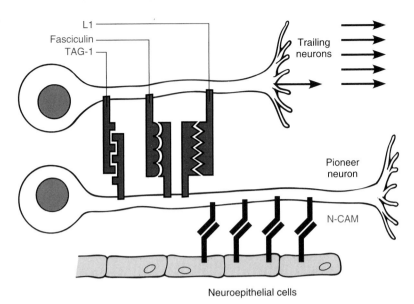

unmyelinated), but once formed, the multilayered
myelin sheath is maintained in a tightly compacted form
to serve its physiologic role (see Chapter 2).

Although the guidance mechanism for attracting
the leading processes of these glial cells to axons is
unknown, a surface transmembrane glycoprotein
unique to myelin-forming cells, termed the *myelin-asso-
ciated glycoprotein* (MAG), is believed to mediate the
initial axon-glial contact before myelin and many of its
characteristic molecules are made (see Fig. 6–17). Dif-

fering from N-CAM, MAG belongs to a subgroup of
the immunoglobulin superfamily of cell adhesion molecules termed *sialoadhesins*. These glycoproteins mediate
close-range interactions between cells by binding specific sialic acid residues, nine-carbon acidic sugars
derived from neuraminic acid, suggesting that an as yet
undefined, sialated glycan (a glycolipid or another glycoprotein) functions as an axonal MAG receptor.

Other adhesion molecules made by myelin-forming
cells participate in homophilic mechanisms to bind the

Figure 6–21. **N-cadherin-mediated neural cell
adhesion requires Ca²⁺.** Astrocytes can serve as
tracks on which axons attach and grow. For
axons of retinal ganglion cells that course through
the optic nerve, this interaction appears to require
N-cadherin in a homophilic adhesion mechanism.

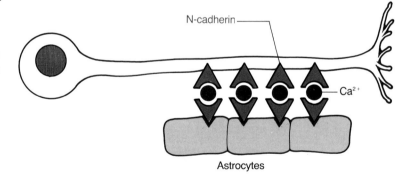

Figure 6–22. **Laminin as a promoter of axonal
growth by heterophilic binding to neuronal
integrins.**

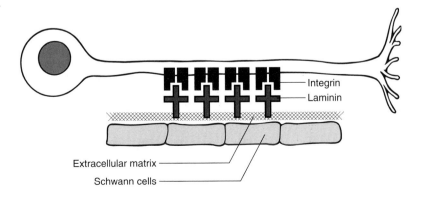

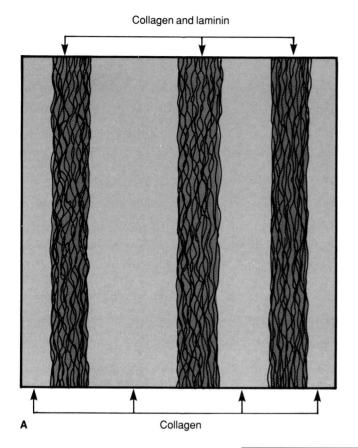

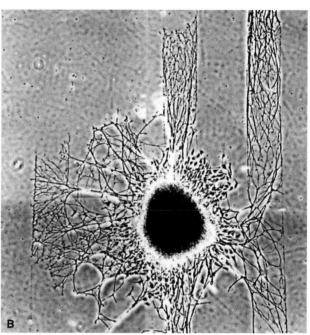

Figure 6–23. Neurites grow preferentially on a substratum of laminin. The flat surface of a tissue culture dish was coated with collagen onto which parallel tracks of purified laminin were painted, as mapped in **A**. The accompanying phase contrast micrograph, in **B**, shows how neurites growing radially from an aggregate of cultured neuronal cell bodies selectively follow the laminin tracks. (bar, 160 μm.) (**B**, Gundersen, RW. *Dev Biol* 1987;121:423, with permission.)

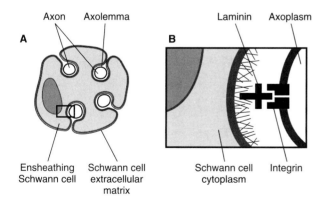

Figure 6–24. The Schwann cell basal lamina promotes axonal growth. **A:** Ensheathing Schwann cells and myelinating Schwann cells (not shown) have a basal lamina (*red*) surrounding their plasma membrane. The basal lamina follows invaginations in the plasma membrane of ensheathing cells and is sandwiched between the Schwann cell and axolemma. **B:** Laminin molecules anchored to the Schwann cell extracellular matrix bind integrin receptors in the axolemma.

apposed leaflets of the myelin sheath together in its functional, compacted state. Under the electron microscope, compacted myelin consists of alternating major dense lines and intraperiod lines (Fig. 6–26). These lines correspond to the respective fused cytoplasmic faces and apposed extracellular faces of the myelinating cell plasma membrane wrapped concentrically around the axon (Fig. 6–27). In myelin of peripheral nerves, an integral transmembrane glycoprotein called P_0 functions in a homophilic manner to bind apposed layers of wrapping to one another across the extracellular space. The P_0 glycoprotein constitutes about 50% of peripheral myelin protein and is one of the simplest members of the immunoglobulin superfamily, having only one immunoglobulin-like repeat (see Fig. 6–17). The functional counterpart of P_0 in CNS myelin is the myelin proteolipid protein (PLP). Unlike P_0, however, PLP is a multipass, nonglycosylated membrane protein unrelated to the immunoglobulin superfamily. Analogous to P_0, it represents about 50% of CNS myelin protein and self-associates across extracellular domains (see Fig. 6–25).

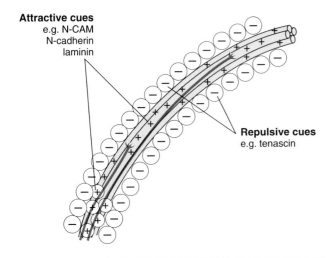

Figure 6–25. Close range interactions that collaborate to guide growth cones along cell surfaces or extracellular matrix deposits are contact-mediated attraction (+) and contract-mediated repulsion or inhibition (–). (Modified from Tessier-Lavigne M, Goodman CS. *Science* 1996;274:1123.)

Clinical Case Discussion

Marfan's syndrome is an autosomal dominant heritable connective tissue disorder characterized by manifestations in the ocular, cardiovascular, and skeletal systems. The prevalence of MFS is 4–6 cases per 100,000 people. Ocular abnormalities include myopia and ectopis lentis (subluxed lenses), which occur in two-thirds of patients. In the infantile form of MFS, 82% have severe cardiac abnormalities. Widening of the aortic root and mitral annulus is the most common cardiac lesion in 80–100% of affected individuals and accounts for 95% of deaths. The skeletal findings include arachnodactyly (long fingers and toes), dolichostenomelia (long limbs), pectus deformities, either excavatum ("funnel chest") or carinatum ("pigeon breast"), scoliosis or kyphosis, tall stature (>95 percentile), high arched palate, and abnormal joint mobility. Hyperextensibility of the hands, feet, elbows, and knees with frequent dislocation of the knees and ankles is often elicited in the medical history. Serum mucoproteins may be decreased and urinary excretion of hydroxyproline increased. Marfan's syndrome can easily

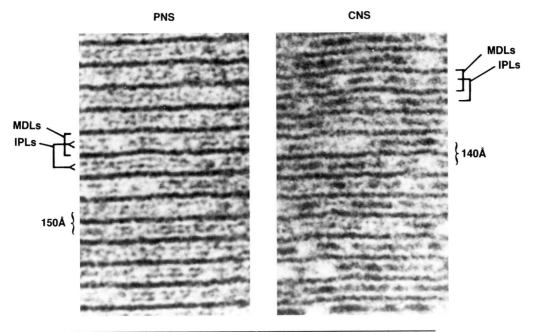

Figure 6–26. High-resolution electron micrographs of compact internodal myelin from the peripheral nervous system (*PNS; left*) and central nervous system (*CNS; right*). The dark-stained major dense lines (MDLs) correspond to fused cytoplasmic faces of each layer of myelin membrane wrapped around the axon (not shown). Lighter-stained intraperiod lines (IPLs) represent apposed extracellular faces of neighboring membranes. The double (PNS) and single (CNS) band of the IPL is probably due to the orientation of P_o, a glycosylated cell adhesion protein of the immunoglobulin superfamily (see Fig. 6–15) that projects farther into the extracellular space between membrane wrappings than does proteolipid protein (PLP), the adhesion molecule used in the CNS. Because of shrinkage during tissue fixation, periodicities measured from the micrographs are slightly smaller than values for unfixed myelin in the PNS (180 Åq) and CNS (156 Åq). (Courtesy of Daniel A. Kirschner.)

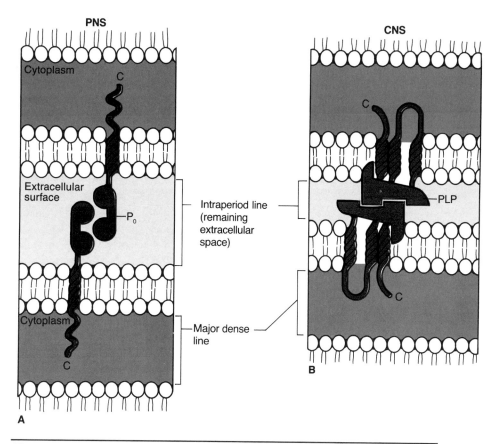

Figure 6–27. Homophilic adhesion molecules in myelin. Protruding into the extra-cellular space between apposed membranes of myelin are the self-association domains of PLP (*CNS*) and the P glycoprotein (*PNS*).

be confused with homocystinuria, as the phenotypic presentation is identical. Absence of homocysteine in the urine is consistent with a diagnosis of MFS.

Marfan's syndrome is caused by molecular defects in the fibrillin gene located on chromosome 15. This gene encodes a 350-kDa glycoprotein, which is an essential component of extracellular matrix microfibrils forming a major component in the suspensory ligaments of the lens and other elastic-containing tissues, including the media of the aorta and other arteries. In 15–35% of patients, the condition appears as a new mutation. Over 30 mutations have been identified. Specific prenatal diagnosis using chorionic villus samples can be provided to families with a previously established mutation.

Suggested Readings

Cell Junctions

Bennet MVL, Barrio LC, Bargiello TA, et al. Gap junctions: new tools, new answers, new questions. *Neuron* 1991;6:305.

Edelman GM. Cell adhesion molecules in the regulation of animal form and tissue pattern. *Annu Rev Cell Biol* 1986;2:81.

Schwartz MA, Owaribe K, Kartenbeck J, et al. Desmosomes and hemidesmosomes: constitutive molecular components. *Annu Rev Cell Biol* 1990;6:461.

The Extracellular Matrix and Its Role in Tissue Formation

Burridge K, Fath K, Kelly T, et al. Focal adhesions: transmembrane junctions between the extracellular matrix and the cytoskeleton. *Annu Rev Cell Biol* 1988;4:487.

Ruoslahi E. Structure and biology of proteoglycans. *Annu Rev Cell Biol* 1988;4:229.

Stevenson BR, Paul DL. The molecular constituents of intracellular junctions. *Curr Opin Cell Biol* 1989;1:884.

Cell Adhesion in the Operation of Muscle

McDonald JW. Extracellular matrix assembly. *Annu Rev Cell Biol* 1988;4:183.

Stevenson BR, Paul DL. The molecular constituents of intracellular junctions. *Curr Opin Cell Biol* 1989;1:884.

Cell Adhesion Mechanisms in Nervous System Development

Kandel ER, Schwartz JH, Jessell TM, eds. *Principles of Neural Science*, 3rd ed. New York: Elsevier, 1989.

Partridge WM, ed. *The Blood–Brain Barrier: Cellular and Molecular Biology*. New York: Raven Press, 1993.

Tessier-Lavigne M, Goodman CS. The molecular biology of axon guidance. *Science* 1996;274:1123.

Review Questions

1. Which adhesive molecule does not engage in homophilic binding?
   a. Po glycoprotein
   b. Laminin
   c. N-CAM (180-kDa isoform)
   d. N-cadherin
   e. N-CAM (120-kDa isoform)

2. Contact-mediated growth cone guidance depends on all of the following except:
   a. Microfilaments
   b. Repulsive contact
   c. Ca^{2+}
   d. Ligands for integrins
   e. All of the above are correct

3. Cell communication that is mediated by gap junctions always involves:
   a. The secretion of neurotransmitter molecules across a synaptic cleft
   b. The diffusion of small molecules through channels that form between the membranes of adjacent cells
   c. The binding of a signaling molecule to a transmembrane receptor
   d. The activation of an elaborate signal transduction pathway
   e. The movement of proteins and ions through a ligand-gated channel

4. Diagnostic criteria for Marfan's syndrome include which of the following:
   a. A molecular defect in the fibrillin gene
   b. Abnormalities of the extracellular matrix microfibrils
   c. Dilatation of the aorta due to abnormalities in the media
   d. a and b
   e. All of the above

5. Which of the following statements concerning the extracellular ground substance is most **CORRECT**?
   a. The ground substance allows connective tissues to resist compressive forces
   b. The ground substance is composed of collagen and elastin
   c. The ground substance inhibits the diffusion of nutrients and wastes between the blood system and cells
   d. The ground substance is composed principally of cadherins and integrins
   e. All of the above are correct

6. Which of following statements about tight junctions is **CORRECT**?
   a. Tight junctions are composed principally of members of the immunoglobulin superfamily
   b. In tight junctions, transmembrane linker proteins bind to the actin cytoskeleton through their cytoplasmic domains
   c. Tight junctions attach cells to the underlying basement membrane
   d. Tight junctions allow metabolic coupling of neighboring cells
   e. Tight junctions encircle the apical-most region of epithelial cells

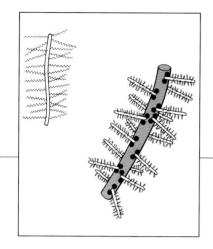

Chapter 7

Intercellular Signaling

Clinical Case

A 25-year-old female, Sophie Smith, presents with complaints of weakness, painful swallowing (dysphagia), and double vision (diplopia) for 3 weeks. Several people on the job have commented that her voice sounds different and that her eyes look droopy. She has been in good health without any recent illnesses. Her physical examination revealed the following: tachypnea, intermittent strabismus, ptosis (droopy eyelids), and a nasal tone to her voice. There was generalized weakness, which was exaggerated by prolonged muscle contractions of the hand when she was asked to squeeze a sphygmomanometer bulb repeatedly. Pertinent negative physical findings included a normal fundoscopic examination and deep tendon reflexes, without muscle atrophy. She was referred to a neurologist for a more complete neurologic examination. In addition to the above findings, the neurologist observed that she had great difficulty chewing. Laboratory tests including a complete blood count and blood chemistries were all normal. While in the neurologist office, she started having great difficulty breathing. He urgently ordered an intravenous Tensilon challenge, which corrected the respiratory difficulty after a few minutes. A serum acetylcholine receptor antibody screen was later found to be positive.

Nerve conduction studies showed a progressive fall in amplitude of the muscle potential with repetitive slow stimulation.

Cells in animals communicate using a variety of molecular messengers. Some of these messengers provide many of the social cues necessary to regulate the growth and differentiation of cells into complex tissues. Others coordinate different physiologic processes and maintain homeostasis in the face of a changing or stressful environment. Most diffusible, extracellular signaling molecules are classified into three general categories of action: endocrine, paracrine, and chemical neurotransmitter (Fig. 7–1), with some overlap in function. More recently, nondiffusible cell adhesion proteins in the plasma membrane and extracellular matrix (Chapter 6) have been identified as signal transducers.

Hormones, the product of islets or glands of endocrine cells are, by definition, released into the blood or lymph so they can circulate to act diffusely at target cells distributed throughout the body. Paracrine mediators, including most polypeptide growth factors, certain amines, and the prostaglandins, derivatives of fatty acids, are not usually transported through the vasculature. Instead, they are released in one form or another by many, if not most, cell types into the extracellular space, where they act on neighboring cells. The remarkable speed and selectivity of cell signaling in the

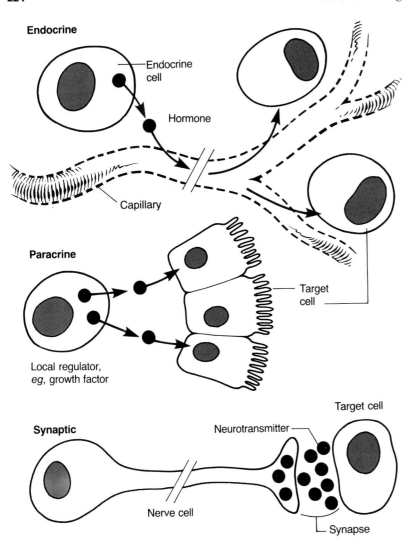

Endocrine

Endocrine cell

Hormone

Capillary

Paracrine

Local regulator,
eg, growth factor

Target cell

Target cell

Synaptic

Neurotransmitter

Target cell

Nerve cell

Synapse

Figure 7–1. Three major modes of intercellular signaling mediated by secreted molecules. Some neurotransmitters can also act as endocrine or paracrine signals to affect multiple target cells in a nondirected fashion. (Modified from Alberts B, Bray D, Lewis J, et al. *Molecular Biology of the Cell.* New York: Garland Publishing, 1989.)

nervous system depends on the transient conversion in the axon terminal of electrical impulses into a discharge of a chemical neurotransmitter. This substance is confined in its action to chemical synapses, highly specialized junctions where neurons communicate with each other, myocytes, and glandular cells across a narrow extracellular cleft. Finally, signaling mediated by adhesion molecules is activated by physical cell–cell contact or cellular interaction with the extracellular matrix. For each type of intercellular-signaling system, the binding of a secreted messenger to a complementary receptor triggers a response in the targeted cell. What differs among these mechanisms is their speed and strategy.

Hormones and Endocrine Signaling

More than 50 hormonal products of the endocrine system have been identified. Each interacts with its own specific receptor protein located in or on the target cell. Hormones are grouped chemically into three categories:

steroid, proteinaceous, and amino acid-related. Table 7–1 gives a brief description of some of the major hormones in each category, in addition to their source and principal action.

Hormones Are Classified Structurally and by Their Intracellular Mechanisms of Action

Hormones act within a relatively slow time frame of up to several minutes. A hormone must diffuse from its site of release through the extracellular space into adjacent capillary beds, depend on blood flow to reach the target area, then leak back out into the extracellular fluid at the cellular site of action. Selectivity in signaling by the endocrine system is based on the types of hormone receptors expressed by target cells, which varies among different cell types (Fig. 7–2). Importantly, receptors for various hormones initiate distinct responses because they are linked to different intracellular signaling pathways (Fig. 7–3). It is not uncommon for a particular hormone receptor to be expressed by multiple cell types, each responding differently.

Hormones are distinguished by their initial action at the cell surface or in the cytoplasm and nucleus. Steroids are small relative molecular mass (M_r) derivatives of cholesterol and are divided according to their physiologic action into six groups: androgens, estrogens, progestins, glucocorticoids, mineralocorticoids, and vitamin D. Because steroids, like the thyroid hormone thyroxine, are lipid-soluble, they freely diffuse across the plasma membrane of target cells. Their receptors are intracellular and contain separate binding sites for the hormone and DNA. The receptor–hormone complex triggers a conformational change in the receptor protein that causes it to bind to certain chromosomal sites and alter genomic activity (see Chapter 8).

Like testosterone, progesterone, and vitamin D, thyroxine requires metabolic transformation for activity. The cellular uptake of T_4 (3,5,3′,5′-tetraiodothyronine (thyroxine) is followed by conversion to the active metabolite T_3 (3,5,3′-triiodothyronine), which interacts with receptors in the cytosol and nucleus. During development, both steroid and thyroid hormones exert effects different from those in adulthood. Nuclear T_3 receptors are expressed in higher amounts by a broad variety of neural and nonneural cell types during development and are particularly important in controlling fetal brain development. The effects are underscored structurally by the stunted differentiation of neurons, with fewer branched dendrites and synaptic contacts in neonatal hypothyroidism, which also results in significant deficits in learning.

Because of their lipophilic nature, steroids and thyroid hormones are insoluble in blood. The problem is surmounted by transporting them throughout the circulatory system as complexes bound reversibly to carrier proteins. In this form, carrier-bound steroids and thyroxine are stabile for much longer intervals (hours to days) than are water-soluble hormones (seconds or less) and, therefore, usually convey longer-lasting effects than do the latter.

Water-soluble hormones, including peptides and those derived from amino acids (other than thyroxine), such as epinephrine, cannot diffuse across the plasma membrane and must interact with cell surface receptors. This may stimulate target cells to release other hormones or trigger rapid electrical activity. These receptors consist of an extracellular domain with a hormone-binding site, a hydrophobic membrane-spanning domain, and an intracellular domain that functions to initiate the cascade of intracellular reactions leading to a biologic response. Many nonsteroidal hormones mediate their effects by activating intracellular second-messenger pathways (see Fig. 7–3) involving cyclic adenosine monophosphate (cAMP), phosphoinositides, or Ca^{2+}–calmodulin in mechanisms that cause the phosphorylation of other proteins and, by doing so, alter their biologic function (see Chapter 8).

Hormonal Secretion Is an Example of Regulated Exocytosis

Selected proteins and other small molecules that are destined for secretion on demand are stored in special secretory vesicles residing within the cytosol (see Chapter 4). The triggering signal for exocytosis is commonly the interaction of a surface receptor on the endocrine cell with its corresponding ligand, either another hor-

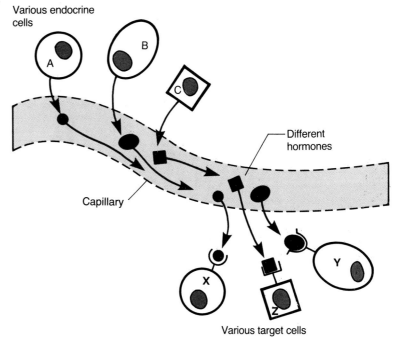

Figure 7–2. Specificity of signaling by the endocrine system. Different endocrine cells secrete varied hormones into the bloodstream, from which they leak back into tissue and selectively bind to only those target cells bearing appropriate receptors. Because hormones are greatly diluted in the blood, their receptor proteins on target cells are extremely sensitive; most activate intracellular responses by binding with high affinity at very low concentration of hormone (usually ≤10^{-8} M). (Modified from Alberts B, Bray D, Lewis J, et al. *Molecular Biology of the Cell.* New York: Garland Publishing, 1989.)

Table 7–1
Some hormones and their properties

Hormones	Site of Origin	Structure	Major Action
Proteins and large polypeptides			
Insulin	β-Cells of pancreas	α chain, 21 amino acids; β chain, 30 amino acids	Utilization of carbohydrate (including uptake of glucose into cells); stimulates protein synthesis; stimulates lipid synthesis in fat cells
Growth hormone-releasing hormone	Hypothalamus	44 amino acids	Stimulates anterior pituitary to secrete somatotropin (growth hormone)
Somatotropin (growth hormone)	Anterior pituitary	191 amino acids	Stimulates liver to produce somatomedin-1, which in turn causes growth of muscle and bone; stimulates fat, muscle, and cartilage cell differentiation
Somatomedin-1 (insulin-like growth factor-1)	Mainly liver	70 amino acids	Growth of bone and muscle; influences metabolism of Ca^{2+}, phosphate carbohydrate, and lipid
Corticotropin-releasing hormone	Hypothalamus	41 amino acids	Stimulates anterior pituitary to secrete ACTH
Corticotropin (adrenocorticotropic hormone; ACTH)	Anterior pituitary	39 amino acids	Stimulates adrenal cortex to produce cortisol; triglyceride breakdown in fat cells
Parathormone	Parathyroid	84 amino acids	Increases bone resorption, thereby increasing blood Ca^{2+} and phosphate; increases resorption of Ca^{2+} and Mg^{2+} and decreases resorption of phosphate in kidney tubules
Erythropoietin	Kidney	Glycoprotein	Stimulates erythrocyte colony-forming cells to differentiate into red blood cells
Luteinizing hormone (LH)	Anterior pituitary	Glycoprotein (α chain, 92 amino acids; β chain, 115 amino acids)	Stimulates oocyte maturation and ovulation and progesterone secretion from ovary stimulates testis to produce testosterone
Follicle-stimulating hormone (FSH)	Anterior pituitary	Glycoprotein (α-chain, 92 amino acids; β-chain, 118 amino acids)	Stimulates ovarian follicles to grow and secrete estradiol; stimulates spermatogenesis in testis
Thyroid-stimulating hormone (TSH)	Anterior pituitary	Glycoprotein (α-chain, 92 amino acids; β-chain, 112 amino acids)	Stimulates thyroid to produce thyroid hormone; fatty acid release from fat cells
Epidermal growth factor (EGF)			
Small peptides			
TSH-Releasing hormone (TRH)	Hypothalamus	3 amino acids	Stimulates anterior pituitary to secrete thyroid-stimulating hormone (TSH)
Somatostatin	Hypothalamus	14 amino acids	Inhibits somatotropin release from anterior pituitary
Vasopressin (antidiuretic hormone; ADH)	Posterior pituitary	9 amino acids	Elevates blood pressure by constricting small blood vessels; increases water resorption in kidney tubules
Oxytocin	Posterior pituitary	9 amino acids	Stimulates uterine contraction and lactation
LH-Releasing hormone	Hypothalamus	10 amino acids	Stimulates anterior pituitary to secrete luteinizing hormone (LH)
Amino acid derivatives (epinephrine)	Adrenal medulla		Increases blood pressure and heart rate; increases glycogenolysis in liver and muscle; fatty acid release from fat cells
Thyroid hormone (thyroxine)	Thyroid		Increases metabolic activity in most cells

(continued)

Table 7–1
(*Continued*)

Hormones	Site of Origin	Structure	Major Action
Steroids			
Cortisol	Adrenal cortex		Affects metabolism of proteins, carbohydrates, and lipids in most tissues; suppresses inflammatory reactions
Progesterone	Ovary (corpus luteum), placenta		Prepares uterus for pregnancy; maintains pregnancy; develops alveolar system in mammary glands
Estradiol	Ovary, placenta		Develops and maintains secondary female sex characteristics; promotes maturation and cyclic function of accessory sex organs; development of duct system in mammary glands
Testosterone	Testis		Develops and maintains secondary male sex characteristics; promotes maturation and normal function of accessory sex organs

mone or a neurotransmitter. A resulting momentary rise in the free cytosolic Ca^{2+} concentration is thought to trigger the fusion of secretory vesicles with the plasma membrane and expulsion of the vesicular contents into the extracellular fluid. Once exocytosis is initiated, vesicular membrane that is newly added to the plasma membrane is endocytosed and recycled into new secretory vesicles. Without a mechanism for membrane retrieval, the surface area of endocrine cells would continuously expand.

Specialized Neurosecretory Cells in the Brain Resemble Both Endocrine Cells and Neurons

In the mammalian hypothalamus, cells with dual properties of endocrine cells and neurons couple the control of the nervous system to endocrine glands (Fig. 7–4). The hypothalamus is linked to the pituitary gland by the infundibulum, or pituitary stalk, and is, in turn, the target of neuronal projections from other brain regions that modulate neuroendocrine activity. When parvicellular (small) hypothalamic neurons in several sites are stimulated, each type secretes a specific peptide hormone or hormone precursor from its axon terminals that diffuses into nearby fenestrated capillaries passing through the pituitary stalk. These hypothalamic hormones include four releasing hormones—thyrotropin releasing hormone (TRH), gonadotropin-releasing hormone (GnRH), corticotropin-releasing hormone (CRH), and growth hormone-releasing hormone (GRH). As these substances reach the pituitary each specifically suppresses or stimulates the release of a secondary hormone, by secretory pituitary cells, into vessels that access the general circulation.

Other peptidergic (peptide-secreting) neuroendocrine cells in the hypothalamus, the magnocellular (large) neurons, extend axons through the pituitary stalk directly into the posterior pituitary, where they terminate near capillaries. On stimulation, separate populations of magnocellular neurons secrete vasopressin and oxytocin, which enter the general circulation through the fenestrated capillary plexus. Whereas vasopressin causes vasoconstriction and increases water resorption by the kidney, oxytocin induces uterine contraction and lactation. Other peptide hormones secreted by hypothalamic neuroendocrine cells are thought to elicit profound behavioral responses, such as thirst (angiotensin II) and eating (cholecystokinin).

Many secondary pituitary hormones effect the release of a third hormone elsewhere in the body. For example, the peptide corticotropin (adrenocorticotropic hormone; ACTH) is stored in secretory vesicles in anterior pituitary cells. Its secretion is stimulated by the hypothalamic CRH. Corticotropin circulates in the main bloodstream and stimulates cells of the adrenal cortex to produce cortisol (Fig. 7–5). If the cortisol concentration in plasma exceeds a narrow physiologic range, the hormone binds to receptors located in neurons of the hypothalamus and pituitary, suppressing the secretion of CRH and ACTH. This sort of feedback inhibition is a fundamental homeostatic mechanism that regulates the production of many other pituitary hormones.

Other neuroendocrine transducers are the chromaffin cells in the adrenal medulla and pinealocytes in the pineal gland. Chromaffin cells are related embryologically to postganglionic neurons of the autonomic nervous system (ANS) that synthesize and secrete norepinephrine but lack axons and dendrites. The cytoplasm

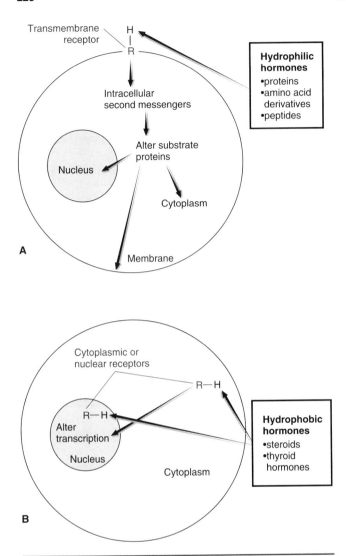

Figure 7–3. Modes of hormonal action in target cells. **A:** Protein, peptide, and certain amino acid-derived hormones (*H*) interact with receptors (*R*) at the cell surface to activate various second-messenger systems. Increased levels of second messengers in the cytosol stimulate the phosphorylation and, thereby, alter the function of substrate proteins, including those that interact with specific nucleotide sequences in the genome. **B:** Steroid hormones and the active thyroid hormone T_3 interact with intracellular receptors that function as transcription factors; second messengers are not utilized. The hormone receptor binding effects a conformational change that exposes a DNA-binding domain on the receptor.

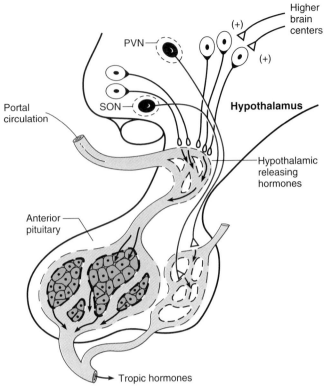

Figure 7–4. Relationship of hypothalamic neuroendocrine cells to the pituitary. Neuroendocrine cells residing in the hypothalamus secrete releasing hormones and release-inhibiting hormones (e.g., somatostatin) from their axon terminals near fenestrated capillaries in the pituitary stalk. These hormones are delivered by the portal vessels to the anterior pituitary where, in turn, they stimulate or inhibit the secretion of other hormones by pituitary cells into the main circulation. (Modified from Hall ZW, ed. *An Introduction to Molecular Neurobiology.* Sunderland, MA: Sinauer Assoc, 1992.)

Photosensory input conveyed along retinal ganglion cell axons by way of the optic nerve to the hypothalamus—specifically, the supraoptic nucleus (location of the biological clock), is relayed via postganglionic sympathetic projections to the pineal gland. The onset of darkness increases the release of the neurotransmitter norepinephrine from sympathetic nerve terminals. Norepinephrine binds to receptors on pinealocytes, which activate the synthesis and secretion of the antigonadotropic hormone, melatonin. Because the pineal gland lies outside the blood–brain barrier, the melatonin it releases can readily enter the general circulation through leaky capillaries.

Paracrine Communication

In multicellular organisms, cells also secrete into extracellular fluid chemical messengers, whose action is largely local and restricted to neighboring target cells

of the adrenal chromaffin cells is packed with large secretory vesicles containing epinephrine, norepinephrine, and peptides such as enkephalin and the chromogranins. When the sympathetic division of the ANS is activated, as in response to emotional stress, sympathetic nerves that innervate the adrenal medullary chromaffin cells stimulate the exocytosis of these products into the bloodstream, which carries them to their target cells in the smooth muscle, heart, and secretory glands.

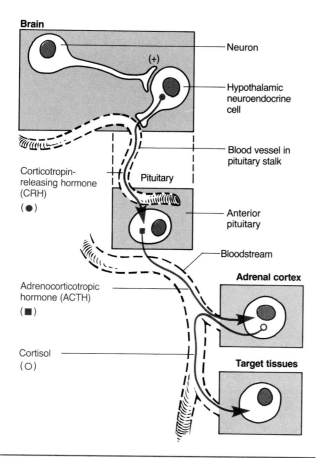

Brain

Neuron

(+)

Hypothalamic
neuroendocrine
cell

Blood vessel in
pituitary stalk

Corticotropin-
releasing hormone
(CRH)
(●)

Pituitary

Anterior
pituitary

Bloodstream

Adrenal cortex

Adrenocorticotropic
hormone (ACTH)
(■)

Cortisol
(○)

Target tissues

Figure 7–5. **Neuroendocrine cells and the regulation of ACTH secretion.** Secretion of ACTH is regulated indirectly by the brain. When stimulated by other centers of the brain, neuroendocrine cells in the hypothalamus secrete corticotropin-releasing hormone (CRH) into capillaries of the pituitary stalk, which deliver the hormone to the pituitary gland. There, CRH stimulates pituitary cells to secrete ACTH into the general circulation, which delivers this hormone to the adrenal gland. ACTH interacts with receptors on cells of the adrenal cortex to release cortisol, which exerts widespread effects on protein, carbohydrate, and lipid metabolism in many tissues. In lieu of excess cortisol in the blood, cortisol suppresses the release of CRH and ACTH in the hypothalamus and pituitary by feedback inhibition, a mechanism by which several other hormones regulate their own blood levels.

(Fig. 7–6). Under normal circumstances, most paracrine signals do not enter the circulation in biologically significant amounts—these substances are secreted or released in small quantities and are rapidly retrieved by cells in the immediate vicinity, degraded by extracellular enzymes, or anchored to components of the extracellular matrix. Although the action of a paracrine signal is highly localized, virtually all cells in the body communicate in this manner to some extent. Unlike endocrine cells, those engaged in paracrine signaling need not be highly specialized for secretion.

Growth Factors Are Typically Pleiotropic Polypeptides

Years of experimentation with tissue culture have revealed that mammalian cells require, in addition to hormonal control, signals by a variety of growth factors, a term usually reserved for soluble polypeptides released by cells into the local environment. Most growth factors are small, typically 10–30 kDa in size, and work by activating surface receptors located on the plasma membrane of target cells. More than 100 growth factors operating in various tissues have been identified, many of which are the products of gene families; novel roles are being ascribed for the more established players. Some prototypical factors and their activities are listed in Table 7–2.

Virtually all cell types in the body make polypeptide growth factors, most of which are released into the surrounding extracellular fluid by regulated exocytosis. Like endocrine signaling, a growth factor's selective effect is initiated because target cells express appropriate receptors. Growth factors that stimulate cell proliferation are termed *mitogens*, whereas others that promote growth and cell survival are termed *trophic factors*. Diffusible gradients of some factors are followed by motile cells. Such factors are termed *chemoattractants*.

Specific names originally given to many growth factors do not always reflect their presently known scope of action. Most of these agents have either multiple growth-promoting effects within the same cell or elicit different responses in different cell types. Consequently, they are said to be *pleiotropic* (Fig. 7–6). This property is exemplified by platelet-derived growth factor (PDGF), which is stored in secretory vesicles of platelets (α-granules) and is known chiefly for its role as one of several local mediators secreted to promote wound repair of damaged blood vessels (Fig. 7–7). As part of the blood-clotting mechanism at the site of vessel injury, platelets exocytose PDGF as a chemoattractant to induce the migration of fibroblasts and macrophages to the site of injury. There, PDGF also stimulates fibroblasts and smooth muscle cells to proliferate and causes the former to increase the production of extracellular matrix components. Under these conditions, macrophages, endothelial cells, and smooth muscle cells may also synthesize and secrete PDGF.

Elsewhere, PDGF acts in other tissues as a mitogen and chemoattractant for different cells of mesenchymal origin. In the developing brain, PDGF is synthesized by neuroectodermally derived cells, including at least one type of glial cell, the astrocyte, and many neurons. The best-studied role of PDGF in neural tissue is in *gliogenesis*, a term given to the generation and maturation of glial cell types during development. Platelet-derived growth factor is thought to act not only as a mitogen for undifferentiated precursor cells that eventually populate the brain with myelin-forming oligodendrocytes (**see**

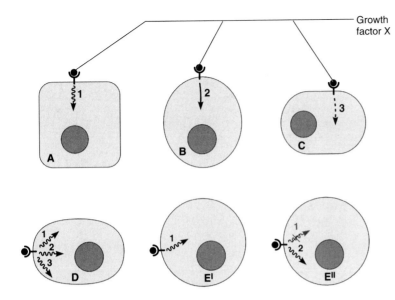

Figure 7–6. Strategies for pleiotropic signaling by growth factors. Identical growth factor–receptor interactions signal different biologic responses in A,B, and C because in each cell type the receptor activates a different signal transduction pathway. Alternatively, the activation of multiple signal transduction pathways in response to ligand-growth factor receptor binding can effect a pleiotropic response within a single cell type. Finally, pleiotropic signaling during development of a cell lineage can reflect a switch in growth factor receptor-mediated activation from one signaling pathway to another. In this way, a single growth factor could switch from being a mitogen for a proliferative blast stage (E′) to a trophic stimulus for mature, postmitotic cells (E″) within a lineage.

Chapter 2), but also as a chemoattractant for the migratory precursor cells to follow as they travel from their germinal origin near the ventricles to their final destination among axonal tracts. Once oligodendrocytes mature and begin to myelinate axons, they stop dividing and cease to express PDGF receptors on the cell surface. Indeed, most cell types during maturation and aging exhibit an altered and typically diminished responsiveness to the growth factors that stimulated their development.

Nerve Growth Factor Is the Prototypical Neurotrophin, a Trophic Factor Required for Neuronal Survival and Differentiation

During the formation of the nervous system, approximately 50% or more of the neurons in many areas of the brain, spinal cord, and peripheral nervous system (PNS) die routinely as part of the normal program of development. Death is thought to result from a process whereby the number of neurons innervating a target tis-

Table 7–2
Prototypical Growth Factors and Their Functions

Factor	Composition	Representative Activities
Platelet-derived growth factor (PDGF)	AA, AB, or BB; A chain, 125 amino acids; B chain, 160 amino acids)	Mitogen for connective tissue cells and immature neuroglia cells
Epidermal growth factor (EGF)	53 amino acids	Mitogen of many cells of ectodermal and mesenchymal origin
Insulin-like growth* factor I (IGF-I)	70 amino acids, 45% homology to insulin	Mediates action of growth hormone
Transforming growth factor β (TGF-β)	Two chains, each 112 amino acids	Potentiates or inhibits response of most cells to other growth factors, depending on the cell type; regulates differentiation of some cell types
Fibroblast growth factor-2 (FGF-2)	146 amino acids	Mitogen for many cell types, including fibroblasts, endothelial cells, and myoblasts; induces embryonic mesoderm
Interleukin-2 (IL-2)	153 amino acids	Mitogen for T lymphocytes
Nerve growth factor (NGF)	Two chains, each 118 amino acids	Promotes axon growth and survival of sympathetic and some sensory and CNS neurons

*Extraneural IGF-I acts as a hormone and is secreted by the liver. Because IGF-I and other polypeptide growth factors do not cross the blood–brain barrier, they are expressed within brain and spinal cord, where they function as paracrine messengers.

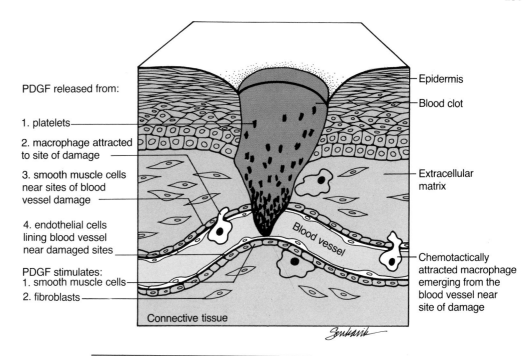

PDGF released from:

1. platelets

2. macrophage attracted to site of damage

3. smooth muscle cells near sites of blood vessel damage

4. endothelial cells lining blood vessel near damaged sites

PDGF stimulates:
1. smooth muscle cells
2. fibroblasts

Epidermis

Blood clot

Extracellular matrix

Chemotactically attracted macrophage emerging from the blood vessel near site of damage

Blood vessel

Connective tissue

Figure 7–7. **PDGF contributes to the process of wound healing at the site of vascular injury by involving many cell types.** Clotting blood causes platelets to secrete PDGF, as do macrophages, recruited chemotactically by PDGF to the damaged area, endothelial cells, and smooth muscle cells near the site of vessel damage. Released PDGF acts as a mitogen for smooth muscle cells and fibroblasts, and stimulates the latter to make more extracellular matrix.

sue is matched appropriately to the number of target cells. By starting with an embryologic excess of neurons, innervation of the entire target field is assured. Death under these circumstances is usually not attributed to an intrinsic neuronal defect. Rather, the loss of neurons is related to the target cells, such as muscle fibers, glandular cells, or other neurons, and their capacity to secrete neurotrophic factors that promote the survival and differentiation of the neurons that innervate them.

Nerve growth factor (NGF) was the first of a small family of neurotrophic factors to be discovered, and its properties are best understood. It is also a prototypical chemoattractant. We have learned how growth cones of axons begin their migration toward target cells by following adhesive pathways (see Chapter 6). As the growth cone approaches its target, it is believed to enter the range of influence of chemoattractant molecules (e.g., NGF) (Fig. 7–8) released specifically by the target cell and to follow a gradient of the substance (chemotaxis) toward the target. Chemoattraction is thought to link a specific type of neuron with its appropriate target cell during a limited developmental period (Fig. 7–9). Axonal growth cones are thought to compete for limiting amounts of the diffusible target cell-derived neurotrophic factor, such as NGF, as they near the target field (Fig. 7–10).

The structure of NGF consists of a homodimer of two 118-amino acid chains coupled by disulfide bonds.

The NGF messenger ribonucleic acid (mRNA) is expressed by only those target cells that receive inputs from NGF-dependent neurons (sympathetic and sensory neurons). Once secreted by the target cell, NGF then binds to NGF receptors on the growth cones of approaching axons. It is internalized by receptor-mediated endocytosis and transported retrogradely to the cell body, where it undergoes lysosomal degradation. The multiple actions of NGF on gene expression, survival,

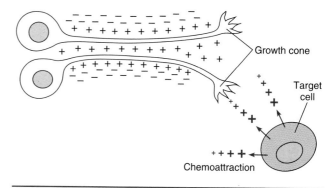

Growth cone

Target cell

Chemoattraction

Figure 7–8. **Growth guidance provided by the force of chemoattraction.** In addition to close range attraction (+) and repulsion (−) provided by contact with nondiffusible adhesion molecules (**Chapter 6**), growth cones are attracted toward gradients of diffusible chemoattractants, e.g., NGF. (Modified from Tessier-Lavigne M, Goodman CS. *Science* 1996;247:1123.)

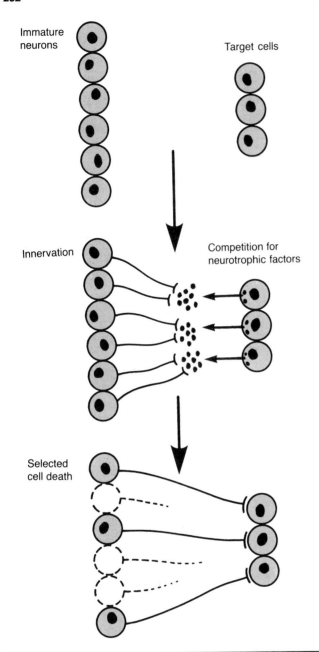

Figure 7–9. Neuronal cell death mediated by target cell-derived neurotrophic factor. Spare neurons formed during embryonic development ensure that corresponding target tissues are fully innervated. During the process of innervation, axons of neurons compete for extremely small numbers of neurotrophic factor molecules secreted by the target cells being innervated. Neurotrophic factor binds to high-affinity receptors on the neuronal surface. Activated signal transduction mechanisms suppress the expression of suicide genes and activate the expression of gene programs for axon elongation and synaptogenesis. Neurons receiving insufficient signal die by apoptosis and are pruned from tissue as part of the developmental program of neurogenesis.

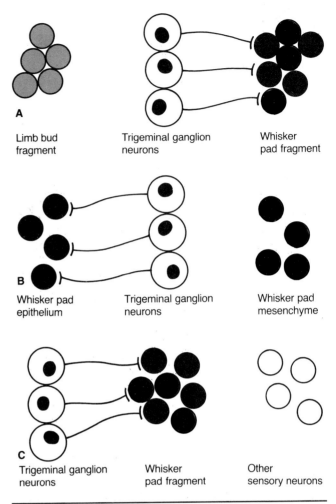

Figure 7–10. Specificity of a chemotactic relationship exemplified between a neurons (trigeminal ganglion) and their target cells of innervation target (whisker pad). From work with tissue culture, Lumsden and Davies provided good evidence with mice that axonal outgrowth of immature trigeminal neurons near—but not touching—potential target tissues will follow a chemical gradient of secreted molecules to contact only the appropriate target. **A:** Trigeminal neurons dissected before finding their contacts *in vivo* extend axons in culture to only the appropriate target (whisker pad fragment) if given an alternative choice (e.g., a piece of cultured limb bud). **B:** The specific source of chemoattraction for axons to the whisker pad is the epithelial cell, which can be dissected from this tissue and grown apart from resident mesenchymal cells. **C:** The chemoattractant provided by the whisker pad epithelial cells is specific for trigeminal ganglion cells; neurite outgrowth is not stimulated from other cultured sensory ganglia that supply other targets.

and neurite outgrowth (Fig. 7–11) are most likely mediated by the activation of multiple signal transduction pathways by NGF receptor (TrkA) autophosphorylation (see Chapter 8).

Although NGF was initially thought to affect only neurons of the PNS, newer evidence points to its role in selected neuronal populations in the brain. For exam-

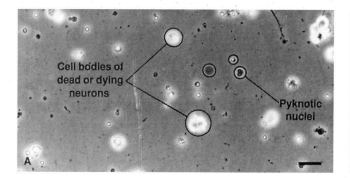

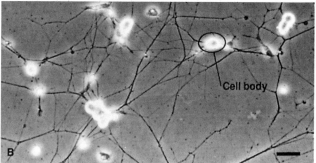

Figure 7–11. Survival and neurite outgrowth of cultured sensory neurons dependent on NGF. In these phase-contrast photomicrographs, sensory neurons that have been dissociated from dorsal root ganglia of embryonic rats are shown 24 hours after culturing in the (A) absence or (B) presence of 50 ng ml^{-1} NGF in the growth medium. A: Without NGF, the neurons die by apoptosis before regenerating neurites shorn away from the cell body by the isolation procedure. B: When maintained in NGF, phase bright neurons survive and exhibit robust neurite outgrowth (bar, 50 μm). (Reichardt F, Weskamp G, Reichardt L. *Neuron* 1991;6:649.)

ple, NGF supports the survival of a small population of cholinergic neurons (those using acetylcholine as their neurotransmitter) in the basal forebrain. These neurons project their axons to specific NGF-producing neurons in the cerebral cortex and hippocampus. Intracerebral injections of NGF in animals rescue dying cholinergic neurons from which axons have been cut. This response may hold some promise for use of trophic factors to treat neurodegenerative disorders such as Alzheimer disease, which causes a specific depletion of similar cholinergic neurons, the function of which relates to memory formation. Other neurotrophins, brain-derived neurotrophic factor (BDNF), neurotrophin-3 (NT-3), and neurotrophin-4 (NT-4), are structurally similar to NGF and are all members of a larger neurotrophic factor gene family. Importantly, different neurotrophins support the survival of distinct but partially overlapping types of neurons, implying that neurotrophic control is widespread during the formation of the nervous system.

Not all neurotrophic factors are provided by target cells of innervation. For example, although ciliary neurotrophic factor (CNTF) is unrelated to the NGF family of neurotrophins but supports the survival of motor neurons supplying skeletal muscle, it is not made by myocytes. Ciliary neurotrophic factor may have clinical relevance for neurodegenerative diseases, such as amyotrophic lateral sclerosis (ALS; Lou Gehrig's disease), that specifically attack motor neurons. Schwann cells in PNS and astrocytes in regions of the CNS produce CNTF and, presumably, secrete it.

One of the most intriguing observations concerning how neurons and many other nonneuronal cells are developmentally programmed to die suggests that such cells are not starved for trophic factors. Surprisingly, inhibitors of protein and RNA synthesis, such as cycloheximide and actinomycin D, can prevent cell death

when administered *in vivo* or in culture. This finding suggests that cells die developmentally by committing suicide; that is, cells evidently synthesize or activate autolytic enzymes. The idea is that at least some neurotrophic factors (e.g., NGF) support cell survival by suppressing the intrinsic expression of cell death genes.

Programmed Cell Death Downsizes Many Different Lineages

Programmed cell death is also a hallmark of developing glial cells and of many nonneural cell types (e.g., hematopoietic lineages). Most affected cells undergo a discernible sequence of degenerative morphologic changes termed *apoptosis*. Characteristics of apoptosis include shrinkage and condensation of the nucleus, which often fragments. A recognition molecule exposed on the surface membrane of apoptotic cells stimulates their rapid phagocytosis within hours by macrophages without spilling cytosolic contents into the surrounding milieu. By contrast, injured cells undergo a different type of death termed *necrosis*. This is a slower process, during which a swelling cytoplasm bursts, releasing cytosol into the extracellular space that triggers an inflammatory response. The emerging picture is that cell numbers are developmentally controlled by the opposing forces of proliferation and programmed death under regulation by a complex variety of growth factors and other cell-derived extracellular signals.

Some Growth Factors Are Bound to Extracellular Matrix for Activity

Not all paracrine growth factors are freely diffusible in the extracellular space. Some families of structurally related growth factors, including the fibroblast growth

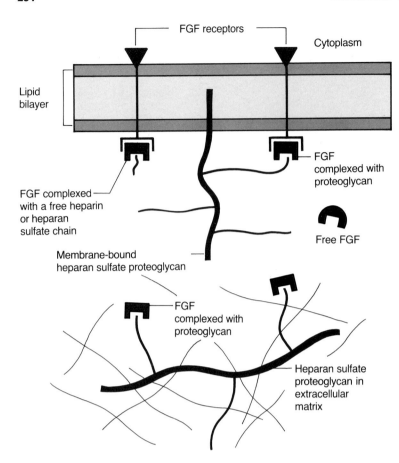

Figure 7–12. Heparin and heparan sulfate proteoglycans as modulators of FGF. FGF binds with low affinity to free heparin, heparan sulfate chains, and heparan sulfate proteoglycans bound to the plasma membrane or extracellular matrix. By acting as low-affinity receptors, these molecules are hypothesized to induce a conformational change in the bound factors, making them able to bind to distinct high-affinity FGF receptors located in the plasmalemma. Only binding to the high-affinity cell surface receptor signals the responding cell. (Modified from Ruoslahti E, Yamaguchi Y. *Cell* 1991;64:867–869.)

factors (FGF) and transforming growth factors (TGFs), bind heparin as well as heparan sulfate-containing proteoglycans attached to the cell membrane or residing in the extracellular matrix (Fig. 7–12). Termed *heparin-binding growth factors* (HBGF), these factors are pleiotropic, acting as mitogens or affecting differentiation in a wide range of biologic events concerning tissue formation, regeneration, and carcinogenesis. Functioning as low-affinity receptors for FGF and TGF, proteoglycans alter the conformation of the bound factors, allowing them to bind high-affinity receptors in the target cell plasma membrane. Anchorage to the cell surface or extracellular matrix also prolongs the resistance of HBGFs to proteolytic degradation and essentially arms the cell with a bound signal for delivery by physical contact.

Cells Under Autocrine Control Respond to Their Own Secreted Local Mediators

Paracrine signaling occurs when growth factors from one set of cells act as diffusible or anchored agents to stimulate an adjacent group of cells. Other situations arise when a cell type responds to one of its own secreted growth factors, a process termed *autocrine signaling*. For example, during the rapid formation of the placenta, clonal proliferation of cytotrophoblasts is thought to occur as the result of autocrine stimulation: these cells both secrete PDGF and express PDGF receptors that relay its mitogenic effect.

Evidence of another autocrine mechanism involves the sequence of signaling events during antigen recognition and the activation of T-helper cells. The proliferation of T-helper cells is necessary to mount a secondary immune response and involves the interleukins, secreted polypeptides that function as local mediators between immune cells. During the process of antigen presentation, T-helper cells are stimulated by interleukin-1 and antigen binding to secrete their own mitogen, interleukin-2 (Fig. 7–13) and to express mitogenically responsive interleukin-2 receptors in their surface membrane. In this way, helper T-cells specific for a given antigen can continue to proliferate, even after losing contact with the antigen itself.

The Mast Cell, Activated During an Allergic Reaction, Engages in Paracrine Signaling By Releasing Histamine

Paracrine signals need not be polypeptides. The mast cell, a resident of connective tissue throughout the body, is activated during an allergic reaction. On its surface are receptors for the IgE class of immunoglobulin. In the cytoplasm are numerous, large secretory

Figure 7–13. Autocrine stimulation of activated T-helper cell proliferation. During the course of the antigen-mediated helper T-lymphocyte response, antigen bound to the surface of an antigen-presenting cell (e.g., a tissue macrophage) also binds to its receptor on a corresponding helper T cell. In response, the T-helper cell stimulates the antigen-presenting cell to secrete the growth factor interleukin-1 (IL-1). This polypeptide binds to IL-1 receptors on the T-helper cell that activate the synthesis and secretion of interleukin-2 (IL-2) and the expression of IL-2 receptors in the T-helper cell plasma membrane. Unlike IL-1, the function of secreted IL-2 appears to be autocrine because it stimulates proliferation of the same cells that secrete it. (Modified from Alberts B, Bray D, Lewis J, et al. *Molecular Biology of the Cell.* New York: Garland Publishing, 1989.)

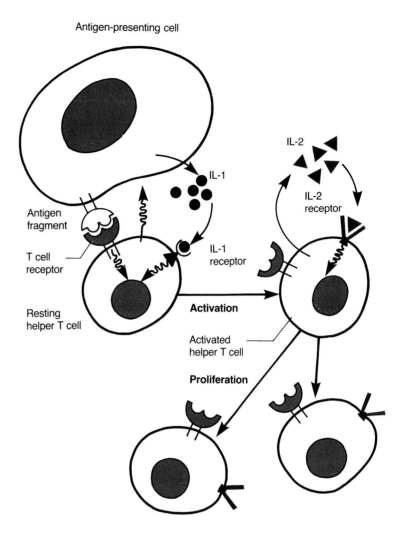

vesicles that have a granular appearance under the electron microscope and house many substances including histamine, a derivative of the amino acid, histidine. During initial exposure to an antigen, such as a venom protein from a bee sting, an immune response is mounted. Antibodies produced by activated B lymphocytes will recognize many antigenic sites (epitopes) on the venom protein and include a small proportion of immunoglobulins of the IgE class. Antivenom IgE molecules bind to specific receptors for IgE located on the mast cell surface, where the complex is retained (see Fig. 7–14). If a subsequent bee sting occurs, new venom protein will bind to the mast cell surface via the IgE antivenom–IgE receptor complex. Cross-linkage of multiple IgE–IgE receptor complexes by a single venom molecule changes the structure of the IgE receptor, which triggers the influx of Ca^{2+} into the cytoplasm and degranulation (exocytosis). Released histamine binds to receptors on endothelial cell surfaces that line postcapillary venules, causing them to leak. As a result, plasma percolates into the extracellular space, and swelling occurs. In addition, the leaky endothelium will allow white blood cells, necessary for neutralization of

the invader and repair of the damaged tissue, to gain ready access to the site of injury. One of many peptides secreted by the mast cells during degranulation is eosinophil chemotactic factor (ECF). This tetrapeptide binds to receptors on the surface of eosinophils and activates a mechanism that enables them to migrate along an ECF chemotactic gradient to mast cells in the immediate vicinity of the bee sting, where the eosinophil begins to repair the region of tissue damage. Other proteins released during mast cell degranulation also contribute to the immediate hypersensitivity response. In sum, these events are the basis for the potentially fatal anaphylactic reaction.

Eicosanoids and Nitric Oxide Are Membrane-Permeant Transcellular Messengers

Some signaling molecules are sufficiently hydrophobic or small enough to freely diffuse through the plasmalemma of cells that make them into neighboring target cells. The generic term *eicosanoids* includes a growing number of oxygenated bioactive derivatives of 20-carbon polyunsaturated fatty acids (prostaglandins,

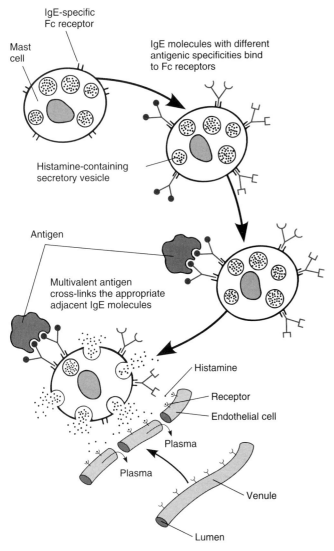

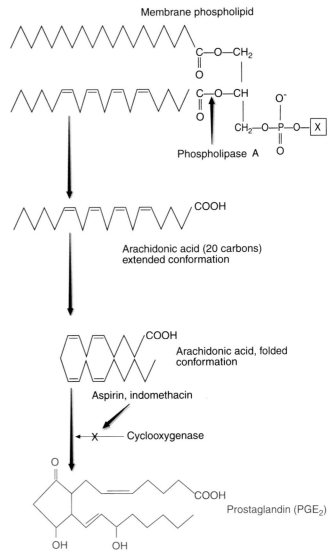

Figure 7–14. Degranulation of mast cells. Mast cells in connective tissue (and eosinophils in blood) are packed with secretory vesicles containing histamine. Exocytosis, or "degranulation," of histamine is triggered by the cross-linking of cell surface receptors that recognize the crystalline fragment (Fc) portion of IgE antibodies. Following a primary immune response, a sampling of IgE antibodies produced by active B lymphocytes remains attached to the cell surface IgE receptor. Degranulation and histamine release are triggered during a secondary immune response when adjacent IgE-bound Fc receptors are cross-linked at the cell surface by the same antigen molecule.

Figure 7–15. Pathways of eicosanoic synthesis. Prostaglandins, thromboxanes, leukotrienes, and other active metabolites are primarily the breakdown products of arachidonic acid. Phospholipase A hydrolyzes phosphoinositol in the plasma membrane. The resulting diffusion of free arachidonic acid into the cytosol initiates a cascade of enzymatic oxidation reactions, giving rise to several types of biologically active metabolites of arachidonate involved in the inflammation response. The mechanism of action of many nonsteroidal antiinflammatory drugs is the inhibition of the cyclooxygenase enzyme that oxidizes arachidonate into prostaglandin, specifically PGE.

thromboxanes, leukotrienes, and other active metabolites). Eicosanoids are formed by virtually all body tissues, and their production is stimulated by activating a variety of cell surface receptors for hormones, growth factors, neurotransmitters, and toxins, in addition to tissue damage resulting, for example, from ischemia, burns, or vascular shock.

Among the eicosanoids, the prostaglandins are an important class that derives from arachidonic acid (Fig. 7–15). An increase in cytosolic free Ca^{2+} is thought to stimulate phospholipase A to hydrolyze fatty acyl esters from membrane phospholipids on the cytosolic side of the plasma membrane. This releases free arachidonic acid from former phospholipids, which is enzymatically oxi-

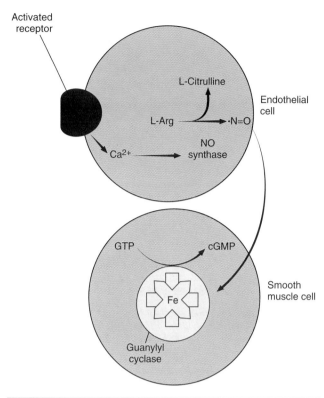

dized in the cytosol into nine major types of prostaglandins (PGA–PGI). Being amphiphilic, prostaglandins diffuse across the lipid bilayer into the extracellular fluid and bind to corresponding surface receptors on neighboring cells. Individual prostaglandins produce a wide variety of diffuse biologic effects in various organs, including contraction of smooth muscle, platelet aggregation, and uterine contraction. The latter activity has led to clinical use of certain prostaglandins as abortifacients.

Prostaglandins released from damaged cells also cause inflammation and activate *nociceptors*—sensory nerve endings specialized for relaying the neural sensation of pain (nociception) to the CNS. Specifically, PGE is released from damaged cells and diffuses to adjacent nociceptors, where it reduces the threshold of nociceptor depolarization necessary to initiate an action potential (see Chapter 2) that relays pain to the spinal cord and brain. The ability of PGE to sensitize nociceptors to subsequent noxious (painful) stimuli is termed *hyperalgesia*. Analgesics, such as aspirin, indomethacin, and nonsteroidal antiinflammatory drugs, inhibit the cyclooxygenase enzyme that forms PGE as a metabolite of arachidonic acid.

Some cell types can also use a gas, nitric oxide (NO), as an extracellular messenger that, like eicosanoids, also diffuses freely across the plasma membrane. An active area of research concerns the function of NO in the brain, where it may contribute to the biochemistry underlying memory (see Chapter 8). However, the role of NO is better understood as a mediator of vasodilation by its action to cause relaxation of smooth muscle cells in the blood vessel wall. Endothelial cells contain NO synthase, the key enzyme that catalyzes the liberation of NO by oxidizing the amino acid, L-arginine to L-citrulline (Fig. 7–16). Activation of the enzyme occurs when acetylcholine, a neurotransmitter released at parasympathetic nerve terminals supplying endothelial cells, binds to specific receptors on the endothelial cell surface. These cholinergic (acetylcholine-binding) receptors are the muscarinic type that, unlike the nicotinic type, do not contain an intrinsic ion channel. Instead, they are associated with a G protein that, when activated, stimulates the phosphoinositide second-messenger pathway leading to a rise in cytosolic Ca^{2+} (see Chapter 8). Elevated intracellular Ca^{2+} activates the Ca^{2+}-sensitive NO synthase, enabling it to catalyze the production of metastable NO. The gas molecules rapidly diffuse from endothelial cells into apposed smooth muscle cells (Fig. 7–16), where they bind iron located at the active site of the enzyme, soluble guanylate cylase. This activates the enzyme and increases the concentration of the second-messenger cGMP in the myocyte. This, in turn, activates cGMP-dependent protein kinase, which phosphorylates proteins (see Chapter 8),

Figure 7–16. Nitric oxide signaling of smooth-muscle cells by endothelial cells in the blood vessel wall. The formation of nitric oxide (NO) in the endothelial cell is driven by the activation of the muscarinic G-protein-linked acetylcholine receptors at the cell surface. This signals second-messenger pathways, leading to a rise in cytosolic Ca^{2+} which, in turn, activates the NO metabolic pathway. The NO freely diffuses into the adjacent smooth muscle cell, where it binds to a heme group associated with soluble guanylyl cyclase. This activates the cyclase to convert guanosine 5'-triphosphate (GTP) into the second-messenger molecule, guanosine 3', 5'-cyclic phosphate (cGMP). In smooth muscle, elevated cGMP stimulates cGMP-dependent protein kinase, resulting in the phosphorylation of key proteins that leads to relaxation of the smooth-muscle wall, causing vasodilation.

leading to a relaxation of the smooth-muscle wall and vasodilation.

The vascular effect of NO explains the mechanism of action of the vasodilator drug, nitroglycerin. Used to treat patients with angina, nitroglycerin is rapidly converted in the blood to NO, which relaxes coronary vessels. Nitric oxide can also mediate toxicity, ranging from the bacteriocidal effect of macrophages to neurotoxicity. For example, excessive glutamate released in response to vascular stroke is thought to cause excessive activation of the NMDA glutamate receptor subtype. The resulting influx of Ca^{2+} through activated NMDA receptors leads to the Ca^{2+}-mediated induction of NO synthase and an overproduction of NO, which binds to and disrupts numerous metabolic enzymes.

Synaptic Transmission

Electrical impulses are conveyed from one nerve cell to the next across specialized regions of intercellular contact, termed *synapses*. Collaborating with the action potential (see Chapter 2), synaptic transmission accounts for the unique speed, precision, and long-distance capability of neural signaling. In the following sections, we will examine the basic mechanisms of synaptic transmission that underlie nervous system function.

Synapses Exist in Two Forms: Chemical and Electrical

The concept of chemicals mediating the action of nerves on their target cells is based on Otto Loewi's discovery in the 1920s that a stimulated vagus nerve slows the heart rate by releasing a diffusible substance, later identified as acetylcholine, into a perfusion fluid. During the following decade, Sir Henry Dale and co-workers found that acetylcholine is also released by stimulated axons of peripheral motor nerves and is responsible for skeletal muscle contraction. The list of chemicals classified as neurotransmitters has grown considerably. Most transmission of electrical signals between neurons, and from nerve fibers to muscle cells or gland cells, is mediated by chemical synapses. In essence, chemical neurotransmission represents a variation of paracrine signaling in which neurons deliver messengers with pinpoint precision to their target cells over great distances. Consequently, the intercellular distance of chemical synaptic signaling is very short, spanning only the width of a synaptic cleft. This is a 20- to 50-nm space that sepa-

rates the plasma membrane of a presynaptic cell from that of a postsynaptic one (Fig. 7–17). Transmission of an electrical signal from the presynaptic cell to the postsynaptic cell occurs indirectly; electrical signals arriving at the presynaptic terminal do not spread to the postsynaptic cell. Instead, they trigger in the presynaptic terminal exocytosis of a chemical neurotransmitter into the synaptic cleft, where it diffuses and binds to specific receptors located on the postsynaptic membrane. Occupation of these receptors induces the opening of ion channels that convert the chemical energy back into an electrical response in the postsynaptic target cell.

Other synapses operate strictly by electrical means, without chemical intervention, and are not restricted to the nervous system. Such electrical synapses are formed between cells by gap junctions acting as ion channels (see Chapter 6). Unlike chemical synapses, the cytoplasm of two cells coupled by electrical synapses is continuous at sites where gap junction-forming ion channels span the lipid bilayers (Fig. 7–18). As a result, signaling across electrical synapses occurs without the brief submillisecond to millisecond delay that typifies chemical synaptic transmission. This gain in speed is advantageous for cells that must function essentially as one large unit within an organ in a relationship called a *syncytium*. Were it not for electrical synapses between cardiac muscle fibers and between smooth muscle cells, for example, the synchronous heartbeat and peristaltic movement of the gut would not be possible (see Chapter 6). By comparison, chemical synapses predominate among nerve cells and provide the versatility and diversification of intercellular signaling necessary for the vast complexity of nervous system function.

Chemical Neurotransmitters Are Classically Defined by Stringent Criteria

To qualify as a neurotransmitter, a substance must be released from a stimulated presynaptic neuron and bind to the membrane of a postsynaptic target cell, where it provokes an excitatory or inhibitory response. Unlike endocrine signaling or paracrine signaling mediated by growth factors, the action of neurotransmitters is highly directed, occurring only at the synapse, and is extremely rapid (less than a few milliseconds). Transmitter release typically occurs from axon terminals that synapse along the dendrite, cell body, or axon of a partner neuron. In the periphery, axon terminals form chemical synapses with myocytes or gland cells to regulate contractility and secretion.

Other criteria for conventional chemical neurotransmission include synthesis and storage near the presynaptic release site of the neurotransmitter molecules in synaptic vesicles. As an exception, peptide transmitters are made in the cell body by the endoplasmic reticulum, packaged into secretory vesicles in the

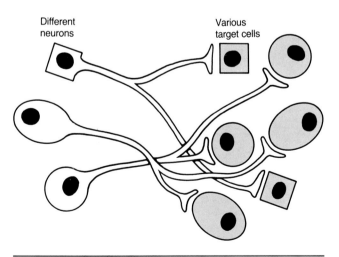

Figure 7–17. Chemical synaptic signaling. Even though many neurons can use the same neurotransmitter, specificity of signaling is achieved in the nervous system because the release of neurotransmitter is highly restricted to target cells at zones of synaptic contact with nerve terminals.

Different neurons

Various target cells

Figure 7–18. Pathways of current flow distinguish chemical and electrical synapses. **A:** In the chemical synapse, current entering the presynaptic terminal during the action potential does not enter the postsynaptic cell. **B:** When two cells are coupled electrically by gap junctions, some of the current entering one cell flows directly into the cytoplasm of the other cell.

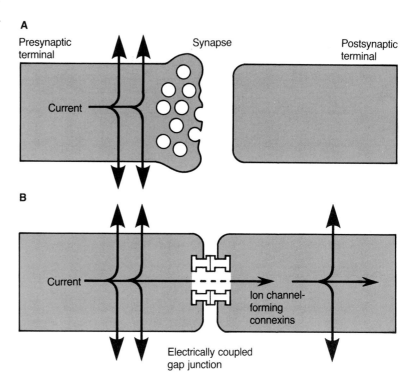

Golgi complex, then transported by the process of fast axonal transport (see Chapters 3 and 4) to the axon terminal. The arrival of an action potential at the presynaptic terminal triggers exocytosis of the neurotransmitter in the synaptic cleft by a Ca^{2+}-regulated process (see Chapter 8). Once released, the transmitter must be recognized and reversibly bound by specific receptor proteins protruding from the postsynaptic membrane surface. When fast-acting neurotransmitters, such as acetylcholine, γ-aminobutyric acid (GABA), glycine, or glutamate, bind to their respective receptors, they cause an ion-selective channel within the receptor itself to open, leading to an influx of a specific ion(s) into the postsynaptic cell (Fig. 7–19). Other, slower-acting neurotransmitters such as the catecholamines (dopamine, norepinephrine, and epinephrine), indoleamines (serotonin, histamine), and various neuropeptides activate receptors without intrinsic ion channels. Instead, these receptors are linked to G proteins and second-messenger systems that indirectly open other ion channels or induce other metabolic changes within the postsynaptic cell.

Finally, a mechanism must exist to rapidly inactivate the effect of the transmitters (Fig. 7–20). For acetylcholine, this is the job of a degradative enzyme, acetylcholinesterase, localized in the synaptic cleft. On the other hand, catecholamines are depleted from the synaptic cleft by extracellular enzymes as well as by a carrier system that takes up transmitter back into the presynaptic terminal. Likewise, amino acid neurotransmitters are taken up into the nerve terminal or surrounding glial cells by other specific carrier systems.

Many Peptides and Some Gases Also Function as Putative Neurotransmitters

New candidates for neurotransmitters in the mammalian nervous system include a burgeoning number of peptides (List 7–1). For most of these messengers, it has not been conclusively shown whether all of the classic criteria of a neurotransmitter are applicable. Nevertheless, neuropeptides are concentrated in distinct subpopulations of neurons, suggesting that peptide release does occur. What is also clear is that many putative neuropeptide transmitters identified in brain are utilized elsewhere in the body. Included here are many peptides that were originally identified as hormones released by specialized neurosecretory cells in the gut. Other transmitter candidates are the endorphins and enkephalins, which function as the brain's own analgesic peptides, and Substance P, the transmitter of primary sensory neurons that convey pain sensation to the CNS. The discovery of NO as a neurotransmitter formed from glutamate acting at NMDA receptors (Chapter 8) has expanded our concept of molecules with transmitter potential. As a gas, NO is neither stored in vesicles nor exocytosed, as are hydrophilic transmitters. Nitric oxide operates retrogradely by diffusing across the postsynaptic terminal membrane and synaptic cleft into the presynaptic terminal. There, it does not bind reversibly to plasma membrane receptors, as most neurotransmitters do, but instead reacts to form covalent linkages with soluble target proteins, e.g., guanylyl cyclases. Recent evidence suggests that carbon monoxide may also function as a transmitter.

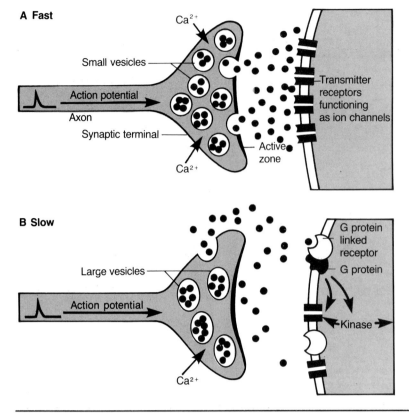

Figure 7–19. Presynaptic and postsynaptic events during fast and slow chemical synaptic transmission. The arrival of an action potential at the presynaptic axon terminal causes an influx of Ca^{2+}, which stimulates the exocytosis of neurotransmitter into the synaptic cleft. **A:** Fast-acting neurotransmitters are released from specialized active zones directly apposed to the postsynaptic membrane and cause the opening of ion channels that form part of their receptors in the postsynaptic cell. Following the removal of neurotransmitter from the cleft, the postsynaptic receptors revert to their previous conformation. **B:** Slower-acting chemical messengers, including peptides and catecholamines, are stored in secretory vesicles that are larger than synaptic vesicles. During Ca^{2+}-mediated exocytosis, the contents of these vesicles are released randomly from different sites around the perimeter of the terminal. These messengers bind to postsynaptic receptors that have no intrinsic ion channels. Instead, they are linked to G proteins, which open other channels indirectly by activating intracellular second-messenger pathways (**see Chapter 9**). (Modified from Scheller RH, Hall ZW, In Hall ZW, ed. *An Introduction to Molecular Neurobiology.* Sunderland, MA: Sinauer Assoc, 1992.)

Different Messengers Localize to Different Types of Vesicles in Nerve Terminals

Individual terminals contain multiple types of transmitter molecules. It is common for a single fast-acting neurotransmitter (e.g., acetylcholine) to colocalize in a nerve terminal with any one of several slower-acting neuropeptides, depending on the specific kind of neuron. Under the electron microscope, presynaptic nerve terminals are typically packed with two types of vesicles. The term *synaptic vesicle* is reserved for small spherical vesicles about 50 nm in diameter (see Fig. 7–21) or ellipsoidal vesicles, both of which are lucent in electron micrographs and are concentrated near the presynaptic terminal membrane. These vesicles contain a fast-acting

transmitter, and when an action potential arrives, many rapidly fuse with the presynaptic membrane to discharge their contents into the synaptic cleft. Discharge occurs at the active zones, specializations on the cytoplasmic face of the presynaptic membrane where synaptic vesicles dock immediately before exocytosis.

Larger vesicles (90–250 nm in diameter), with an electron-dense core, are more randomly distributed within the terminal (see Figs. 7–19, 7–21). These are referred to as *secretory vesicles* and usually contain peptides or amines serving as neurotransmitters. Multiple peptides are often contained within a single secretory vesicle. Secretory vesicles release their contents from random sites around the perimeter of the axon terminal (see Fig. 7–20).

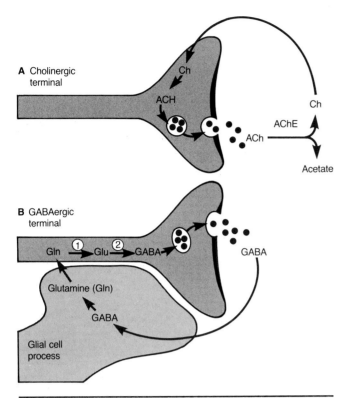

Figure 7–20. Neurotransmitter reuptake and recycling into synaptic vesicles. **A:** Acetylcholine (*ACh*) released from cholinergic terminals is rapidly hydrolyzed by acetylcholinesterase (*AChE*) into choline and acetate in the synaptic cleft. Choline (*Ch*) is retrieved by a choline transporter in the presynaptic terminal membrane, enzymatically converted to acetylcholine by choline acetyltransferase, and repackaged into newly formed synaptic vesicles. **B:** In contrast, the inhibitory amino acid transmitter GABA, released from GABAergic nerve terminals, is retrieved by astroglial cells, the plasma membranes of which contain a specific GABA transporter. In the glial cytoplasm, GABA is converted to glutamine, which reenters the nerve ending. Only neurons containing the enzymes glutaminase (*1*) and glutamate decarboxylase (*2*) can convert glutamine back into GABA in the terminal.

The Skeletal Neuromuscular Junction: A Prototypical Chemical Synapse

Many of the unifying principles that apply to synaptic transmission in the brain and spinal cord of the central nervous system have come from examining cells in the periphery with relatively simple patterns of innervation. The synapse subject to the greatest scrutiny has been the neuromuscular junction formed between axon terminals of cranial or spinal motor neurons and skeletal muscle cells which, like neurons, are electrically excitable. Figure 7–22 follows the course of an axon as it nears a target muscle fiber and typically branches to innervate hundreds of muscle fibers within a specific muscle. Each muscle fiber (myocyte) forms

List 7–1

Classes of established and putative neurotransmitter molecules

Gut–brain peptides

Vasoactive intestinal peptide (VIP)
Cholecystokinin (CCK)
Gastrin
Motilin
Pancreatic polypeptide
Secretin
Substance P
Substance K
Bombesin
Neurotensin
Gastrin-releasing peptide (GRP)

Opioid peptides

Dynorphin
β-Endorphin
Met-enkephalin
Leu-enkephalin

Amino acids

Excitatory

Glutamate
Aspartate

Inhibitory

γ-Aminobutyric acid (GABA)
Glycine

Biogenic Amines

Catecholamines

Norepinephrine
Epinephrine
Dopamine

Indoleamines

Serotonin (5-HT)
Histamine

Ester

Acetylcholine

Other

Nitric oxide
Carbon monoxide

one synapse with the terminal of a single axon branch that has lost its myelin but remains ensheathed by Schwann cell plasma membrane. Innervation of the muscle fiber occurs only at a special region of the muscle membrane surface, termed the *endplate*, where an axon branch, in turn, splits into several finer terminal

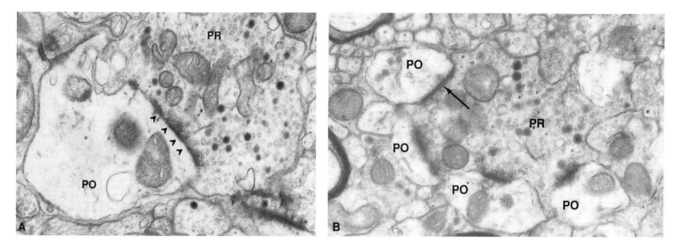

Figure 7–21. Colocalization of synaptic vesicles and secretory vesicles in the presynaptic terminal. **A:** In this electron micrograph of the dorsal horn of the spinal cord, the small electron-translucent vesicles within the presynaptic terminal (PR) contain the excitatory neurotransmitter glutamate, and the larger electron-dense secretory vesicles contain neuropeptides. Note the dark electron-dense "thickening" of the postsynaptic terminal (PO) membrane (*arrowheads*) that is thought to represent a concentration of postsynaptic transmembrane receptors. **B:** In some synapses, similar thickenings (*arrow*) are observed along the presynaptic terminal and are referred to as *active zones*, special sites where synaptic vesicles dock to facilitate exocytosis. (Courtesy of H. J. Ralston, III.)

branches. Overlying the endplate, the terminal axon ends as varicosities, or synaptic boutons, the specific sites at which the neurotransmitter acetylcholine is released. Examination of the neuromuscular junction by electron microscopy (see Fig. 7–22) reveals a high degree of organization at both the presynaptic and postsynaptic sides of these synaptic contacts. Every synaptic bouton overlies an invagination of the muscle membrane, called a *junctional fold*. A high density of acetylcholine receptors is located in the postsynaptic membrane where it begins to infold.

When an action potential arrives at the presynaptic terminal of a neuromuscular junction, the resulting depolarization triggers a cycle of events that begins with the opening of voltage-sensitive Ca^{2+} channels. Ca^{2+} influx triggers the mobilization of many synaptic vesicles (see Chapter 8) that are free to fuse with the presynaptic membrane, where exocytosis of these vesicles and others predocked by spectrin II, and then SNAREs, at special active zones releases acetylcholine into the synaptic cleft. At the postsynaptic membrane, acetylcholine binds nicotinic receptors to open ion channels, leading chiefly to the influx of Na^+ down its steep electrochemical gradient into the muscle fiber (see Chapter 2). If the resulting depolarization exceeds a threshold value, voltage-dependent Na^+ channels open, and an action potential spreads throughout the muscle membrane to initiate the chain of events causing contraction (see Chapter 3).

The most common disease affecting neuromuscular transmission is myasthenia gravis, an autoimmune disorder. Autoantibodies made against the nicotinic cholinergic receptor cause the disease by blocking the receptor and promoting accelerated endocytosis of the receptor from the motor endplate.

Synaptic Vesicles Package Neurotransmitters Into Single Quanta, Unit-Doses for Exocytotic Release

When a single action potential depolarizes an axon terminal at the neuromuscular junction, several hundred synaptic vesicles release their contents into the synaptic cleft, whereas thousands more remain conserved within the terminal. This ensures that a repetitive discharge of acetylcholine will result from a succession of high-frequency impulses arriving at the nerve ending. Even at rest, brief subthreshold depolarizations randomly occur at the muscle fiber membrane. Termed *miniature synaptic potentials* (MEPs), each of these postsynaptic MEPs results from the release of acetylcholine by a single synaptic vesicle. The amplitude of each MEP is strikingly similar (about 1 mV) (Fig. 7–23) because each synaptic vesicle contains roughly the same number (~5000) of acetylcholine molecules. The contents of a single vesicle correspond to a single quantum (minimum amount) of releasable transmitter. Depending on the nerve impulse, the amount of Ca^{2+} that enters the nerve terminal (see Chapter 8) transiently increases the num-

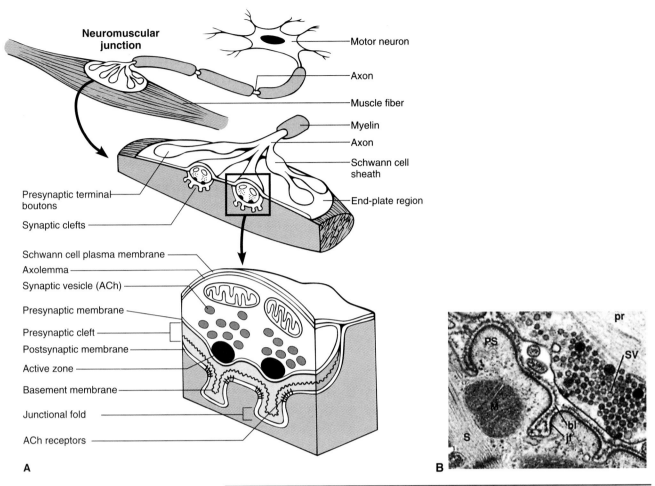

Neuromuscular junction

Motor neuron

Axon

Muscle fiber

Myelin

Axon

Schwann cell sheath

End-plate region

Presynaptic terminal boutons

Synaptic clefts

Schwann cell plasma membrane

Axolemma

Synaptic vesicle (ACh)

Presynaptic membrane

Presynaptic cleft

Postsynaptic membrane

Active zone

Basement membrane

Junctional fold

ACh receptors

A

B

***Figure 7–22.* A neuromuscular junction. A:** Progressive enlargements of a neuro-muscular junction are depicted in a drawing of a single axon (multiple branches are not shown) terminal synapsing at the endplate of the muscle fiber membrane. The presynaptic terminal (*pr*) consists of a splay of terminal varicosities, also termed *synaptic boutons.* Although the terminal loses its myelin, a thin Schwann cell membrane covering remains. Separating the presynaptic membrane from the postsynaptic (*Ps*) cell is a narrow synaptic cleft about 50 nm wide. Within the cleft is a prominent basal lamina. Terminal boutons overlie regions at which the postsy-naptic junctional folds (*jf*) occur extensively and concentrate acetylcholine recep-tors. Within each bouton of the presynaptic terminal are clusters of synaptic vesi-cles (*SV*). Many of these vesicles are docked at active zones, appearing as electron densities. **B:** An electron micrograph of a similar neuromuscular junction reveals many of these characteristic features. Although active zones are not prominent, note the distinct basal lamina (*bl*) in the synaptic cleft. Acetylcholinesterase, which degrades discharged acetylcholine, is secreted by the muscle fiber and bound mainly to heparan sulfate proteoglycans in the basal lamina. Agrin is deposited by the nerve terminal into the basal lamina. (*m*, mitochondria; *s*, sarcomere) (**A,** modi-fied from McMahon UJ, Kuffler SW. *Proc R Soc Lond (Biol)* 1971;177:485; **B,** courtesy of John Heuser.)

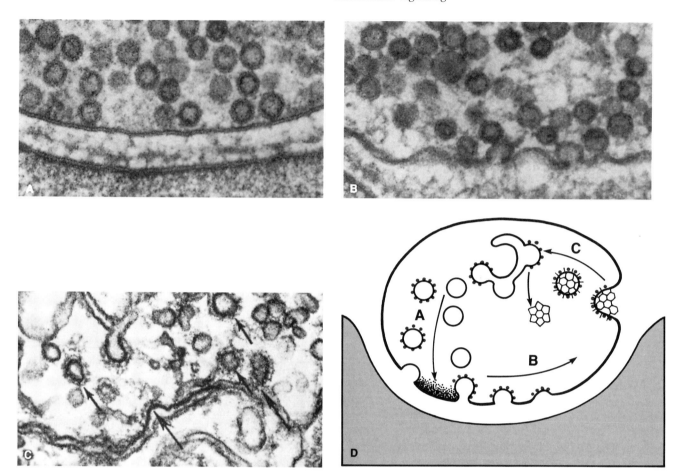

Figure 7–23. Exocytosis and recycling of synaptic vesicles at the presynaptic terminal. For this series of electron micrographs (**A–C**), the frog neuromuscular junction was frozen at successive time intervals following depolarization of the presynaptic terminal. Within 5 msec after the depolarizing stimulus, synaptic vesicles docked at the presynaptic terminal membrane (**A**) discharge their contents into the synaptic cleft and form "omega figures" (**B**), a term for the transient shape of the vesicles fusing with the plasmalemma. As early as 10 seconds after depolarization (**C**), the first examples of endocytosed vesicles (*long arrows*) budding from the presynaptic terminal membrane are identified by a clathrin coat (*short arrows*). The process of exocytosis and recycling at the presynaptic terminal of the neuromuscular junction is summarized schematically in **D**. (A–C, Heuser JF, Reese TS. *J Cell Biol* 88:564,1981; D, modified from Heuser JE, Reese TS. *J Cell Biol* 57:315,1973.)

ber of vesicles undergoing exocytosis by as much as 10^4-fold above the level in quiescent terminals. Following exocytosis, synaptic vesicle membrane proteins are believed to remain concentrated at sites of the presynaptic membrane and are rapidly retrieved by endocytosis to form a recycled vesicle (see Fig. 7–23).

Multiple Synaptic Inputs Converge on a Single Neuron to Decide the Output of an Action Potential

Quite unlike the single synaptic input to a muscle fiber is the common situation where thousands of presynaptic terminals converge to synapse on a single neuron (Fig. 7–24). The postsynaptic neuron must integrate all of these inputs to determine whether an output is further relayed in the form of an action potential. At each synapse, the opening of ion-selective channels and resulting ion influx causes a characteristic local change in the resting membrane potential of the postsynaptic cell, termed a *postsynaptic potential* (PSP). This is analogous to the endplate potential formed at the postsynaptic skeletal muscle membrane, except that the result can be excitatory, resulting from depolarization (e.g., channels open to Na^+), or inhibitory, due to hyperpolarization (e.g., channels open specifically to Cl^-). Excitatory postsynaptic potentials (EPSPs) and inhibitory postsynaptic potentials (IPSPs) vary in size and duration for a given target

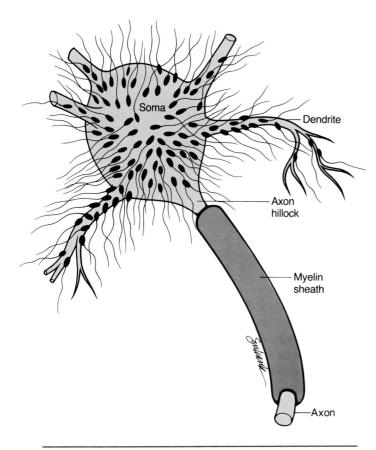

where the net effect of all synaptic inputs determines whether the output of an action potential will occur, because this is where voltage-sensitive Na+ channels are first encountered in sufficient quantity as electrical signals move toward the output pole of the neuron.

Adhesion Molecules Have Signaling Function

Over the past decade, it has become clear that many adhesion molecules are multifunctional, activating through physical cell–cell and cell–matrix interactions signal transduction pathways affecting gene transcription (Fig. 7–26). For example, members of the Src family of nonreceptor tyrosine kinases (Chapter 8) associate with the cytoplasmic domains of specific cell adhesion molecules (CAMs) in the immunoglobulin superfamily. The activation of such kinases appears instrumental in mediating key recognition events in neural development, e.g., CAM-mediated neurite outgrowth. Likewise, integrin receptors for extracellular matrix components have also been associated with tyrosine kinase-mediated signal transduction. Signaling mediated by CAMs may also converge on pathways regulated by growth factors. Both CAM- and integrin-mediated signaling has been linked to the influx of Ca^{2+} functioning as a potential second messenger.

Figure 7–24. Axon terminals synapsing on a motor neuron. Distal portions of axonal terminals and their boutons appear (*black*) as they synapse on the cell body of a single large motor neuron, the dendrites and axon of which are partially visible. (Modified from Hagger RA, Barr ML. *J Comp Neurol* 1950;93:17.)

Clinical Case Discussion

Myasthenia gravis is characterized by easy fatigability of the muscles innervated by the brain stem, in particular, the extraocular muscles and those of mastication, swallowing, and respiration. Girls are affected more often than boys (3.5:1). The age of onset is usually greater than 10 years of age. There also exist both neonatal (transient; infant born to a myasthenic mother) and congenital (persistent) forms. Myasthenia gravis is an autoimmune disorder involving thymic dysfunction and the production of abnormal T cells. A circulating antibody is produced in the serum that binds to the acetylcholine receptor protein and reduces the number of motor end plates available for binding acetylcholine. This receptor controls a ligand-gated ion channel at the neuromuscular junction that controls muscle contractions.

The most prominent clinical symptoms are difficulty with chewing, dysphagia, and a nasal voice. The most common presenting symptom is ptosis. Patients can also have difficulty with respirations and handling their secretions. A consistent finding on physical examination is fatigue of muscle groups after repetitive stimulation, such as squeezing the examiner's hand or raising and lowering the eyelids for a prolonged period of time, which causes an exaggeration of the ptosis. Both of these abnormalities are corrected by the administra-

neuron and are said to be "graded" according to the strength of presynaptic stimulus. Because the dendrites and cell body of the neuron are essentially devoid of voltage-sensitive Na+ channels, these structures will not generate an action potential. Instead, all of the PSPs arising from synaptic inputs spread as local potentials by current diffusion (see Chapter 2) to the cell body, where they are summed to form a grand postsynaptic potential (GPSP). Because local potentials decay in amplitude over distance, unlike action potentials, the contribution of a given PSP to the GPSP becomes progressively greater with increasing proximity of the synapse that generates it to the cell body. In addition, a high frequency of action potentials arriving at the presynaptic terminal will also increase the open time of postsynaptic receptor channels and, thereby, the magnitude of the corresponding PSP. In general, when excitatory inputs predominate, the net effect is depolarization at the cell body. Alternatively, an overbalance of inhibitory inputs will cause hyperpolarization and will inhibit the occurrence of an action potential (Fig. 7–25). The base of the axon, or axon hillock, is

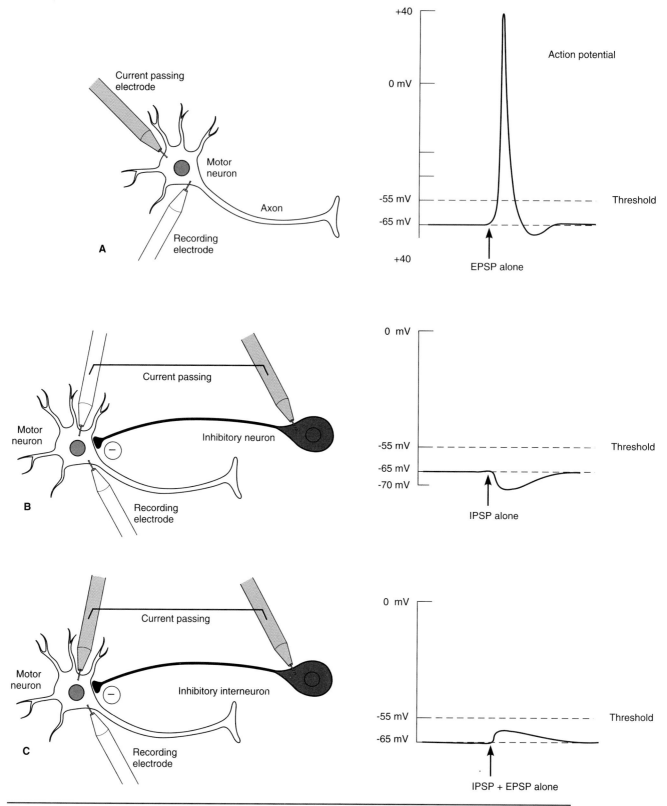

Figure 7–25. The action of an inhibitory chemical synapse counteracts depolarization leading to an action potential. **A:** A recording electrode is used to measure the membrane potential of the postsynaptic neuron artificially depolarized with a current-passing electrode. If an EPSP surpasses threshold in the depolarized cell, an action potential is triggered. **B:** An IPSP produced in the motor neuron by an inhibitory interneuron; the inhibitory interneuron can hyperpolarize the motor neuron, rendering it incapable of initiating an action potential. **C:** Simultaneous depolarization of the inhibitory interneuron and artificial depolarization of the motor neuron result in an additive IPSP and EPSP that, in essence, cancel each other. (Modified from Kandel ER, Schwartz JH. In: Kandel ER, Schwartz JH, Jessell TM, eds. *Principles of Neural Science*, 3rd ed. Norwalk, CT: Appleton & Lange, 1991.)

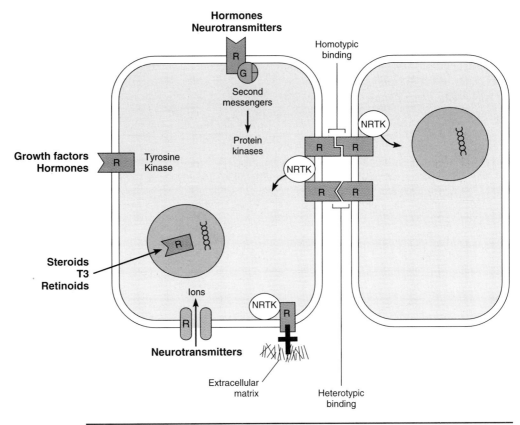

Figure 7–26. Adhesion molecules included among the major categories of receptors (R) engaged in intercellular signaling. Homotypic and heterotypic cell–cell adhesion mediated by CAMs and cell–matrix adhesion mediated by integrins can activate associated nonreceptor tyrosine kinases (NRTK) in the cytoplasm (see chapter 8).

tion of edrophonium (Tensilon, Enlon, Reversol) in 92% of patients with a diagnosis of myasthenia gravis. Nerve conduction studies are usually normal. By contrast, repetitive stimulation of motor nerves at a slow rate shows progressive decrease in the muscle potential in myasthenia patients. Treatment includes the administration of the acetylcholinesterase antagonists pyridostigmine (Mestinon, Regonol) or neostigmine (Prostigmin), immunosuppression, and plasma exchange to remove the circulating antibodies. Thymectomy results in clinical improvement.

Suggested Readings

Hormones and Endocrine Signaling

Norman AW, Litwack G. *Hormones*. New York: Academic Press, 1987.

Synder SH. The molecular basis of communication between cells. *Sci Am* 1985;253(4):132.

Paracrine Communication

Barde Y-A. Trophic factor and neuronal survival. *Neuron* 1989;2:1525.

Loughlin SE, Fallon JH, eds. *Neurotrophic Factors*. San Diego: Raven Press, 1993.

Synaptic Transmission

Alberts B, Bray J, Lewis J, et al. *Molecular Biology of the Cell*, 3rd ed. New York: Garland Publishing, 1994.

Siegel G, Agranoff B, Alberts RW, et al., eds. *Basic Neurochemistry*, 5th ed. New York: Raven Press, 1994.

Review Questions

1. Which is not a property of steroids?
   a. Released from endocrine cells by exocytosis
   b. Bound to carrier protein in plasma and cytosolic receptors
   c. Released directly from the hypothalamus

    d. Diffuse across the plasma membrane of target cells

    e. Derived from cholesterol

2. Neuroendocrine cells:

    a. Do not synapse with neurons

    b. In the pineal gland secrete melatonin in response to activation of catecholamine receptors

    c. Release hormones near capillaries distinguished by the lack of tight junctions

    d. Classified as magnocellular neurons project directly into the anterior pituitary

    e. b and c

3. Nerve growth factor:

    a. Activates a receptor tyrosine kinase

    b. Is produces by target cells of innervation

    c. Is not neurotrophic for all neurons

    d. Suppresses programmed cell death

    e. All of the above are correct

4. Chemical synaptic transmission:

    a. Mediated by GABA produces EPSPs

    b. Is faster than electrical synaptic transmission

    c. Involves the exocytosis of Substance P from secretory vesicles

    d. Is mediated by norepinephrine-gated ion channel receptors

    e. Mediated by glutamate and GABA is inactivated by enzymatic degradation of transmitters in the synaptic cleft

5. Nitric oxide:

    a. Is a retrograde neurotransmitter

    b. Evokes vasodilation by inactivating soluble guanylyl cyclases

    c. Is exocytosed

    d. Causing vasodilation is produced by nitric oxide synthase present in endothelial cells

    e. a and d

6. All of the following processes occur at the neuromuscular junction **EXCEPT**:

    a. Depolarizations of the postsynaptic membrane at rest

    b. Endocytosis of synaptic vesicle membrane at the presynaptic terminal

    c. Inhibitory postsynaptic potentials

    d. a and c

    e. b and c

7. All of the following are true regarding myasthenia gravis **EXCEPT**:

    a. Involves abnormal T cell function

    b. It is caused by autoantibodies to the acetylcholine receptor protein

    c. The symptoms can be treated with acetylcholinesterase antagonists

    d. Nerve conduction velocity is abnormal

    e. Acetylcholine is unable to bind to its receptor and stimulate muscle contraction

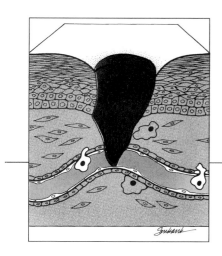

Chapter 8
Signal Transduction Events

Clinical Case

A 4-month-old infant, Linus Powell, developed fever, cough, and difficulty feeding for about 2 weeks. Over the last 2 days, the cough has become more severe, and the infant is unable to catch his breath at the end of a coughing episode. He also turns blue while coughing and sometimes vomits. He has received his first set of routine immunizations, which included pertussis. Dad had a mild cold about 1 week ago, which is better now. On examination, the infant appears very agitated and moderately dehydrated. Auscultation of the chest revealed diffuse wheezes and rales; a respiratory rate of 50 and cyanosis after a coughing paroxysm was observed. The transcutaneous oxygen saturation was 92%. He was unable to feed, due to frequent coughing. His white blood cell count was 30,000/mm³ with 80% lymphocytes. A nasopharyngeal swab was obtained to culture for *Bordetella pertussis*. A chest x-ray showed perihilar infiltrates and patchy atelectasis. The infant was admitted to the hospital for intravenous hydration and antibiotic therapy.

Cells can communicate with each other through physical contact, secreted molecules, or gap junctions, using mechanisms discussed in previous chapters. With the exception of direct communication by gap junctions, cell communication requires that a signaling molecule interact with a specific receptor molecule.

The major intercellular signaling mechanisms—endocrine, paracrine, and chemical synaptic—are classified according to the distances over which the messenger molecule must travel between source and target cells (see Chapter 7). To receive and respond to these signals, all target cells for a given hormone, growth factor, or neurotransmitter contain specific receptors that bind the messenger to provoke a biologic effect inside the cell. Most often, these receptors are transmembrane proteins, although steroid hormones enter directly into the target cells and bind to intracellular protein receptors. The binding to a receptor by a *ligand*, a generic term for the bound messenger, causes the protein receptor molecule to undergo a conformational change. This sets off a complex series of reactions inside the cell that results in cellular activation. Importantly, many drugs exert their effects in the body by acting as specific ligands that activate or block the same receptors that are activated by endogenous signals. In this chapter, the various mechanisms by which an activated receptor converts a signal into cellular activity will be considered.

Steroid Hormone Receptor Mechanisms

Steroid Hormones and Other Lipophilic Signaling Molecules Act by Binding to Intracellular Receptors

A few hormones exert their activity by diffusing across the plasma membrane and binding to receptors that are located either in the cytoplasm or nucleus of the target cells. After binding to the receptors, the activated hormone–receptor complex then binds to specific regions of DNA to regulate the transcription of various genes. Hormones that work in this fashion are relatively small, hydrophobic molecules that can readily diffuse through lipid bilayers. The most intensively studied examples are the steroid hormones, which are all derivatives of cholesterol (see Chapter 7). Although not steroidlike in structure, the thyroid hormones and retinoids also diffuse through the plasma membrane and bind to intracellular receptors during cell signaling.

Like all signaling processes, steroid hormone activation requires the presence of appropriate receptors in target cells. Molecular cloning studies have demonstrated that steroid hormone receptors all have very similar structures. Steroid hormone receptors are composed of three domains: a COOH-terminus domain that contains the hormone-binding site; an NH₂-terminus domain that is involved in the activation of gene transcription; and a middle domain that contains the DNA-binding site (Fig. 8–1). According to current thought, the resting, or inactive, steroid hormone receptor contains an inhibitor protein, called *Hsp90*, bound to the DNA-binding site of the molecule. When a steroid hormone binds at the COOH-terminal hormone-binding region, the steroid receptor undergoes a conformational change, and the Hsp90 inhibitor protein dissociates from the DNA-binding domain. The activated receptor then dimerizes and the complex binds to specific DNA nucleotide sequences and regulates the transcription of the adjacent genes. The activity of steroid hormones is relatively long-lived, generally lasting from hours to days.

Signaling Mechanisms Utilizing Water-Soluble Molecules

Water-Soluble Signaling Molecules Bind to Cell Surface Receptors

Steroids and steroid-like molecules are special classes of signaling molecules in that they are able to cross lipid bilayers and enter the cytosol. Most signaling molecules, such as hormones, growth factors, and neurotransmitters, must bind to surface receptor proteins on their appropriate target cells to modify the behavior of those

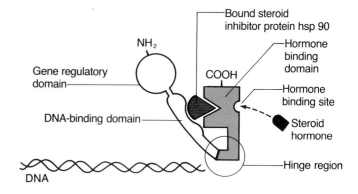

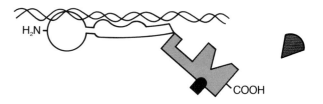

Figure 8–1. A schematic representation of a steroid hormone receptor and receptor activation. The steroid hormone receptor is composed of three domains: a COOH-terminal hormone-binding region, an NH₂-terminal gene regulatory domain, and a domain in the middle of the molecule that binds to DNA. In the inactive state, an inhibitor protein called *Hsp 90* is bound at the DNA-binding region. Hormone binding induces a conformational change in the receptor that causes the inhibitor protein to disassociate from the DNA-binding region. This exposes the nucleotide-specific DNA-binding site and allows receptor dimerization to occur. The hormone–receptor complex then binds to a specific nucleotide sequence of the DNA and regulates the transcription of the adjacent genes. (Modified from Alberts B, Bray D, Lewis J, et al. *Molecular Biology of the Cell*, 2nd ed. New York: Garland Publishing, 1989.)

cells. Therefore, most signaling molecules do not modulate cellular behavior by the direct regulation of gene expression, as do the steroid hormones. Instead, cellular activation by water-soluble signaling molecules occurs by transmission of the signal across the plasma membrane. In many instances, the signaling molecule induces the production or release of secondary messenger molecules inside the cell. These secondary messengers then act on various cytoplasmic proteins, eventually resulting in the activation or repression of the target cell.

Cell surface receptors generally are divided into one of three classes. The first are the ion channel-linked receptors, an example of which is the acetylcholine receptor described previously. This type of receptor functions by changing the ion permeability of the cell membrane following ligand binding. The second type is the *enzymatic,* or *catalytic,* receptor. Most catalytic receptors are transmembrane proteins that contain a tyrosine kinase domain on the cytoplasmic surface of the plasma membrane. This type of receptor activates a cell directly by phosphorylating intracellular proteins on tyrosine residues following activation of the receptor by ligand. Included in this class are the insulin receptor and the receptors for several growth factors. The final class of receptor is the G protein-coupled receptor. After ligand binding, the activated receptor binds to a protein on the cytoplasmic surface of the plasma membrane called a *G protein*. G proteins are so called because these proteins bind to and cleave GTP following receptor activation. The binding of GTP allows the G protein to initiate a cascade of intracellular enzymatic events that result in the modulation of intracellular activity. The remainder of this section will be devoted to describing how these processes occur.

Ion Channel-Linked Receptors

This class of receptors is designed simplistically for the rapid conversion of the chemical energy of fast-acting neurotransmitter molecules back into electrical signals at the postsynaptic cell. An ion-selective aqueous channel that spans the lipid bilayer is formed by the polypeptide subunits of the transmembrane receptor molecule itself. The receptor operates by binding, at an external site, a ligand neurotransmitter that causes a transient allosteric change and opening of the ion channel (Fig. 8–2). As a result, and depending on the specificity of the receptor, selected cations (K^+, Na^+, Ca^{2+}) or the Cl^- anion will flow through the opened channel down their individual electrochemical gradients to alter the membrane potential and, thus, the excitability of the cell. Most often identified by their location on dendrites and at the neuromuscular junction, ion channel-linked receptors are responsible for the local potentials that, if sufficient, open voltage-gated ion channels along the axon and muscle fiber to trigger an action potential (Fig. 8–3).

Many Ion Channel-Linked Receptors Share a Common Structure

Ion channel-linked receptors respond either to the major, fast-acting excitatory neurotransmitters, glutamate and acetylcholine, or to the principal inhibitory neurotransmitters, γ-aminobutyric acid (GABA) and glycine (Fig. 8–4). Of these, the receptors for acetylcholine, GABA, and glycine are closely related structurally and have evolved from a common ancestral gene. Each receptor consists of a tetramer or pentamer with different types of subunits that vary from 50 to 60 kDa in size. Of these, the prototype and best-understood example is the acetylcholine receptor, located in the postsynaptic folds of the skeletal neuromuscular junction (see Chapters 2 and 7). These receptors mediate the fast excitatory action of spinal nerves and cer-

Figure 8–2. Gating of the acetylcholine receptor-linked ion channel by acetylcholine released at the neuromuscular junction. The acetylcholine receptor, which is located in the plasma membrane of the muscle cell end plate at the neuromuscular junction, is gated by binding acetylcholine released from the presynaptic terminal.

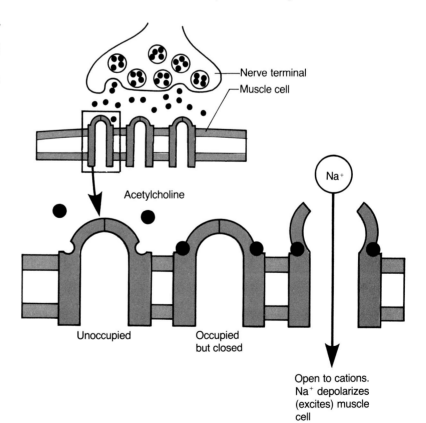

Nerve terminal
Muscle cell

Acetylcholine

Na⁺

Unoccupied

Occupied but closed

Open to cations. Na⁺ depolarizes (excites) muscle cell

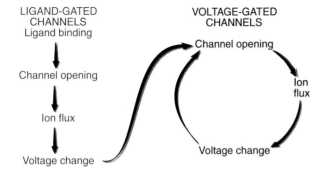

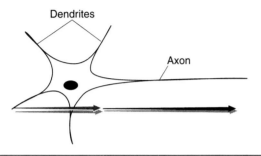

Figure 8–3. Flow of information initiated by opening ligand-gated ion channels. The binding of a neurotransmitter to ligand-gated channels at the postsynaptic membrane initiates a passive local potential that, if sufficient, triggers voltage-gated channel opening at the axon hillock and the regenerative action potential.

tain cranial nerves, and are termed *nicotinic* because they bind nicotine, a potent receptor-activating drug. Nicotinic acetylcholine receptors differ from a second class of slow-acting *muscarinic* acetylcholine receptors (they selectively bind the drug muscarine) that can be either inhibitory or excitatory, depending on their loca-

tion. Indeed, most receptors for neurotransmitters come in several subtypes to mediate different effects of a common messenger released at different locations.

The binding of one acetylcholine molecule to each of two identical α-subunits of the receptor protein shifts the nicotinic receptor from a closed state to an open one that lasts as long as the ligand remains bound (Fig. 8–5). When open, the channel is nonselectively permeable to all cations (Na^+, K^+, Ca^{2+}) but excludes anions. Despite the indiscriminate permeability to cations, opened acetylcholine receptor channels cause membrane depolarization primarily by enhanced Na^+ influx. To appreciate why, recall that the voltage gradient and concentration gradient both act in the same direction to force Na^+ into the cell. On the other hand, the voltage gradient essentially counterbalances the concentration gradient of K^+ across the membrane, resulting in a net driving force for K^+ influx of near zero. Although the situation for Ca^{2+} resembles that of Na^+, Ca^{2+} contributes little toward the inward current because its extracellular concentration is significantly less than that of Na^+. Increased cytosolic Ca^{2+} functions in a nonelectrical manner as an important second messenger to couple surface receptor activation with a variety of Ca^{2+}-sensitive biochemical responses.

Ion Channel-Linked Receptors Are Affected by Many Neurotoxins and Psychoactive Drugs

Ion channel-linked receptors are often the site of action for neurotoxins and drugs that exert their effect by binding to and modifying the allosteric control site of an ion channel. The term *agonist* is applied to a ligand that biases a receptor to open state, as nicotine and muscarine do to their respective acetylcholine receptors. Receptor blockers, termed *antagonists*, can also differ-

Figure 8–4. Chemical structures of the principal excitatory and inhibitory neurotransmitters that open ion channel-linked receptors.

EXCITATORY

$$H_3C - \overset{\overset{\displaystyle O}{\|}}{C} - O - CH_2 - CH_2 - \overset{+}{N} - (CH_3)_3$$

acetylcholine

$$^+H_3N - \underset{\underset{\underset{\underset{COO^-}{|}}{CH_2}}{|}}{\overset{|}{CH}} - COO^-$$

glutamate

INHIBITORY

$$^+H_3N - CH_2 - CH_2 - CH_2 - COO^-$$

γ-aminobutyrate (GABA)

$$^+H_3N - CH_2 - COO^-$$

glycine

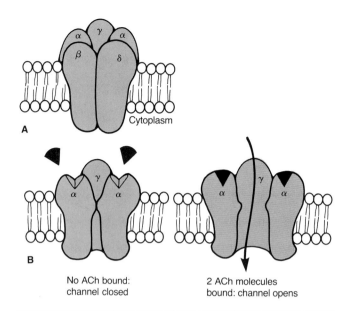

A

Cytoplasm

B

No ACh bound: channel closed

2 ACh molecules bound: channel opens

Figure 8–5. **Model of the nicotinic acetylcholine receptor in three dimensions. A:** The receptor channel consists of five transmembrane subunits ($\alpha_2;\beta_2\gamma$) surrounding a central pore. **B:** The cut-away models show how the binding of two acetylcholine molecules, one to each α-subunit, changes the receptor conformation to open the ion pore. (Modified from Kandel E, Schwartz J, Jessell T, eds. *Principles of Neural Science*, 3rd ed. Norwalk, CT: Appleton & Lange, 1991.)

these, the kainate and quisqualate-A receptors account for the major excitatory action of glutamate on motor neurons; they derive their names from the experimental drugs that selectively bind and activate them. Both receptors directly gate a channel that is selectively permeable to Na^+ and K^+, but not to Ca^{2+} (Fig. 8–7). A different quisqualate receptor (quisqualate-B) is not directly gated to an ion channel but activates a phosphoinositide-linked second-messenger pathway (discussed later).

Glutamate binds to yet another directly gated receptor type, the NMDA receptor (named for its selective activation by the agonist drug N-methyl-D-aspartate). The NMDA receptor is unique among glutamate receptor types because it is permeable to Ca^{2+} as well as to K^+ and Na^+. Moreover, the NMDA receptor is plugged by extracellular Mg^{2+} at the normal membrane resting potential. Membrane depolarization must be sufficiently large to displace Mg^{2+} from these channels, thereby allowing Na^+ and Ca^{2+} to enter the cell (Fig. 8–7). The NMDA receptor type is also distinguished pharmacologically as the only glutamate receptor that is inhibited by the hallucinogenic drug, phencyclidine (PCP, "angel dust").

The NMDA-activated channel is the only known example of a receptor-gated channel that is maximally opened by the ligand (glutamate) binding combined with cell depolarization (i.e., voltage gating). By allowing Ca^{2+} entry, activated NMDA receptors are believed

entiate between receptor subtypes. For example, the paralytic effects of α-bungarotoxin in cobra venom and curare at the tip of a poison dart are directly attributed to the blockage of nicotinic acetylcholine receptors at the skeletal neuromuscular junction. On the other hand, the nerve gas antidote, atropine, selectively acts as a muscarinic acetylcholine receptor blocker.

Receptors that bind GABA, a major inhibitory neurotransmitter in brain and spinal cord, are also sites of significant drug action. A conformational change resulting from the binding of GABA at the α-subunit of the receptor protein opens a Cl^--selective channel centered within the receptor. The well-characterized $GABA_A$ receptor subtype also has separate extracellular binding sites for benzodiazepines, a class of antianxiety medications, skeletal muscle relaxant drugs such as diazepam (Valium) and chlordiazepoxide (Librium), and barbiturates (Fig. 8–6). The binding of either GABA, benzodiazepine, or barbiturate to the $GABA_A$ receptor will cooperatively increase the receptor affinity for the other two agents. Accordingly, these drugs are thought to act by decreasing the amount of GABA needed to open the receptor Cl^- channel, thereby enhancing many of GABA's inhibitory properties and resulting in overt changes in behavior.

Glutamate, a major excitatory neurotransmitter in the mammalian CNS, binds to four different glutamate receptors, three of which are gated ion channels. Of

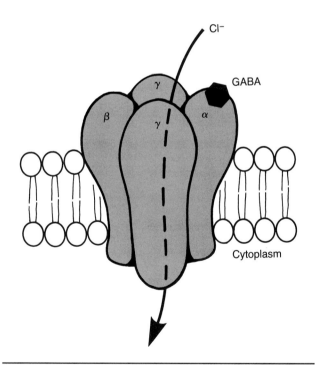

Figure 8–6. **Three-dimensional model of the GABA receptor.** The channel consists of four subunits ($\alpha\beta\gamma_2$) surrounding a ligand-gated channel that is selective for Cl^-. GABA binds all subunits, but with greatest affinity to the α-subunit. Separate binding sites exist for barbiturates (the α- and β-subunits) and benzodiazepines (both γ-subunits).

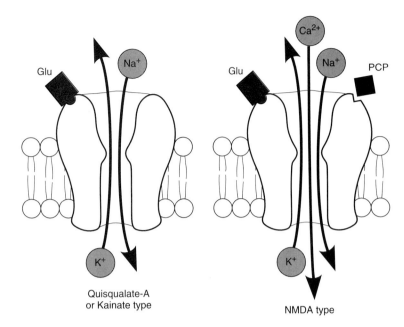

Quisqualate-A
or Kainate type

NMDA type

Figure 8–7. Glutamate receptor types. Kainate and quisqualate-A receptors bind the glutamate agonist AMPA and open a channel permeable to Na^+ and K^+. The NMDA receptor binds PCP and gates a channel permeable to Ca^{2+}, as well as to Na^+ and K^+. The quisqualate-B receptor (not shown) is not directly linked to an ion channel. Instead, it stimulates the phosphoinositide second-messenger pathway and the formation of inositol 1, 4, 5-trisphosphate (IP_3) and diacylglycerol (DAG), two second messengers discussed later in this chapter. (Modified from Kandel E, Schwartz J, Jessell T, eds. *Principles of Neural Science*, 3rd ed. Norwalk, CT: Appleton & Lange, 1991.)

to initiate intracellular Ca^{2+}-dependent signal transduction events that contribute not only to long-term changes in synaptic function but also to glutamate toxicity when Ca^{2+} influx is excessive. This predicament is hypothesized to promote the neuronal degeneration that occurs following continuous seizure activity.

Catalytic Receptors/Tyrosine Kinases

Catalytic Receptors Contain Three Distinct Subregions

Catalytic receptors are particularly important for sensing environmental signals that influence cell growth and tumor formation. They are single-pass transmembrane proteins that usually become enzymatic when activated by binding an extracellular signal. Serving as receptors for peptide hormones (e.g., insulin) and many growth factors, such as fibroblast growth factor, platelet-derived growth factor, and epidermal growth factor, catalytic receptors have three functional regions: (a) a regulatory ligand-binding portion containing the NH_2-terminus that is exposed to the extracellular fluid; (b) an α-helical domain that spans the lipid bilayer; and (c) an effector region containing the COOH-terminus that protrudes into the cytosol and includes the catalytic domain in which the enzymatic activity resides (Fig. 8–8). Activation of this type of receptor often sets off a signal transduction cascade inside of the target cell that eventually leads to the nucleus and results in either activation or repression of gene transcription. The major class of catalytic receptor is the tyrosine kinase family receptors, although other examples of catalytic receptors are known (Fig. 8–8).

The Activity of a Cellular Protein Is Regulated by Phosphorylation of Its Serine, Threonine, or Tyrosine Residues

Before considering the methods of signal transduction employed by catalytic receptors, it first will be necessary to describe the regulation of protein activity by phosphorylation and dephosphorylation. One of the principal mechanisms that cells use for modulating the activity of a protein is to regulate the phosphorylation state of the protein. For example, the activity of a protein can be turned on (or off) by attaching a highly charged phosphate group to the polypeptide; the level of activity of that protein then can be returned to its resting state by removing that phosphate group (Fig. 8–9). The enzymes that are responsible for the catalytic transfer of a phosphate group from ATP to a target protein are called *kinases*. In most instances (>99%), a kinase will couple a phosphate group to either a serine or a threonine on a target protein. Hence, these types of enzymes are called *serine-threonine kinases*. The remainder of the intracellular phosphorylation events are due to the catalytic coupling of a phosphate group to a tyrosine residue of a target substrate by enzymes called *tyrosine kinases*. As will be discussed in the next section, tyrosine kinases are critical mediators of intracellular signaling. As stated previously, the phosphorylation of a target protein is a transient state. Members of another class of enzymes, the protein phosphatases, remove phosphate groups from phosphorylated substrates and return the level of activity of a protein to its resting state. Proteins whose activities are regulated by phosphorylation and dephosphorylation are involved in numerous physiologic processes in cells and include cytoskeletal com-

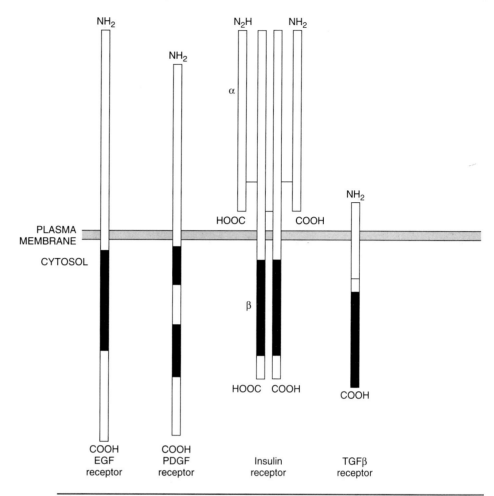

Figure 8–8. **Examples of catalytic receptors and their transmembrane orientation.** The tyrosine kinase receptors for epidermal growth factor (EGF), platelet-derived growth factor (PDGF), and insulin are represented. The catalytic domains for each receptor (*black*) reside in the cytosol and represent homologous sequences of approximately 250 amino acids to suggest that all have evolved from a single ancestral gene. (Modified from Alberts B, Bray D, Lewis J, et al. *Molecular Biology of the Cell*, 3rd ed. New York: Garland Publishing, 1994.)

ponents, receptors, ion channels, transcriptional regulators, and other enzymes (List 8–1).

Protein Kinases Have Regulatory and Catalytic Regions

Although broadly classified as being either serine-threonine or tyrosine kinases, all protein kinases are related by certain structural and functional properties. Each contains one or more regulatory and catalytic domains, and the modulation of the regulatory domain is essential for enzymatic activity at the catalytic site. However, the molecular mechanism leading to activation differs among various kinase classes. For example, the cyclic adenosine monophosphate (cAMP)-dependent protein kinase, or A-kinase, is an inactive holoenzyme that exists as a tetramer comprising two heterodimers, each heterodimer containing a regulatory and a catalytic subunit (Fig. 8–10). The activating ligand, cAMP, triggers the kinase by binding to the regulatory subunits and altering their conformation. This allosteric change releases the catalytic subunits in active form. By comparison, other cytoplasmic protein kinases contain a single regulatory and catalytic domain as part of the same polypeptide chain (see Fig. 8–10). In these enzymes, the binding of other molecules [e.g., cyclic guanosine monophosphate (cGMP), diacylglycerol, Ca^{2+}–calmodulin] to the regulatory domain is believed to unmask and activate the catalytic region by unfolding this portion of the molecule. Receptor tyrosine kinases also require activation and, in this example, the regulatory domains are extracellular and bind to signaling molecules. The binding of a ligand to the extracellular domain of the receptor tyrosine kinase results in activation of the receptor protein at its intracellular, catalytic domain to carry out phosphorylation in the

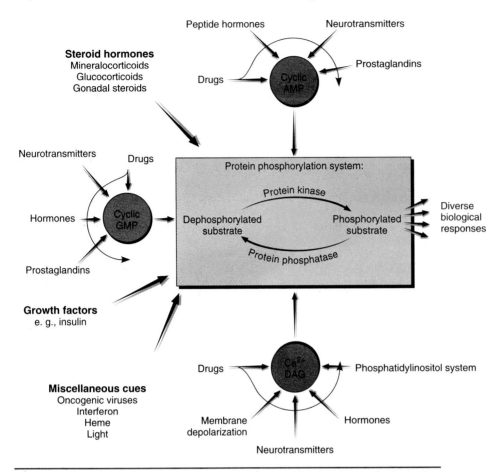

Figure 8–9. Pathways used by cells to convert extracellular signals into biological responses converge on the protein phosphorylation system. In this schematic diagram, various regulatory agents modulate protein phosphorylation by altering levels of a second messenger (*red*), though other agents act directly to regulate protein phosphorylation without second messengers. Most drugs affecting protein phosphorylation (*curved arrows*) influence the capacity of a primary messenger (e.g., neurotransmitter hormone, growth factor, or nerve impulse) to change second-messenger levels. A few drugs [e.g., Ca^{2+}-channel blockers and phosphodiesterase inhibitors, such as caffeine and theophylline (Marax, Quibron, Mudrane, Sio-Phyllin, Theolate)] directly alter second-messenger levels to regulate protein phosphorylation. (Modified from Nestler EJ, Greengard P. *Protein Phosphorylation in the Nervous System*. New York: John Wiley & Sons, 1984.)

cytosol. Finally, in other instances, a kinase may be activated by being phosphorylated by another kinase during a cascade of enzymatic reactions.

Specific Amino Acid Sequences Identify the Phosphorylation Sites of Proteins Targeted by Different Kinases

Catalytic subunits of the major families of protein kinases contain two different binding sites, one for ATP and one that recognizes sites on target substrates. Termed a *phosphorylation sequence*, this region of a target protein differs for each class of protein kinase and includes both residues to be phosphorylated and a characteristic pattern of surrounding arginine residues. For example, a substrate protein containing the sequence Arg-Arg-X-Ser-X (where *X* is any other amino acid) is subject to phosphorylation on the serine residue by cAMP-dependent protein kinase but cannot bind sufficiently to cGMP-dependent protein kinase, which requires the sequence Arg-X-X-Ser-Arg-X-. This selectivity directs a specific kinase not only to the appropriate substrate protein but also to preferred sites among the many phosphorylation sequences that can exist within a given polypeptide.

Regulatory domains of many kinases often contain a pseudosubstrate sequence that binds to and inhibits the catalytic domain of the enzyme. This region of

List 8–1

Neuronal proteins regulated by phosphorylation[a]

Enzymes involved in neurotransmitter biosynthesis

Tyrosine hydroxylase
Tryptophan hydroxylase

Enzymes involved in cyclic nucleotide metabolism

Adenylate cyclase
Guanylate cyclase
Phosphodiesterase

Autophosphorylated protein kinases

cAMP-dependent protein kinase
cGMP-dependent protein kinase
Ca^{2+}-calmodulin-dependent protein kinases
Ca^{2+}-diacylglycerol-dependent protein kinase
    (protein kinase C)
Casein kinases
Protein tyrosine kinases
Double-stranded RNA-dependent protein kinase
Rhodopsin kinase

Phosphate inhibitors

Inhibitor 1
Inhibitor 2
DARPP-32
G substrate

Proteins involved in transcription and translation regulation

RNA polymerase
Histones
Nonhistone nuclear proteins
Ribosomal protein S6
Other ribosomal proteins

Cytoskeletal proteins

MAP-2
Tau
Other microtubule-associated proteins
Neurofilaments

Calspectin
Myosin light chain
Actin
Tubulin

Synaptic vesicle-associated proteins

Synapsin I
Protein III
Clathrin
Synaptophysin

Neurotransmitter receptors

Nicotinic acetylcholine receptor
Muscarinic acetylcholine receptor
β-Adrenergic receptor
α-Adrenergic receptor
GABA-benzodiazepine receptor

Ion channels[b]

Voltage-dependent
    Na^+ channel
    K^+ channel
    Ca^{2+} channel
Ca^{2+}-dependent K^+ channel
Neurotransmitter-dependent
    Nicotinic acetylcholine receptor
    Serotonin-regulated K^+ channel in *Aplysia*
      sensory neurons
    Serotonin-regulated (anomalously rectifying) K^+
      channel in *Aplysia* neuron R15
Na^+ channel in rod outer segments

Miscellaneous

B-50
Rhodopsin
Myelin basic protein
87-kDa protein
100-kDa protein
G proteins

[a]Some of the proteins included are specific to neurons. The others are present in many cell types in addition to neurons and are included because among their multiple functions in the nervous system is the regulation of neuron-specific phenomena. Not included are the many phosphoproteins present in diverse tissues (including brain).
[b]Several of the ion channels listed are physiologically regulated by protein phosphorylation reactions, although it is not known whether such regulation is achieved directly through the phosphorylation of the ion channel or indirectly through the phosphorylation of a modulatory protein that is not part of the ion channel molecule.
(Modified from Nestler EJ, Walaas SI, Greengard P. *Science* 1984;225:1357.)

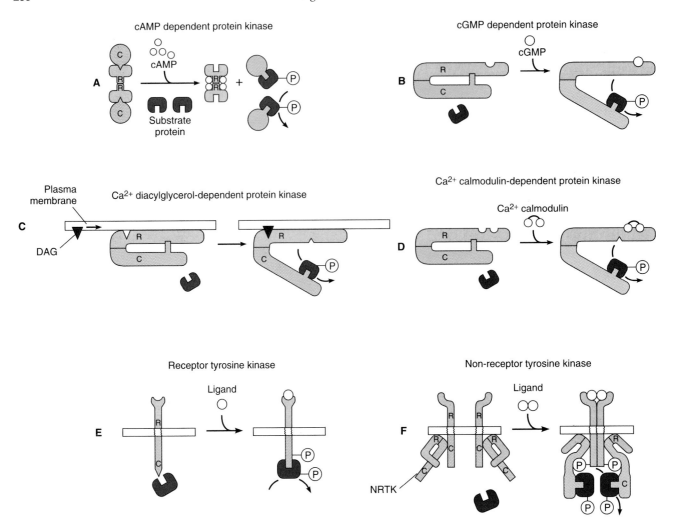

Figure 8–10. **Allosteric changes activate the major protein kinases.** When inactive, the catalytic (C) domains of (**A**) cAMP-dependent protein kinase; (**B**) cGMP-dependent protein kinase; (**C**) Ca^{2+}-/DAG-dependent protein kinase; and (**D**) Ca^{2+}–calmodulin-dependent protein kinase are occupied by a nonphosphorylated region of the enzymes regulatory (R) domains whose amino acid sequences are homologous to the phosphorylation sequences of other substrate proteins. cAMP bound to the two R subunits of A-kinase causes them to dissociate from two C subunits, thereby activating the latter. (**E**) Ligand bound to the extracellular (regulatory) domain of receptor tyrosine kinases causes an allosteric shift that activates the C domain. (**F**) Ligand-induced activation of noncatalytic receptors secondarily activates associated nonreceptor tyrosine kinases (NRTK). Activated kinases transiently associate with the substrate proteins they phosphorylate.

10–20 amino acids resembles the phosphorylation sequence recognized by the catalytic domain of the kinase but contains nonhydroxylated amino acids substituted for serine and threonine residues. Although it cannot be phosphorylated, this region of the kinase can still bind to its own catalytic domain and function as a competitive inhibitor for protein kinase activity (Fig. 8–10). For example, this relation enables the regulatory subunits of cAMP-dependent protein kinase to bind the catalytic subunits, thereby inactivating the enzyme.

Phosphoprotein Phosphatases Inactivate Phosphoproteins

As stated previously, at any instant, the functional status of a phosphoprotein is determined by the proportion of it that exists in phospho and dephospho forms. The steady-state level of protein phosphorylation attained in the cell is balanced by the activities of protein kinases as well as those of phosphoprotein phosphatases that reverse the phosphorylation step (see Fig.

Figure 8–11. The cAMP-dependent protein phosphatase regulation in skeletal muscle. **A:** In the absence of cAMP, the constitutively active cAMP-dependent protein phosphatase shifts the equilibrium of phosphoprotein metabolism to the left, in favor of the dephospho form. **B:** Increased intracellular cAMP shifts the equilibrium far to the right by activating cAMP-dependent protein kinase. The latter, in turn, activates—through phosphorylation—not only key glycogenolytic enzymes (represented as P-protein), but also a phosphatase inhibitor protein.

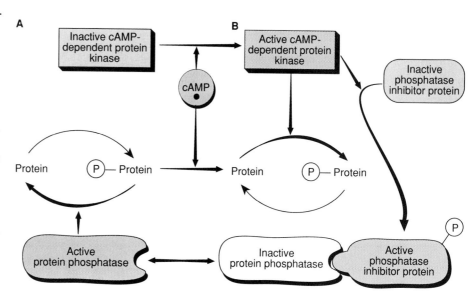

8–9). Protein phosphatases generally have a much broader substrate specificity than do protein kinases. Two major classes of phosphatase are those that dephosphorylate either serine or threonine residues and those that dephosphorylate tyrosine residues, and protein phosphatase activity can be regulated either directly or indirectly. For example, calcineurin, one of four phosphatases now identified in the brain, is directly activated by Ca^{2+} in association with calmodulin. When activated, it dephosphorylates several protein substrates phosphorylated by cAMP-dependent protein kinase. In skeletal muscle, however, phosphatase activity is regulated very differently. There, the phosphatase is constitutively active (Fig. 8–11). However, when the myocyte is activated, a protein called *phosphatase inhibitor protein* is stimulated to bind and inhibit the protein phosphatase. This mechanism will be described in more detail later.

Signal Transduction for a Catalytic Receptor Generally Is Accomplished by Intrinsic Tyrosine Kinase Activity

With this introduction to the modulation of protein activity via phosphorylation, we are now ready to consider in detail the mechanism of signal transduction that occurs following activation of a receptor tyrosine kinase. To convert an extracellular signal to an intracellular one, the binding of a ligand to the regulatory domain of a catalytic receptor must induce activation of the cytoplasmic domain of the receptor molecule. The catalytic receptors achieve this by dimerizing following ligand binding. This dimerization brings the cytoplasmic domains of the catalytic receptor into an arrangement that allows the cytoplasmic portions of the receptor subunits to phosphorylate one another on multiple tyrosine residues (Fig. 8–12). The phosphory-

Figure 8–12. Catalytic receptor-mediated signal transduction across the plasmalemma. When a peptide growth factor binds to the extracellular side of its tyrosine kinase receptor, the activated receptor phosphorylates its own tyrosine residues (*black reaction*). Phosphotyrosinated sites on the receptor serve as docking sites for small SH adaptor proteins via their SH2 domains. SH3 domains allow small SH adaptor proteins to interact with each other.

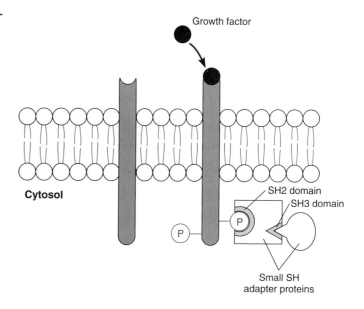

lation of the cytoplasmic subregions of the receptor activates the cell in a novel manner. Specifically, the phosphorylated tyrosines on the cytoplasmic portion of the receptor serve as docking sites for intracellular proteins that play a key role in the subsequent relay of the activating signal inside of the cell. Different docking proteins recognize different phosphorylated sites on the activated receptor tyrosine kinase. In this manner, multiple signal transduction pathways can be activated inside of a cell, adding to the complexity of the cellular response.

Molecular studies have demonstrated that all docking proteins contain conserved regions, called *SH2* and *SH3 domains*, that are essential for recognition of phosphotyrosines on the activated receptor tyrosine kinase and allow binding of the docking protein to the cytoplasmic portion of the receptor. In many instances,

the docking protein itself is activated via phosphorylation by the kinase activity of the receptor, resulting in signal transduction. In other instances, however, the docking proteins serve as anchoring sites that allow the formation of protein complexes on the cytoplasmic side of the activated receptor, thereby allowing either phosphorylation of the protein that binds to the docking protein by the receptor tyrosine kinase or activation of the recruited protein in some other manner. An example of how this might occur is the activation of an important signal relay molecule called *ras*. Ras is a GTPase that helps to relay signals from the activated receptor tyrosine kinase to the nucleus to regulate gene transcription following binding of a growth factor to the receptor. Ras is unlike the other signaling GTPases, called *G proteins* (described later in this chapter), in that ras is a monomer rather than a trimer. However,

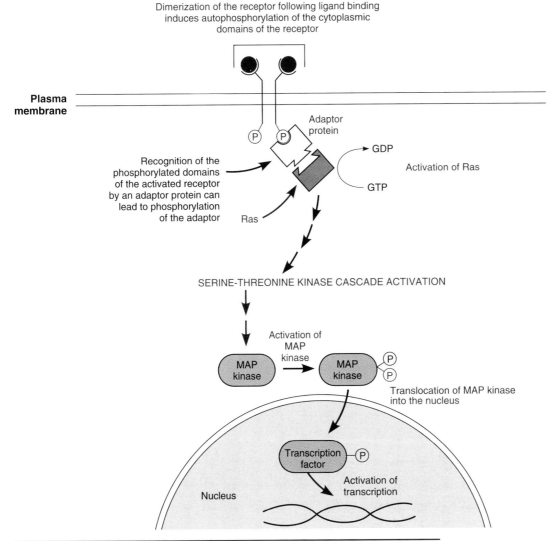

Figure 8–13. Ras activity triggers an enzymatic pathway to the nucleus. A schematic showing the activation of transcription that occurs following activation of a catalytic receptor. See text for details.

ras is similar to the trimeric G proteins in that it serves as a molecular switch that is inactive when bound to GDP but becomes activated upon binding to GTP. The binding of ras to an activated docking protein triggers ras to exchange GTP for bound GDP at its guanyl nucleotide binding site, resulting in the activation of ras. Once activated, ras initiates a cascade of kinase reactions inside the cytoplasm that eventually results in gene activation.

Figure 8–13 summarizes the events that result in ras activation and the signaling events leading to the nucleus that occur under the direction of ras. As stated previously, the tyrosine kinases dimerize and undergo an autophosphorylation process following ligand binding. The phosphotyrosine residues on the cytoplasmic domain of the activated receptor are recognized by a number of intracellular proteins, and the proteins that bind to the phosphotyrosine residues vary from cell to cell. Regardless, the phosphotyrosines on the activated receptor allow the transient formation of a protein complex on the cytoplasmic surface of the plasma membrane. Once this complex forms, some of the protein subunits of the complex will be phosphorylated on tyrosine residues by the tyrosine kinase receptor and become activated. The activated proteins then trigger the exchange of GDP for GTP by the ras protein. Ras then sets off a signal cascade to the nucleus by activating a family of serine-threonine kinases. The final enzyme involved in this cascade is a protein called *mitogen-activated protein kinase* (MAPK). Upon phosphorylation by an upstream kinase, MAP kinase translocates to the nucleus, where it phosphorylates various transcription factors. The transcription factors then transcribe the genes that will allow either cell proliferation or cell differentiation, depending on the nature of the signaling molecule that is acting at the cell surface.

The Src and Jak-Tyk Subfamilies of Nonreceptor Tyrosine Kinases Propagate Signals Mediated by Cytokines and Some Hormones

In addition to their critical role as receptor tyrosine kinases, a number of additional cytosolic tyrosine kinases have been described. Two major classes of nonreceptor tyrosine kinases are the Src and Jak (Janus kinase)/Tyk kinases. Members of the Src tyrosine kinase family localize to the cytoplasmic face of the plasma membrane, where they interact with activated receptors via SH2 and SH3 domains located within the protein. The Src proteins are then activated by phosphorylation, and the activated kinase then phosphorylates various target substrates on tyrosine residues. Similar mechanisms of activation and action are visualized for the cytosolic Jak kinases.

The Requirement of Multiple Growth Factor Families for Cellular Development Reflects the Activation of Different Signal Transduction Pathways

Intensive investigation concludes that multiple growth factors employing distinct receptor-mediated signal transduction pathways collaborate to promote cellular development. This is exemplified in the nervous system by comparing the signaling pathways utilized by the prototypic nerve growth factors, NGF and CNTF (see Chapter 7). Signaling by NGF (Fig. 8–14) and related neurotrophins is initiated by activation of the specific tyrosine kinase NGF receptor (Trk), which dimerizes and autophosphorylates. This indirectly activates the serine-threonine kinase, Raf-1 (also termed *MAPK kinase–kinase*), by a ras-mediated mechanism. Activated Raf-1 then initiates the cytosolic MAPK cascade (discussed earlier). Activated MAPK then translocates to the nucleus, where it phosphorylates a protein transcription factor, Elk-1, allowing it to bind regulatory regions of target genes and to activate transcription. Although the name reflects its regulation of cell cycle progression, other aspects of growth, including differentiation and apoptosis, are regulated by NGF via the MAPK cascade.

Another neurotrophic factor unrelated to NGF, ciliary neurotrophic factor (CNTF), uses different receptor components employing the Jak-STAT (signal transducers and activator of transcription) pathway (Fig. 8–14). When activated, the CNTF receptor causes the Jak component of the transmembrane receptor heterodimer to phosphorylate itself as well as the CNTF receptor. The resulting tyrosinated motifs serve as receptor docking sites for SH2 domain-containing members of the STAT family of transcription factors, which are constitutively present in the cytoplasm. When docked, STATs are phosphorylated by the activated Jaks and rapidly translocate to the nucleus, where they bind to regulatory regions of genes targeted by CNTF to activate transcription. The Jak-STAT pathway represents one of the fastest transduction mechanisms affecting genomic regulation. Moreover, these two pathways demonstrate the complexity of growth factor regulation in the development and homeostasis of a single cell type.

Some Oncogenes Code for Proteins Resembling Dysfunctional Receptor Tyrosine Kinases

Cell transformation resulting from the expression of cancer-causing genes can often be attributed to proteins that mimic the action of a specific growth factor or its receptor (see Chapter 9). One example is the viral oncogene v-*erb-B* expressed by the avian erythroblastosis virus. The coded v-*erb-B* protein is strik-

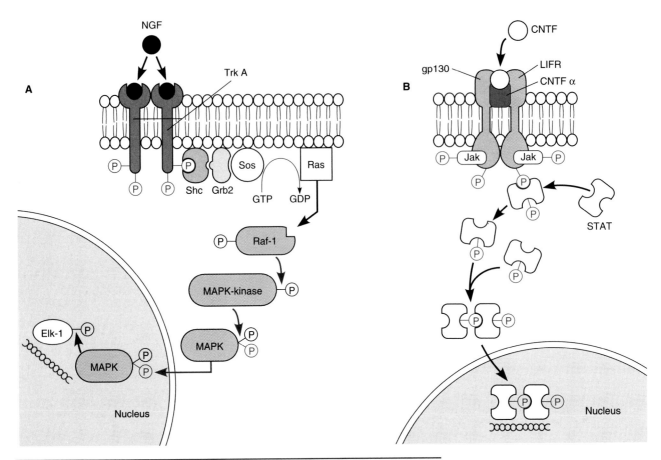

Figure 8–14. Neurotrophin-mediated signaling via the MAPK cascade and Jak-STAT pathways. **A:** Trk A receptors bound by NGF dimerize and autophosphorylate. A sequence of SH adaptor proteins bind phosphotyrosinated Trk A and each other. The guanine nucleotide factor Sos activates the G protein, ras, which, in turn, binds and activates Raf-1. This serine-threonine kinase activates MAPK-kinase, which activates MAPK. Activated MAPK translocates to the nucleus, where it phosphorylates transcription factors (e.g., Elk-1) that regulate transcription of immediate early response genes in regulatory regions of target delayed response genes. **B:** CNTF bound to its tripartite receptor protein complex, initially to CNTFα, followed by gp130 and LIFR, causes associated nonreceptor tyrosine Jak kinases to autophosphorylate and phosphorylate gp130 and LIFR. STAT protein binds transiently to the phosphotyrosinated receptor, where Jaks phosphorylate them to an active form. Upon dissociation from the receptor, activated STATs dimerize via their SH2 domain–phosphotyrosine interactions, migrate to the nucleus, and bind CNTF-response elements on CNTF-responsive genes to initiate transcription.

ingly similar to amino acid residues 551–1154 of the human receptor for epidermal growth factor (EGF) but lacks almost all the receptor's extracellular ligand-binding domain (Fig. 8–15). Normally, the EGF receptor has only minimal tyrosine kinase activity, but this is greatly stimulated by ligand binding at the cell surface. As with many growth factor receptors, the tyrosine kinase activity of the EGF receptor ultimately signals mitosis by activating a signal transduction cascade

to the nucleus. However, the shorter version of the EGF receptor produced by the v-*erb-B* oncogene has constitutively high tyrosine kinase activity. This most likely is due to the lack of a regulatory (EGF-binding) domain that, when unoccupied by EGF in the normal receptor, inhibits its tyrosine kinase activity. Consequently, cells expressing the truncated, oncogenic form of the receptor proliferate continuously, as though EGF was always present.

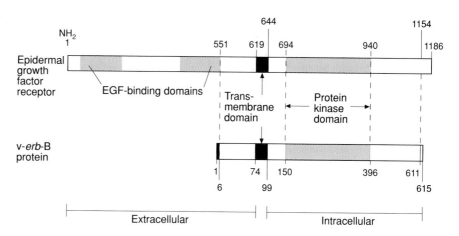

Figure 8–15. The v-*erb*-B protein as a variant of the EGF receptor. The carboxyl half of the EGF receptor (**top**) is almost identical with the v-*erb*-B protein (**bottom**). Capped by six amino acids from the virally encoded *gag* gene, the amino half of the v-*erb*-B protein is considerably shorter and lacks EGF-binding domains. (Modified from Hunter T. *Trends Biochem Sci* 1985;10:275.)

G Protein-Linked Receptors

Several Different Types of G Proteins Exist in Cells

The final type of signal transduction mechanism to be considered is the pathway that utilizes any one of several different relay proteins called *trimeric G proteins*. Unlike the monomeric G protein ras (described earlier), the major trimeric G proteins that have been characterized are composed of three polypeptide subunits: α, β, and γ. The α-polypeptide is the most variable subunit of G proteins, and it is the functionally important portion of the protein. During the signal relay process, the α-subunit binds to GTP, carries out its specific function, then hydrolyzes the GTP to GDP to help turn off the signal. At least 20 different α subunits have now been described. Some of these are cell-specific, whereas others appear to be expressed ubiquitously. The β- and γ-subunits function as a complex. Multiple isoforms of the β and γ-subunits also have been identified, and one of the major functions of the β- and γ-chains is to bind the α-polypeptide and to anchor the α-subunit to the cytoplasmic surface of the plasma membrane.

The G proteins are classified according to their subunit composition and function. Some G proteins activate the enzyme adenylyl cyclase and, as a result, are called *stimulatory G proteins* (G_s). G_s is composed of a distinct α-subunit, $G_{s\alpha}$, and β- and γ-chains. Inhibitory G proteins (G_i) perform a variety of cellular roles, including the inhibition of the enzymatic activity of adenylyl cyclase, and perform other functions. The G_i proteins may be composed of the same or different β- and γ-chains as is the G_s protein but contain a specific α-polypeptide called $G_{i\alpha}$. The third major class of G proteins (G_q) does not interact with adenylyl cyclase but instead activates the enzyme phospholipase C. Again, it is thought that this G protein has a specific α-subunit. In addition to the major G proteins, others have also been identified. These specialized G proteins, which

have been identified in sensory cells, will be described in a later section.

$G_{s\alpha}$ Turns on the Production of cAMP by the Activation of Adenylyl Cyclase

The G proteins act as intermediates in the relay of signaling messages across the plasma membrane. A working model that summarizes how this occurs for G_s is illustrated in Figure 8–16. The G protein-linked hormone receptors form a large family of transmembrane proteins, and each member of this family is a seven-pass transmembrane protein that contains an extracellular hormone-binding domain and a cytoplasmic domain that can interact with G_s. After hormone binding, the receptor undergoes a conformational change that enables the cytoplasmic domain of the receptor to interact with G_s. Receptor binding induces a conformational change in the nucleotide-binding site on $G_{s\alpha}$, and GDP is replaced by GTP at the guanyl nucleotide-binding site of the $G_{s\alpha}$ subunit. This induces $G_{s\alpha}$ to dissociate from the β- and γ-subunits of G_s and allows $G_{s\alpha}$ to bind and activate adenylyl cyclase, an enzyme that catalyzes the conversion of ATP to cAMP. Cyclic AMP is then able to diffuse through the cytoplasm to target the appropriate intracellular enzymes. $G_{s\alpha}$ cleaves GTP to GDP in less than a minute, which induces $G_{s\alpha}$ to dissociate from adenylyl cyclase. Adenylyl cyclase then ceases to produce cAMP, and $G_{s\alpha}$ reassociates with the β- and γ-subunits to form an inactive G_s complex.

The $\beta\gamma$ complex of the activated G_s also is involved in the signal transduction process. Whereas $G_{s\alpha}$ acts on adenylyl cyclase in most cell types, the $\beta\gamma$ complex acts upon various targets in different cell types. In some cell types, the $\beta\gamma$ complex apparently works along with the $G_{s\alpha}$ subunit to increase stimulation of adenylyl cyclase, whereas in other cells it has been shown that the $\beta\gamma$ complex acts on either K$^+$ channels or one of several phospholipases or kinases. The $\beta\gamma$ complex is inactivated when the $G_{s\alpha}$ subunit cleaves GTP to GDP,

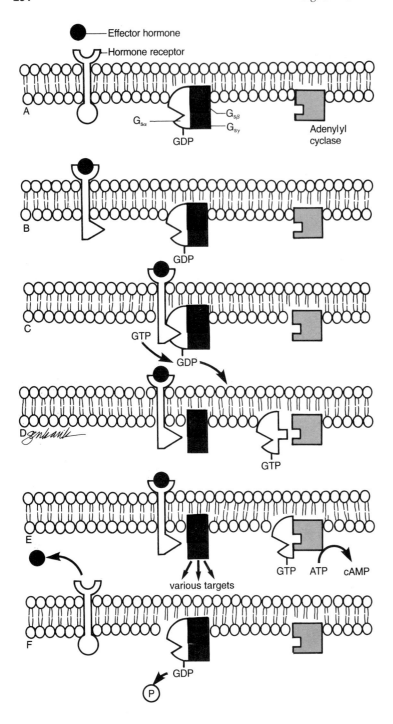

— Effector hormone

— Hormone receptor

$G_{s\alpha}$

$G_{s\beta}$

$G_{s\gamma}$

GDP

Adenylyl cyclase

GDP

GTP

GDP

GTP

various targets

GTP   ATP   cAMP

GDP

P

Figure 8–16. Events of receptor activation, G_s activity, and cAMP production. **A:** Nonsteroid hormone receptors are transmembrane proteins. **B:** Hormone binding causes a conformational change in the receptor, and (**C**) this allows the activated receptor to bind to the G_s complex. Binding of the receptor allows $G_{s\alpha}$ to replace bound GDP with GTP. **D:** GTP binding causes the dissociation of $G_{s\alpha}$ from the G_s complex, and (**E**) the activated $G_{s\alpha}$ subunit then binds to and turns on the activity of membrane-associated adenylyl cyclase. Activated adenylyl cyclase catalyzes the conversion of ATP to cAMP; cAMP then serves as the secondary messenger. The $\beta\gamma$ complex targets different effector molecules in different cell types. **F:** Dissociation of hormone from the receptor and cleavage of GTP to GDP by $G_{s\alpha}$ result in the signal being turned off. (Modified from Alberts B, Bray D, Lewis J, et al. *Molecular Biology of the Cell*, 2nd ed. New York: Garland Publishing, 1989.)

thereby allowing the α subunit to reassociate with the $\beta\gamma$ complex to turn off the activated G protein.

cAMP Acts on a Family of Regulatory Enzymes

Cyclic AMP fulfills its role as a secondary messenger by acting on a handful of enzymes, with one of the enzymes being activated by cAMP, whereas others are inhibited. The enzyme that is activated by cAMP is called *cAMP-dependent protein kinase* (A-kinase). The binding of cAMP to A-kinase causes the A-kinase to phosphorylate

serine and threonine residues on various target substrate proteins. The substrates of the A-kinase differ from cell to cell, which, in part, explains why the effects of increased levels of cAMP vary from cell type to cell type.

Increased levels of cAMP also cause the inhibition of a specific phosphatase. Because the effects of hormonal activation are transient, cells must have a mechanism for reversing the phosphorylations by A-kinase. This is accomplished by a small family of cytoplasmic phosphatases, one of which is regulated by cAMP. The cAMP-dependent protein phosphatase (protein phosphatase I) is active in the absence of cAMP. One of the substrates of activated A-kinase is a protein called *phosphatase inhibitor protein* (mentioned earlier). When cAMP levels rise, phosphatase inhibitor protein is activated and binds to the cAMP-dependent protein phosphatase, blocking the activity of the phosphatase. This allows a hormonally-stimulated cell to overcome the opposing action of the phosphatase.

Epinephrine Induces Muscle Cells to Break Down Glycogen in a cAMP-Dependent Reaction

A well-studied example of the effects of increased cAMP levels on a cell is the regulation of glycogen metabolism in skeletal muscle. Epinephrine (sometimes called *adrenaline*) is the hormone released from the adrenal gland that induces the "fight-or-flight" mechanism in humans. When epinephrine binds to surface receptors on skeletal muscle cells, G_s is activated, and $G_{s\alpha}$ turns on the activity of adenylyl cyclase. Adenylyl cyclase catalyzes the production of cAMP, which binds to and allosterically activates A-kinase. A-kinase then phosphorylates and activates an enzyme called *phosphorylase kinase*. Phosphorylase kinase phosphorylates an enzyme, termed *glycogen phosphorylase*, which is the final enzyme in this reaction cascade. Glycogen phosphorylase then catalyzes the breakdown of glycogen to glucose-1-phosphate, and glycolysis ensues (Fig. 8–17).

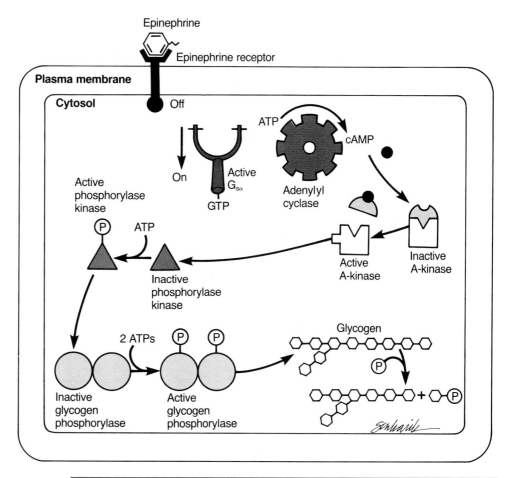

Figure 8–17. The induction of glycogen breakdown in muscle cells by epinephrine. When epinephrine binds to a transmembrane epinephrine receptor, the receptor is activated. The activated receptor induces the $G_{s\alpha}$ subunit to bind GTP and to disassociate from the $G_{s\beta\gamma}$ complex. $G_{s\alpha}$ turns on the activity of adenylyl cyclase, which catalyzes the conversion of ATP to cAMP. The cAMP binds to A-kinase and induces the disassociation of the molecule's regulatory portion from the catalytic subunits. One of the substrates of A-kinase is the enzyme phosphorylase kinase. Phosphorylation activates phosphorylase kinase, which then catalyzes the phosphorylation of glycogen phosphorylase. Activated glycogen phosphorylase catalyzes the breakdown of glycogen to glucose-1-phosphate, and glycolysis begins. (In this schematic, A-kinase is depicted as only a dimer; in reality, it is a tetramer; see text for details.)

The Reaction Cascade Allows Signal Amplification

The obvious question when considering the mechanism of signal transduction by G protein activation and cAMP production is: Why did such a complex pathway evolve? As seen earlier in this section, direct signaling mechanisms, such as steroid hormone activation and tyrosine kinases, work very well in the cells that use these types of hormone-receptor pathways. Why, then, have an elaborate activation sequence such as the one that was just outlined for epinephrine action? Complex reaction cascades appear to have evolved because this form of signaling mechanism allows the amplification of an extracellular signal. Consider how this might work. Suppose that hormone X binds to receptor Y on the surface of a cell somewhere in the body. The receptor Y would be activated by binding to the hormone and could, for the sake of this discussion, activate five G proteins. Each of the $G_{s\alpha}$ subunits could then bind to and stimulate an adenylyl cyclase molecule. Each adenylyl cyclase molecule could produce many cAMP molecules. For this hypothetical example, each adenylyl cyclase enzyme will produce 50 cAMPs, resulting in an overall amplification of the original signal 250 times. Each cAMP could then bind to and allosterically activate a subunit of the A-kinase enzyme. As A-kinase has two cAMP-binding sites, this would result in the activation of 125 A-kinases. Now, each of the 125 A-kinases will phosphorylate and activate five target enzymes, resulting in a 625-fold amplification of the original hormonal signal. Then, in turn, each of the 625-activated target enzymes could catalyze five additional enzymatic reactions, with the net result of the original signal of hormone X being a 3125-fold amplification within the cytoplasm.

The Inhibitory G Protein G_i Blocks the Activity of Adenylyl Cyclase and Regulates Ion Channels

Recent studies have shown that the inhibitory G protein G_i can modulate cellular activity in one of two ways. In one instance, the inhibitory G protein subunits act on adenylyl cyclase to block the activity of the enzyme, while in some cell types the $G_{i\alpha}$ subunit regulates the activity of ion channels. Each of these pathways will be considered.

The first several steps in the signal transduction pathway are the same, whether G_s or G_i is activated; that is, ligand binds to receptor, and the cytoplasmic portion of the activated receptor interacts with a G protein. In this instance, the specific G protein is G_i, and receptor binding causes an exchange of GTP for GDP at the nucleotide-binding site on $G_{i\alpha}$. The G_i protein subunits then dissociate. When acting on adenylyl cyclase, both the $G_{i\alpha}$ and β–γ complex appear to be able to inhibit the activity of the enzyme. Moreover, excess β–γ complex can bind to free $G_{s\alpha}$ subunits, leading to the formation of inactive G_s proteins, which further represses the activity of adenylyl cyclase.

A somewhat different mechanism is used during the regulation of ion channels by $G_{i\alpha}$. Following activation of the G_i trimeric protein complex by ligand-bound receptor, the $G_{i\alpha}$ subunit binds GTP and dissociates from the subunits. The $G_{i\alpha}$ then can bind directly to a channel protein to modulate the activity of the channel. An example of this type of regulation is the muscarinic acetylcholine receptors on cardiac muscle cells that were discussed previously. When acetylcholine stimulates the receptor on cardiac myocytes, G_i is activated. The $G_{i\alpha}$ subunit then binds to K^+ channels in the plasma membrane, causing the K^+ channels to open. This makes it more difficult to depolarize the cardiac muscle cells, resulting in a reduction in both the rate and strength of cardiac muscle contraction.

Specialized G Proteins Play Key Roles in Sensory Reception Processes

Studies have shown that G proteins play a key role in sensory perception for vision, smell, and taste. In each of these systems, specialized G proteins have evolved that perform key functions during sensory stimulation. In photoreceptor cells of the retina, a modified G protein called *transducin* (G_t) is involved in visual transduction, while in the olfactory epithelium an olfactory-specific G protein (G_{olf}) plays a key role in the sense of smell. Two different G proteins have been shown to be involved in taste sensing; G_s is involved in the sensation of sweets, and a gustatory-specific G protein (G_{gust}) plays a role in bitter sensation. Each of these specialized G proteins is composed of an α, β, and γ subunit, although each has a unique α subunit. Moreover, in the specialized sensory cells, activation of the G protein results in the generation of an electrical signal, rather than a second-messenger molecule.

During visual perception, the activating substance is a photon of light, rather than of a classical ligand. The light receptor in the photoreceptor cells of the retina is rhodopsin. Rhodopsin is a seven-pass transmembrane protein like other G protein-linked receptors. Stimulation of rhodopsin by a photon of light leads to activation of G_t. $G_{t\alpha}$ does not target adenylyl cyclase. Instead, $G_{t\alpha}$ activates cyclic GMP phosphodiesterase. Cyclic GMP is essential for the opening of Na^+ channels in the plasma membrane of the photoreceptor cells (i.e., the membrane contains cyclic GMP-gated Na^+ channels), and cyclic GMP phosphodiesterase is an enzyme that catalyzes the breakdown of cyclic GMP. The decrease in cyclic GMP levels caused by the activation of G_t results in a closing of the Na^+ channels. This causes a hyperpolarization of the receptor cell that is transmitted to the optic nerve, resulting in the perception of light. Cyclic GMP levels are rapidly returned to resting levels in the photoreceptor cells, due to the activity of the enzyme guanylyl cyclase.

Specialized cells exist in the olfactory epithelium of the nose that allow the detection of various odors. Olfactory receptor molecules activate the G_{olf} protein following odorant stimulation. It is presumed that hundreds of different odorant receptors exist that allow the perception of different smells. At present, it is not known whether receptor cells express only a single type of odorant receptor or whether several different types of olfactory receptors are expressed on a single olfactory receptor cell. Regardless, the $G_{olf\alpha}$ subunit activates adenylyl cyclase, causing an increase in cAMP levels. The cAMP that is produced binds to cAMP-gated cation channels in the plasma membrane of the receptor cells, causing the channels to open and allowing an influx of Na^+. This induces a depolarization of the cell and triggers an action potential that travels to the brain.

Several different mechanisms are used for taste perception by the specialized taste receptor cells in the taste buds of the tongue. Although the sensation of salty and sour substances appears to be due to simple ionic movements across the plasma membrane of the receptor cells, resulting in a depolarization of those receptors, the detection of bitter and sweet substances appears to involve the activity of G proteins that are regulated by transmembrane receptors. Sweet substances activate G_s in the receptor cells following receptor binding. The activated $G_{s\alpha}$ subunit then turns on adenylyl cyclase, and cAMP levels increase. This triggers a closing of K^+ channels and an eventual depolarization of the receptor cell. Bitter compounds are sensed due to the activity of a specialized G protein called *gustducin* (G_{gust}). The $G_{gust\alpha}$ subunit apparently activates cyclic GMP phosphodiesterase in a manner similar to that described for $G_{t\alpha}$ in the retina. This causes a decrease in cyclic GMP levels, the closure of Na^+ channels, hyperpolarization of the cell, and the sensation of taste. To date, receptors for taste stimuli have not been identified but, presumably, different receptor cells express different types of taste receptors.

Aberrant Regulation of Adenylyl Cyclase Is the Molecular Error that Underlies Certain Diseases

By altering the behavior of a G protein, a cell can be either hyperstimulated or inhibited from responding to extracellular signals. This is demonstrated by the activity of certain toxins. Cholera toxin modifies the α-subunit of G_s in intestinal epithelial cells, thereby inhibiting the α-polypeptide from cleaving GTP to GDP. This results in the overproduction of cAMP in the intestinal cells, and the intestinal epithelium continuously transports Na^+ and water into the lumen of the gut. This is the cause of diarrhea and dehydration associated with cholera. In a somewhat similar manner, pertussis toxin

disrupts the normal activity of G_i. This toxin, which is made by *Bordetella pertussis*, the bacterium responsible for whooping cough, inhibits the functioning of $G_{i\alpha}$. As a result, adenylyl cyclase activity is not turned off in response to certain signals.

Calcium as a Second Messenger

Calcium Ions Exert Their Effects by Allosterically Modifying Intracellular Target Proteins

Calcium is the other important secondary messenger that cells can use to respond to an extracellular signal. As for fluxes in cAMP levels, the effects on cellular activity initiated by changes in the cytoplasmic free Ca^{2+} concentration can be dramatic. Like cAMP, Ca^{2+} exerts its effects by binding to certain target proteins and allosterically modifying the activity of those targets. The principal Ca^{2+}-binding protein in the cytoplasm is calmodulin. Calcium-binding causes a conformational change in calmodulin that allows calmodulin to bind to any one of several target proteins in a cell. This can result in either the activation or inhibition of the particular Ca^{2+}–calmodulin target.

In addition to calmodulin, several other Ca^{2-}-binding proteins have been identified and characterized. Perhaps the best-known example is troponin C. Troponin C is found in skeletal muscle cells, and binding of Ca^{2+} to troponin C allows skeletal muscle contraction to occur. Another example of a protein whose activity is regulated by Ca^{2+}-binding is the proteolytic enzyme calpain. In its inactive state, calpain is free of bound Ca^{2+}. Calcium binding causes a conformational change that allows calpain to bind to its substrate proteins and to degrade the molecules proteolytically.

Changes in Intracellular Calcium Levels Usually Are Due to the Activity of a G Protein

Intracellular Ca^{2+} levels normally are very low (usually near 10^{-7} M), whereas the extracellular free Ca^{2+} concentration is in the millimolar range. Different cells use various mechanisms to keep cytosolic free Ca^{2+} levels low. The most important of these is a Ca^{2+} pump in the plasma membrane. This Ca^{2+}-ATPase uses the energy of ATP hydrolysis to transport Ca^{2+} ions from the cytoplasm to outside the cell. Cells also can sequester Ca^{2+} inside internal membranous spaces. For example, Ca^{2+}-ATPases that are present in the membranes of organelles, including the mitochondria and smooth endoplasmic reticulum, can be used to pump Ca^{2+} ions from the cytosol into the lumen of one of these membrane-bounded organelles. In addition, certain specialized cells that are very dependent on changes in intracellular Ca^{2+} levels for their functioning, such as nerve and muscle cells, have an accessory Ca^{2+} pump in their plasma membranes. This specialized Ca^{2+} pump couples the efflux of Ca^{2+} to the influx of Na^+ ions in these cells.

The binding of signaling molecules to receptors on the surface of cells can result in rapid changes in the level of Ca^{2+} ions in the cytoplasm. In some instances, such as neurotransmission, increases in cytoplasmic Ca^{2+} levels are due to the influx of Ca^{2+} ions from the extracellular environment through gated ion channels. However, increases in cytoplasmic Ca^{2+} ion levels most often are caused by the release of these ions from intracellular stores. The release of Ca^{2+} from internal membranes that is triggered by hormone binding is due to the activity of a specialized G protein (G_q).

The steps that result in the stimulation of G_q are the same as those for the activation of G_s and G_i. That is, a signaling molecule binds to a membranous receptor, and the ligand-binding induces a conformational change in the receptor molecule. The activated receptor molecule then binds to G_q, which can then bind to GTP. This binding induces the dissociation of the α-subunit from the β–γ complex (Fig. 8–18). However, the α-subunit does not stimulate cells through binding to adenylyl cyclase; instead, it exerts its effects by interacting with the membrane-associated enzyme phospholipase C-β. Phospholipase C-β then catalyzes the reactions that eventually result in the release of Ca^{2+} from intracellular stores.

The Catalytic Activity of Phospholipase C-β Produces Inositol Trisphosphate, Which Triggers the Release of Calcium From Intracellular Stores

Phospholipase C-β catalytically cleaves members of the membrane phospholipid family called the *phosphatidylinositols* after activation by $G_{q\alpha}$. The most important member of the inositol phospholipid family for the signal transduction pathway is a phosphorylated

derivative of phosphatidylinositol, phosphoinositol bisphosphate (PIP_2). PIP_2 is principally located in the inner leaflet of the lipid bilayer of the plasma membrane, and, during signal transduction events, PIP_2 is cleaved into two products, inositol trisphosphate (IP_3) and diacylglycerol, by the enzymatic activity of phospholipase C-β. Each of these molecules plays important roles in the signal transduction pathway.

Inositol trisphosphate is the water-soluble portion of the membrane phospholipid that is released into the cytosol after PIP_2 cleavage. It diffuses through the cytoplasm, then induces the release of Ca^{2+} from intracellular storage compartments. Although the details are unknown, an IP_3 receptor has been identified in these membranes (Fig. 8–19), and it is presumed that this receptor is a gated Ca^{2+} channel. The binding of IP_3 to this receptor would induce a conformational change in the receptor that would allow Ca^{2+} to be released into the cytosol. The Ca^{2+} ions can then bind to and activate cytoplasmic receptor molecules such as calmodulin, thereby allowing the cell to respond in the appropriate fashion to the signaling molecule that is bound at the cell surface. Clearly, the reason for using such an involved pathway is the same as the rationale for using cAMP as a secondary messenger: the activity of IP_3 results in a considerable amplification of the original signal. In theory, the binding of a ligand to the cell surface could result in a substantial change in the level of intracellular Ca^{2+} ions, with the subsequent activation of numerous calmodulin molecules.

The increase in cytosolic Ca^{2+} concentration induced by ligand binding is transient. The cell uses two mechanisms for returning the cytosolic Ca^{2+} levels to normal. First, Ca^{2+}-ATPases either pump Ca^{2+} out of the cell or resequester Ca^{2+} in internal storage compartments. Second, IP_3 is dephosphorylated by a cytosolic phosphatase. The duration of the Ca^{2+} signal, however, can last several seconds.

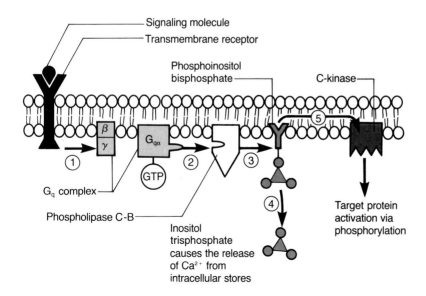

Figure 8–18. A schematic showing the activation of phospholipase C-β. A signal molecule binds to a transmembrane receptor, inducing a conformational change in the receptor. (*1*) The activated receptor binds to G_q, and allows GTP to bind to the $G_{q\alpha}$ subunit. The GTP-binding causes the $G_{q\alpha}$ subunit to dissociate from the $G_{q\beta\gamma}$ complex, and (*2*) the $G_{q\alpha}$ molecule then binds to and activates phospholipase C-β. (*3*) Phospholipase C-β catalyzes the cleavage of PIP_2 to yield IP_3 and diacylglycerol. (*4*) IP_3 diffuses through the cytoplasm and induces the release of Ca^{2+} from intracellular storage sites. (*5*) Diacylglycerol remains in the lipid bilayer and binds to C-kinase and promotes the activation of C-kinase. (Modified from Matthews CK, van Holde KE. *Biochemistry.* Redwood City, CA: Benjamin/Cummings Publishing, 1990.)

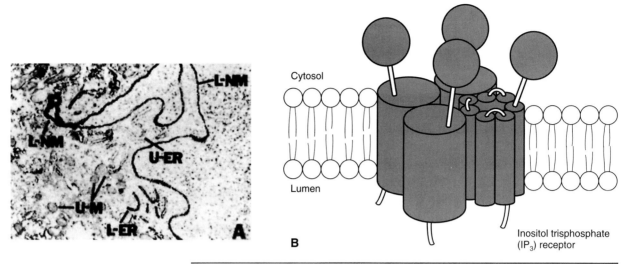

Figure 8–19. The localization and structure of the IP₃ receptor. **A:** Antibodies were generated against the 260-kDa polypeptide subunit of the IP₃ receptor. The protein antigen was localized in Purkinje neurons by immunoelectron microscopy. Regions of the nuclear membrane were labeled (*L-NM*), as were many regions of the endoplasmic reticulum (*L-ER*). Mitochondria (*U-M*) and some regions of endoplasmic reticulum (*U-ER*) were unlabeled. **B:** A schematic showing the putative structure of the IP₃ receptor. The IP₃ receptor consists of a tetramer of subunits, each containing seven transmembrane domains, and a region that protrudes into the cytosol and binds IP₃. (**A,** Ross et al. *Nature* 1989;339:468–470; **B,** modified from Hall Z. *An Introduction to Molecular Neurobiology.* Sunderland, MA: Sinauer Assoc, 1992.)

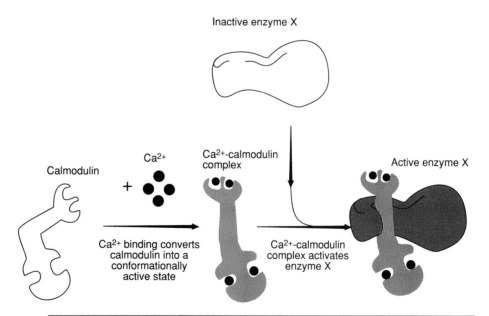

Figure 8–20. Calmodulin relays the signal of increased cytosolic Ca²⁺ to a conformationally sensitive target protein. The binding of Ca²⁺ evokes an allosteric change that enables calmodulin to bind and convert an inactive substrate protein into a conformationally active state. Calmodulin binds a myriad of substrate proteins with different affinities. As more Ca²⁺ enters the cell, more Ca²⁺–calmodulin complexes become available to activate substrate proteins with progressively lower binding affinities. Accordingly, the amount of Ca²⁺ entry into the cytosol may determine which target proteins are activated first. (Modified from Alberts B, Bray D, Lewis J, et al. *Molecular Biology of the Cell*, 2nd ed. New York: Garland Publishing, 1989.)

Calcium–Calmodulin Activates Multiple Target Proteins

Free Ca^{2+} released into the cytosol rapidly binds to calmodulin to cause an allosteric change in the protein (Fig. 8–20), which enables Ca^{2+}–calmodulin to regulate other calmodulin-sensitive proteins downstream in the signal–response pathway. Target proteins include several enzymes, as well as receptors and transport proteins. A family of Ca^{2+}–calmodulin-dependent (CaM) kinases exists that includes the myosin light chain kinase responsible for phosphorylating myosin in smooth muscle (see Chapter 3). Other members of this family, such as the type I CaM kinase, phosphorylate a single recognized substrate protein. Specifically, CaM kinase I phosphorylates the protein synapsin that binds synaptic vesicles to F-actin or spectrin II. The phosphorylation of a much broader variety of proteins in the cell is carried out by CaM kinase II. When activated, this kinase becomes autophosphorylated by a mechanism that keeps the enzyme catalytically active for a short time, even after Ca^{2+}–calmodulin has dissociated (Fig. 8–21). This modification probably allows

the enzyme to maintain relatively high kinase activity for a short time after the cytosolic concentration of Ca^{2+} returns to a basal level. The Ca^{2+}–calmodulin complex also interacts with other second-messenger pathways. In particular, isozymes of adenylyl cyclase and cAMP phosphodiesterases associated with cAMP metabolism can be affected by Ca^{2+}–calmodulin. The interplay of the various signaling pathways is shown in Fig. 8–22.

Diacylglycerol Activates Protein Kinase C

The other product of PIP_2 hydrolysis, diacylglycerol, also contributes to the response of a cell to a signaling molecule. Whereas IP_3 is released into the cytosol following PIP_2 cleavage, diacylglycerol remains embedded in the lipid bilayer of the plasma membrane (see Fig. 8–18). Diacylglycerol and phosphatidylserine, which is also present in the lipid bilayer, bind to protein kinase C and activate the enzyme. The C-kinase has several substrates, including various membrane pumps and channel proteins. It also appears to modulate the rate of

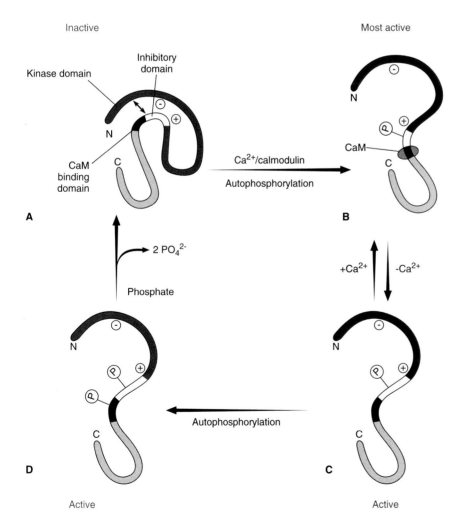

A Inactive

Kinase domain
Inhibitory domain
N
CaM binding domain
C

Ca^{2+}/calmodulin
Autophosphorylation

B Most active

CaM
N
C

Figure 8–21. Autophosphorylation leaves CaM II kinase in a transient switched-on state. **A:** When the enzyme is inactive, the inhibitory domain of CaM II kinase binds and, thereby, inhibits its own kinase (catalytic) domain. **B:** The binding of Ca^{2+}–calmodulin (CaM) to a separate domain on CaM kinase II causes an allosteric change that frees the catalytic domain to phosphorylate other substrates, as well as the enzyme itself. **C:** When the Ca^{2+} level drops in the cytosol, the autophosphorylation step prevents inactivation of the kinase, which continues to autophosphorylate itself at another site (**D**) until phosphatases return the enzyme to its inactive form. (Modified from Hall Z. In: Hall, Z. ed. *Introduction to Molecular Neurobiology.* Sunderland, MA: Sinauer Assoc, 1992.)

2 PO_4^{2-}
Phosphate

$+Ca^{2+}$ $-Ca^{2+}$

D Active
N
C
Autophosphorylation

C Active
N
C

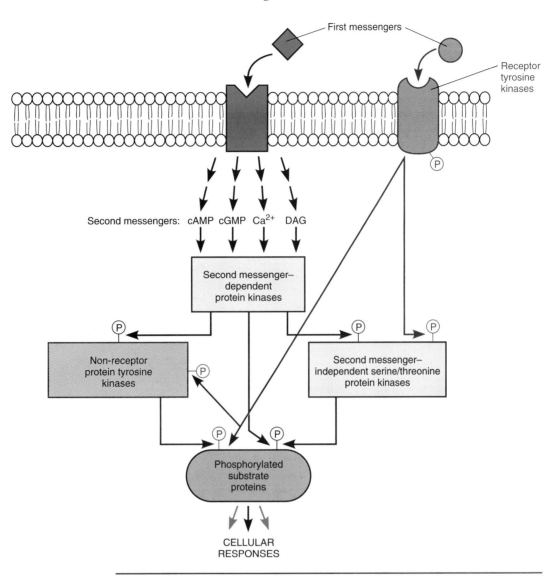

Figure 8–22. Second messenger-dependent and -independent protein kinases and their interplay in protein phosphorylation. Some extracellular ligands (first messengers) bind and activate transmembrane receptors, leading to the production of second messengers (cAMP, cGMP, Ca²⁺, or DAG), which activate the corresponding second messenger-dependent protein kinases. These kinases can, in turn, activate by phosphorylation second messenger-independent kinases (serine-threonine and non-receptor protein tyrosine kinases). Other first messengers directly activate second messenger-independent receptor tyrosine kinases, which autophosphorylate and phosphorylate substrate proteins, including both nonreceptor protein tyrosine kinases and second messenger-independent serine-threonine protein kinases. In this way, the cellular response to first messengers can activate a network of multiple kinase families with partially overlapping and differing substrate specificities.

transcription of various genes by regulating the activity of proteins that bind to promotor regions of the genes.

Diacylglycerol can exert its effects on a cell in a second fashion. In some instances, diacylglycerol can be converted to arachidonic acid, which is further metabolized to produce prostaglandins (see earlier discussion).

The Activities of Inositol Trisphosphate, Diacylglycerol, and Prostaglandins Can Be Regulated Pharmacologically

Many of the activities that are induced by the signaling molecules that are produced following G$_q$ stimulation can be mimicked pharmacologically. Drugs called *phor-*

Table 8–1
G proteins, α-subunits, and functions

G Protein	α-Subunit	Function
G_s	sα	Activates adenylyl cyclase
G_i	iα	Inhibits adenylyl cyclase and regulates many ion channels
G_q	qα	Activates phospholipase C
G_t	tα	Regulates cyclic GMP phosphodiesterase in photoreceptors
G_{olf}	olfα	Activates adenylyl cyclase
G_{gust}	gustα	Regulates cyclic GMP phosphodiesterase in taste receptors

bol esters are able to bind to C-kinase and to trigger the activity of the enzyme. Because phorbol esters cause stimulation of cellular activity, they are commonly used as tumor promoters in carcinogen studies. The Ca^{2+} ionophore A23187 is able to mimic the behavior of IP_3. *Ionophores* are small hydrophobic molecules that dissolve into membranes and increase the ion permeability of the membranes. When added to cells, A23187 increases the permeability of the plasma membrane to Ca^{2+}, and Ca^{2+} enters the cytosol from the surrounding extracellular fluid. Finally, it is thought that aspirin and ibuprofen can act as antiinflammatory agents by inhibiting prostaglandin biosynthesis. The various types of G proteins that have been described in this chapter are summarized in Table 8–1.

Calcium Influx Is Required for Neurotransmitter Release

Whereas the previous sections dealt with cellular activity that occurs by the release of Ca^{2+} ions from intracellular stores, cellular behavior also can be modified by an influx of Ca^{2+} ions from the extracellular environment. In this section, we will focus on how an influx of Ca^{2+} ions into neurons leads to synaptic transmission events. Although many of the same key molecules, such as the various kinases and the Ca^{2+}–calmodulin complex, are involved, the key difference in this type of system is that Ca^{2+} enters the cell from the extracellular space.

Mechanisms underlying even the most sophisticated neural functions, such as learning and memory, are thought to depend on the fundamental properties of synaptic chemical transmission. The rapid conduction of electrical signals within a given neuron is made possible by the action potential, which results from the sequential opening of voltage-dependent Na^+ and K^+ channels in the plasma membrane (see Chapter 2). Although the arrival of an action potential at the presy-

naptic terminal increases the likelihood that synaptic transmission will occur, neither Na^+ influx nor K^+ efflux actually controls neuronal output. Instead, the opening of voltage-sensitive Ca^{2+} channels, caused by the incoming wave of depolarization at the presynaptic terminal, couples action potentials to biochemical reactions that trigger neurotransmitter release.

In pioneering experiments demonstrating this principle, Bernard Katz and his co-workers used selective ion channel-blocking agents to disrupt the electrophysiologic events leading to neurosecretion in giant synapses obtained from the squid superior cervical ganglion. The presynaptic neuron was either bathed in tetrodotoxin (TTX)—a toxin produced by puffer fish that specifically binds to and blocks only voltage-gated Na^+ channels—or injected with tetraethylammonium (TEA)—a chemical that selectively inactivates voltage-gated K^+ channels. With one or both channel types rendered nonfunctional by these neuropharmacologic tools, Katz and Miled showed that transmitter release was still possible by artificially depolarizing the presynaptic cell with a current-passing electrode. This suggested that depolarization normally requires current in the form of some other unidentified cation. Subsequently, it was found that when Na^+ and K^+ channels are fully blocked with TTX and TEA, all residual current across the presynaptic membrane is caused by Ca^{2+} ions. A decrease in the extracellular Ca^{2+} concentration reduces, and ultimately prevents, synaptic transmission. Conversely, transmission in neurons deprived of extracellular Ca^{2+} is restored by pipetting Ca^{2+} ions near the presynaptic terminal.

Calcium Channels Responsible for Secretory Potentials Are Restricted to Active Zones of the Presynaptic Terminal

The brief delay (only a few milliseconds) between the onset of an action potential at the presynaptic terminal and a corresponding postsynaptic potential is largely due to the time required for depolarization to trigger the opening of voltage-sensitive Ca^{2+} channels, allowing Ca^{2+} influx into the presynaptic cell (Fig. 8–23). The resulting inward current of Ca^{2+} ions moving down their steep concentration gradient is termed the *secretory potential* because it is essential for neurosecretion by the presynaptic cell. Secretory potentials are restricted to the axon terminal. Only at this site of the neuronal plasma membrane are voltage-dependent Ca^{2+} channels sufficiently concentrated to pass a relevant amount of current. In contrast to voltage-gated Na^+ channels, which are inactivated rapidly, voltage-gated Ca^{2+} channels inactivate slowly once they have been opened by depolarization (in squid neurons, at least). Consequently, Ca^{2+} continues to enter the terminal as long as it remains depolarized.

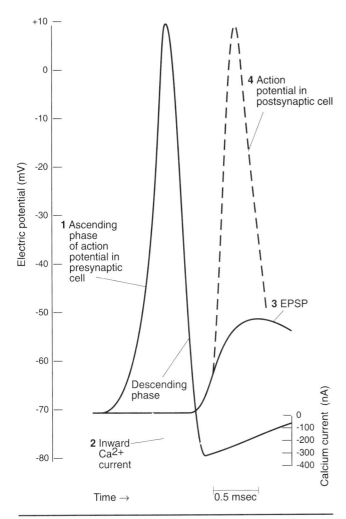

Figure 8–23. Sequence and timing of four successive electrical events of synaptic transmission. (*1*) Depolarization of the presynaptic terminal by an incoming action potential opens presynaptic voltage-gated Ca²⁺ channels, resulting in (*2*) the flow of current caused by the influx of Ca²⁺ ions into the cell. The subsequent release of neurotransmitter leads to a postsynaptic depolarization (*3*) that must surpass a threshold value to trigger an action potential (*4*) in the postsynaptic cell. (Modified from Linás RR. *Sci Am* 1982;247(4):56.)

the site where synaptic vesicles are positioned for release. This is evident in axon terminals injected with fura-2, an indicator dye that fluoresces upon binding to intracellular Ca²⁺. Under view in the fluorescence microscope, Ca²⁺ influx is observed to occur at a rate tenfold greater near the active zone than elsewhere in the terminal of a depolarized neuron. Consistent with this observation, microscopic studies of the frog neuromuscular junction show that voltage-sensitive Ca²⁺ channels marked with a fluorescent form of the channel-blocker ω-connotoxin are also confined to the active zone directly opposing the postsynaptic membrane. Under the electron microscope, these channel proteins probably correspond in number and size to rows of 10-nm intramembranous particles flanking the active zone, the topography of which is revealed by the method of freeze-fracture (Fig. 8–24).

By concentrating Ca²⁺ channels within active zones, neurons differ from other exocrine and endocrine secretory cells that disperse them throughout their plasma membrane. This specialized arrangement speeds the release of the transmitter following depolarization by minimizing not only the distance required for Ca²⁺ diffusion to the synaptic vesicle release site, but also the number of Ca²⁺ channels that are needed. Following the arrival of an action potential, a transient increase of internal Ca²⁺ from a basal level of 0.1 μM to as much as 100 μM is rapidly achieved in the active zone because the influx is local and readily terminated by diffusion throughout the remainder of the presynaptic terminal. All of these factors ensure that transmitter release can occur quickly and repetitively to match a high-frequency train of incoming electric signals.

Calcium Influx Mobilizes Synaptic Vesicles for Docking at the Active Zone and Promotes Exocytosis

Only a small subpopulation of the synaptic vesicles housed in the presynaptic terminal are immediately positioned at sites for quick release (see Chapter 7); most are anchored nearby to a cytoskeletal lattice composed of actin and brain spectrin filaments. Increased Ca²⁺ leads to the detachment of synaptic vesicles from this framework and allows more of them to dock at the active zone. The mechanism for detachment involves the regulation of synapsin I, a fibrous 86-kDa phosphoprotein that links synaptic vesicles to the actin cytoskeleton (Fig. 8–25). Facing the cytosol, synapsin I is attached to the external surface of the vesicle membrane, where it is a substrate for cAMP-dependent protein kinase and CaM kinases I and II at three phosphorylation sites (Fig. 8–26). However, only in its dephosphorylated form can synapsin I attain the proper conformation to bind to both the vesicle and actin or spectrin filaments. The influx of Ca²⁺

An action potential of longer duration will thereby increase the amount of Ca²⁺ that flows into the terminal and, ultimately, the quantal release of transmitter, which increases the size of the postsynaptic potential. However, nerve cells in species other than squid contain two classes of voltage-sensitive Ca²⁺ channels, and one of them (the N-type channel) is, in fact, rapidly inactivating and is primarily responsible for neurosecretion.

The discharge of neurotransmitter into the synaptic cleft occurs as soon as 200 μsec after the secretory potential forms because voltage-sensitive Ca²⁺ channels in the presynaptic terminal are positioned very close to

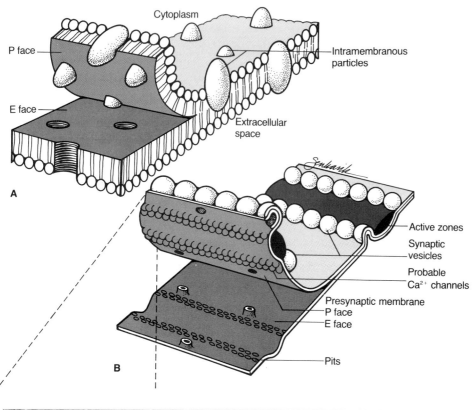

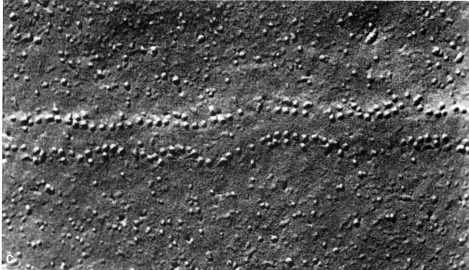

Figure 8–24. Location of Ca²⁺ channels at the presynaptic terminal viewed by freeze-fracture and electron microscopy. **A:** The process of freeze-fracture splits the lipid bilayer of the plasma membrane to reveal the topographic features of both the interior half facing the cytoplasm, termed the *protoplasmic* (*P*) face, and the exterior half bordering the extracellular space, termed the *external* (*E*) face. **B:** When applied to the presynaptic terminal membrane, this technique reveals on the P face rows of intramembranous particles that flank active zones where docked synaptic vesicles are released. **C:** Linear arrays of putative Ca²⁺ channels as they actually appear by electron microscopy, showing the protruding particles of the presynaptic terminal (P face) at the frog neuromuscular junction. (Courtesy of J. Heuser.)

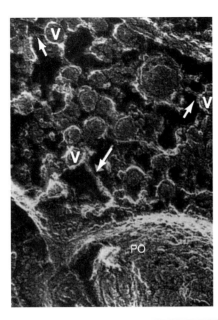

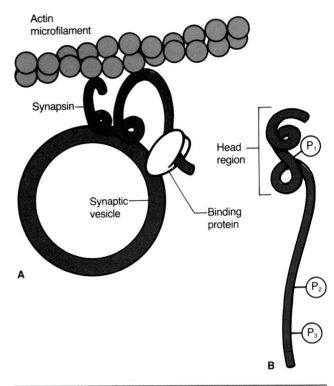

Figure 8–25. Cytoskeletal anchorage of synaptic vesicles in the presynaptic terminal. For this electron micrograph, tissue was freeze-fractured and etched to show presynaptic structures in a three-dimensional view. Synaptic vesicles (*V*) are linked by short rods (20–30 nm), thought to represent individual synapsin molecules (*short arrows*) that attach to longer filaments consisting of actin or spectrin II (*long arrow*) PO, postsynaptic cell. (Hirokawa N, et al. *J Cell Biol* 1989;108:111–126.)

Figure 8–26. Anchorage of synaptic vesicles to actin regulated by Ca^{2+}-dependent and Ca^{2+}-independent phosphorylation of synapsin I. **A:** In association with a synaptic vesicle and actin, dephosphorylated synapsin contains a globular head group buried in the vesicle membrane and a helical tail attached to an accessory binding protein. **B:** Vesicle detachment from actin is mediated by the phosphorylation of synapsin I at three sites. Cyclic AMP-dependent protein kinase and CaM kinase I both recognize phosphorylation sequence 1, located at the head group. Sites 2 and 3, positioned in the tail region, are recognized by CaM kinase II. (Modified from Hall Z. *Introduction to Molecular Neurobiology.* Sunderland, MA: Sinauer Assoc, 1992.)

provided by an action potential at the axon terminal is thought to stimulate both kinases to phosphorylate synapsin I, thereby altering its shape and allowing synaptic vesicles to release from the restraint of the actin filaments. This allows the vesicles to move from the reserve pool into the releasable pool of vesicles that are docked at the active zone of the presynaptic plasma membrane.

By functioning in a second neurosecretory role, Ca^{2+} appears to directly mediate transmitter release from vesicles docked at the active zone. As detailed in Chapter 4, Ca^{2+} binds to the major vesicle protein, synaptotagmin. The resulting conformational change in the protein causes fusion of the vesicle and terminal membranes, presumably through the active participation of acidic membrane phospholipids. A fusion pore, which resembles in dimension an ion channel, connects the vesicle lumen with the extracellular space and mimics a gap junction in its electrical properties (Fig. 8–27).

Hyperactivity and Axoaxonic Synapses Modulate Transmitter Release by Altering Presynaptic Calcium Influx

Short-term changes in synaptic transmission are brought about by mechanisms within and outside the presynaptic neuron that further modulate the concen-

tration of intracellular free Ca^{2+}. For transmitter secretion to accurately reflect the cadence of electrical signals that produce it, free Ca^{2+} levels at the vesicle release site in the presynaptic terminal must rapidly rise and diminish. The rapid decrease in the rate of transmitter release following a single action potential predicts that the increased Ca^{2+} concentration at the vesicle release site dissipates within 1 msec. Mechanisms immediately responsible for rapid decay include (a) diffusion of Ca^{2+} to the remainder of the terminal, (b) binding of free Ca^{2+} to cytoplasmic proteins, and (c) Ca^{2+} uptake into temporary storage vesicles and mitochondria (Fig. 8–28). When the intracellular Ca^{2+} level begins to fall following a single action potential, free Ca^{2+} is then slowly released from these temporary storage sites and is transported out of the presynaptic terminal into the extracellular fluid by a Ca^{2+}-ATPase and a Na^{+}–Ca^{2+} exchange pump located in the plasma membrane (see

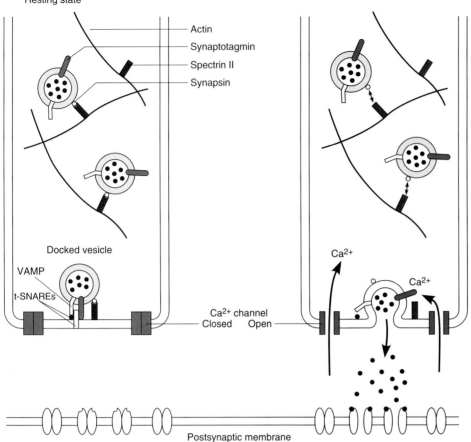

Figure 8–27. Calcium plays a dual role in neurosecretion at the presynaptic terminal. During the resting state (*left panel*), voltage-sensitive Ca²⁺ channels are closed, and few vesicles are bound by fusogenic proteins to release sites in the active zone. The linker protein, synapsin, immobilizes most vesicles to actin filaments and spectrin II in the cytosol, and binds some of them to the active zone of the presynaptic plasma membrane by spectrin II. After the arrival of an action potential (*right panel*), voltage-gated Ca²⁺ channels open. Calcium influx promotes the mobilization of synaptic vesicles from the actin-based cytoskeleton by stimulating the phosphorylation of synapsin I. For vesicles docked at the presynaptic terminal membrane by SNARE protein complexes (**see Chapter 4**), Ca²⁺ binds synaptotagmin to change the conformation of the protein and trigger exocytosis.

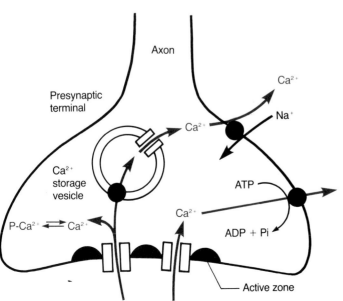

Figure 8–28. Schematic diagram of Ca²⁺ localization following the arrival of an action potential at the presynaptic terminal. Opened Ca²⁺ channels allow Ca²⁺ to enter near the active zone. Calcium ions diffuse through the terminal, where they bind to proteins (*P*) and are taken up into Ca²⁺ storage vesicles. When its cytosolic concentration decreases, Ca²⁺ is slowly released from storage and transported out of the terminal by a Ca²⁺-ATPase and a Na⁺–Ca²⁺ exchange pump.

Figure 8–29. Axoaxonic synapses as modulators of Ca²⁺ influx and neurotransmitter release. **A:** Presynaptic inhibition transpires when the axon terminal of an active inhibitory neuron synapses on the presynaptic terminal of another neuron. Fewer voltage-gated Ca²⁺ channels open because the presynaptic terminal membrane of neuron "a" is hyperpolarized. When an action potential arrives at the presynaptic terminal of neuron "a," the inward Ca²⁺ current is diminished, causing a reduction in the amount of neurotransmitter released and the magnitude of the excitatory potential recorded in the postsynaptic neuron. **B:** Presynaptic facilitation requires an active facilitating neuron that depresses the K⁺ current flowing out of another presynaptic neuron. This prolongs the duration of an action potential arriving at the presynaptic terminal and, as a result, the inward Ca²⁺ current is prolonged. Consequently, an increase in the amount of transmitter released by the presynaptic cell enhances the excitatory postsynaptic potential.

A. Presynaptic inhibition

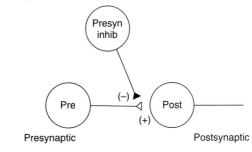

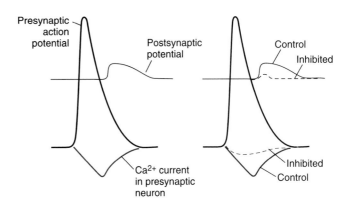

B. Presynaptic facilitation

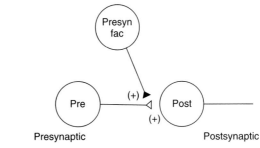

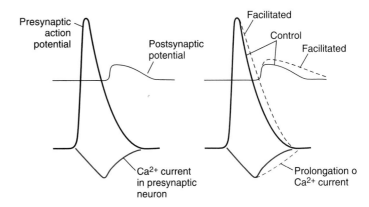

Chapter 2). It is emphasized that the complete process of Ca²⁺ reequilibration in the nerve terminal is relatively slow, taking 0.1–1.0 second after a single action potential. Consequently, continuous depolarization, resulting from a rapid succession of action potentials impinging on the terminal, can overload these local mechanisms for Ca²⁺ removal, thereby causing the accumulation of intracellular free Ca²⁺ and prolonging synaptic transmission for minutes or longer.

The level of intracellular free Ca²⁺ available for neurosecretory function can also be altered by the processes of presynaptic inhibition or presynaptic facilitation. Axoaxonic synapses allow one neuron to modulate the neurotransmitter release of another when the axon of one neuron synapses on the presynaptic terminal of a second neuron (Fig. 8–29). In presynaptic inhibition, the release of an inhibitory neurotransmitter, such as GABA, at the axoaxonic synapse evokes a receptor-mediated opening of Cl⁻ channels located on the postsynaptic membrane; that is, the presynaptic terminal of the second neuron. As a result, the influx of Cl⁻ ions into the second neuron

counteracts the depolarization caused by an action potential arriving at the same terminal. The net effect reduces both the opening of voltage-sensitive Ca^{2+} channels and neurosecretion by the modulated neuron. On the other hand, precedence for presynaptic facilitation exists in certain molluscan neurons that use serotonin as a facilitory neurotransmitter at axoaxonic synapses. There, serotonin released by one neuron binds to receptors on the presynaptic terminal of the target neuron and acts through cAMP-dependent pro-

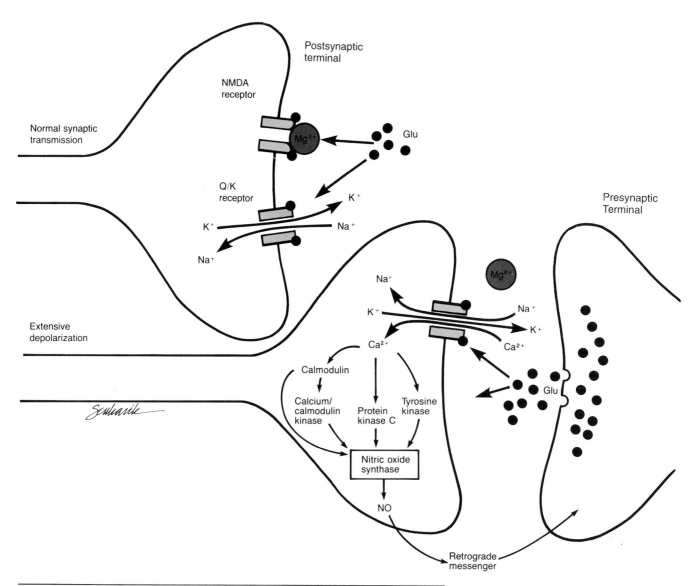

Figure 8–30. The role of the NMDA receptors in triggering LTP at the postsynaptic terminal. During normal synaptic transmission (*left panel*), glutamate released from the presynaptic terminal can bind to both NMDA and quisqualate kainate (Q/K) receptor subtypes. However, in this state only Q/K receptor channels are opened, allowing the flow of Na⁺ and K; NMDA receptor channels remain blocked by Mg^{2+}. Only in response to high-frequency stimulation, leading to significant depolarization of the postsynaptic terminal (*right panel*), is the Mg^{2+} block removed, thereby allowing flow of Ca^{2+}, K⁺, and Na⁺ through NMDA receptor channels. Increased intracellular Ca^{2+} at the dendritic spine is only the first step in a series of subsequent Ca^{2+}-mediated biochemical reactions responsible for inducing LTP. (Modified from Hall Z. *Introduction to Molecular Neurobiology.* Sunderland, MA: Sinauer Assoc, 1992.)

tein phosphorylation to close voltage-dependent K$^+$ channels. The reduction in the rectifying K$^+$ current across the plasma membrane prolongs depolarization brought about by an action potential at the terminal. Consequently, Ca^{2+} influx is enhanced, as is the amount of transmitter released.

In other instances, receptors located on the presynaptic terminal itself, termed *autoreceptors*, bind some of the parent neuron's own released transmitter to downregulate secretion further. For example, some of the norepinephrine released by sympathetic nerves binds to α$_2$-adrenergic receptors located on the presynaptic nerve terminal. This autoreceptor-binding activates a G protein to reduce the inward Ca^{2+} current through altered Ca^{2+} channels, thereby decreasing transmitter release.

Memory Formation Is Linked to Calcium Influx Stimulated by the Activation of N-Methyl-D-Aspartate Receptors

This text has concentrated on the function of Ca^{2+} and voltage-gated Ca^{2+} receptors in presynaptic release mechanisms. Calcium ions also play a central role in postsynaptic events mediated by ligand-gated Ca^{2+} channels in the dendrites of neurons in brain regions, such as the hippocampus, whose function is strongly implicated with memory formation. Most neuroscientists espouse the theory that the storage of memory ultimately results from the encoding of detailed sensory information as intricate structural changes in synaptic connections. Cell ablation studies further suggest that specific memory is attributed not to one neuron, but to many of them interconnected by considerable distances within and among different brain regions.

At the subcellular level, one of the neurochemical events suggested to contribute substantially to memory formation is long-term potentiation (LTP). This phenomenon was originally described in 1966 with the observation that a brief (≤1 second), high-frequency electrical stimulation of certain hippocampal neurons leads to prolonged synaptic transmission, lasting for as long as several weeks. It is now known that the activation of a subclass of postsynaptic glutamate receptors, specifically, the *N*-methyl-D-aspartate (NMDA) subtype, is essential for LTP (Fig. 8–30). These receptors are found in particularly high concentration in the hippocampus. When the hippocampus of a mammal is infused with NMDA receptor antagonists, memory skills are selectively impaired, leaving normal unrelated tasks of neurotransmission unaffected. The NMDA receptors are unique (see Chapter 7) in that both the binding of glutamate and strong depolarization of the postsynaptic cell must occur simultaneously for NMDA receptor channels to open, whereupon they are rendered freely permeable to Ca^{2+} as well as to Na$^+$ and K$^+$. The influx of Ca^{2+} into the dendrites of the postsynaptic cell is thought to stimulate CaM kinases

that cause the induction of LTP. One of the substrates for these kinases is nitric oxide (NO) synthase, which, in its active state, generates NO (see Chapter 7) in the postsynaptic neuron. Because NO can freely dissociate through the plasmalemma, it is thought to act in the presynaptic neuron to prolong neurotransmitter release, as occurs during LTP.

Clinical Case Discussion

Pertussis is an acute, highly contagious infection of the respiratory tract caused by *Bordetella pertussis*, characterized by severe bronchitis. The disease is most common and most severe in early infancy. Fifty percent of children affected will be less than 1 year of age. The onset of pertussis, or "whooping cough," is insidious, with catarrhal upper respiratory tract symptoms; slight fever, rhinitis, and sneezing; followed by a paroxysmal cough characterized by 10–30 forceful coughs and ending with a loud inspiration (the "whoop"). Vomiting and cyanosis commonly follows a paroxysm. White blood cell counts of 20,000–30,000/mm^3 with 70–80% lymphocytes typically appear near the end of the catarrhal stage. The chest x-ray reveals thickened bronchi and sometimes shows a "shaggy" heart border, indicative of bronchopneumonia and atelectasis. *Bordetella pertussis* can usually be cultured from the nasopharynx.

The pertussis toxin produced by *Bordetella pertussis* catalyzes the ADP-ribosylation of the G protein, blocking inhibition of adenylate cyclase by G$_i$. This defect produces the symptoms of whooping cough, including hypersensitivity. Erythromycin is the antibiotic of choice to treat both the index case and the exposed family members and hospital contacts. Active immunization with pertussis vaccine in combination with diphtheria and tetanus toxoids (DPT) is routinely performed in early infancy. More recently, an acellular pertussis vaccine has become available.

Suggested Readings

Steroid Receptor Mechanisms

Gehring V. Steroid hormone receptors: biochemistry, genetics, and molecular biology. *Trends Biochem Sci* 1987;12:399.

Yamamoto KR. Steroid receptor regulated transcription of specific genes and gene networks. *Annu Rev Genet* 1985;19:209.

Ligand-Gated Receptors

Cooper JR, Bloom FE, Roth RH. *The Biochemical Basis of Neuropharmacology*, 6th ed. New York: Oxford University Press, 1991.

Hall ZW. *An Introduction to Molecular Neurobiology*. Sunderland, MA: Sinauer Associates, 1992.

Kinases and Regulation

Krebs EG. Role of the cyclic AMP-dependent protein kinase in signal transduction. *JAMA* 1989;262:1815.

Ingebritsen TS, Cohen P. Protein phosphatases: properties and role in cellular regulation. *Science* 1983;221:331.

Tyrosine Kinases

Kahn P, Graf T, eds. *Oncogenes and Growth Control.* Springer-Verlag, Berlin, 1986.

Pelech SL. Networking with protein kinases. *Curr Biol* 1993; 3:513.

Schlessinger J, Ullrich A. Growth factor signaling by receptor tyrosine kinases. *Neuron* 1993;9:383.

G Protein-Linked Receptors

Gilman AG. G-proteins: transducers of receptor-generated signals. *Annu Rev Biochem* 1987;56:615.

Neer EJ. Heterotrimeric G proteins: organizers of transmembrane signals. *Cell* 1995;80:249.

Stryer L, Bourne HR. G-proteins: a family of signal transducers. *Annu Rev Cell Biol* 1986;2:391.

Calcium as a Second Messenger

Gerday C, Bolis L, Gilles R, eds. *Calcium and Calcium Binding Proteins.* Springer-Verlag, Berlin, 1988.

Schulman, H. The multifunctional Ca^{2+}/calmodulin-dependent protein kinases. *Curr Opin Cell Biol* 1993;5:247.

Review Questions

1. The function of calmodulin is:
   a. To produce IP_3 from PIP_2
   b. To serve as an intracellular Ca^{2+} binding protein that mediates many Ca^{2+}-regulated intracellular processes
   c. To bind to receptors on intracellular compartments, thereby triggering the release of Ca^{2+} from these intracellular storage sites
   d. To associate with cadherins to regulate cell-matrix adhesion
   e. To remove tropomyosin from actin during skeletal muscle contraction

2. Steroid hormone receptors:
   a. Are multipass transmembrane proteins
   b. Stimulate the dissociation of the alpha subunit of a G protein from the β and γ subunits of the protein
   c. Act directly on adenylyl cyclase to regulate cAMP production
   d. Have separate domains for ligand binding, DNA recognition, and transcriptional activation
   e. Activate the MAP kinase pathway that results in the regulation of various transcription factors

3. Phospholipase C-β carries out which of the following reactions?
   a. The cleavage of GTP to GDP+phosphate
   b. The conversion of ATP to cAMP
   c. The cleavage of PIP_2 to IP_3 and DAG
   d. The removal of phosphate groups from phosphorylated proteins
   e. The breakdown of LDL and HDL in adipocytes

4. Which of the following statements about steroid hormones is correct?
   a. All steroid hormones are DNA-binding proteins.
   b. Steroid hormones are synthesized from cholesterol.
   c. Steroid hormones generally are polypeptides.
   d. Steroid hormones control cellular activity by regulating the phosphorylation of numerous tyrosine kinases.
   e. Steroid hormones bind to cell surface receptors.

5. Catalytic receptors on the cell surface:
   a. Activate cells by producing Ca^{2+}
   b. Convert cAMP to ATP
   c. Often activate cells by phosphorylating intracellular target proteins on tyrosine residues
   d. Convert PIP_2 to IP_3 and DAG
   e. Are activated by steroid hormones and subsequently bind to DNA to regulate transcription

6. Unlike other ligand-gated receptors, NMDA receptors:
   a. Pass Mg^{2+} current
   b. Pass Ca^{2+} current
   c. Require a depolarized membrane and the combination of agonist binding for maximal activation
   d. a and c are correct
   e. b and c are correct

7. Ca^{2+} taken up into the presynaptic terminal from the extracellular space
   a. Gives calmodulin its enzymatic activity
   b. Is triggered by an increase in the resting membrane potential
   c. Is subject to vesicular storage and release by IP_3 receptor activation
   d. Stimulates phosphorylation of synapsin by CaM kinase
   e. Cannot be released from the terminal back into the extracellular fluid

8. A 6-month-old child was diagnosed with whooping cough. The mechanism for the effects of *Bordetella pertussis* in the respiratory tract is:
   a. Lymphocyte cellular reaction
   b. Release of pertussis toxin
   c. Inhibition of G-protein activity
   d. All of the above
   e. None of the above

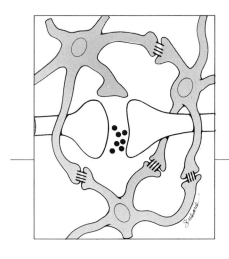

Chapter 9
Cell Division

Clinical Case

A 50-year-old man, Solomon Goldman, developed fatigue, weakness, night sweats, weight loss, and bone pain over the last 3 months. On further questioning, he has had intermittent episodes of prolonged erections (priapism) for approximately 2 weeks. On physical examination, he has a low-grade fever, pallor, ecchymoses, and hepatosplenomegaly. Papilledema of the optic nerve was seen in both fundi. Laboratory tests showed a hemoglobin of 10 gm/dl, leukocyte count of 100,000/mm^3 with 15% blast, 6% metamyelocytes, 9% bands, 55% neutrophils, and 15% lymphocytes; and a platelet count of 500,000/mm^3. The serum uric acid and lactic dehydrogenase levels were markedly elevated. Histochemical stain showed reduced leukocyte alkaline phosphatase activity.

Cellular proliferation is essential for normal growth and development. The most obvious example of the importance of regulated cell division events is embryogenesis. After fertilization, the zygote nucleus is formed, and the process of embryonic development begins. Over a 40-week gestation-period, millions of mitotic cell divisions occur, cellular differentiation takes place, and a human being is formed. However, there are numerous other examples for which rapid cellular division is necessary for human survival. For example, cellular proliferation must occur if humans are to repair wounds and mount effective immune responses. In addition, every day the body must replace millions of cells that are lost from normal wear and tear. If these normal repair mechanisms are blocked, by radiation damage, for example, a person will die within a few days.

Cells divide by the process of mitosis. However, mitosis, or M phase, occupies only a very small segment of the cell cycle. Most of a cell's existence is spent outside M phase in a period known as *interphase*. It is during interphase that cell growth, DNA synthesis, and the duplication of other important cellular constituents occur. Understanding the molecular mechanisms responsible for driving cells through the cell cycle is now one of the most rapidly advancing areas in the field of cell biology.

For a human to survive, the proliferation of cells must be tightly regulated. If a cell is able to evade the normal machanisms that control cell growth a cancer is formed, and the effects on the person's health are devastating. Therefore, the regulation of cellular growth must be unerring. Complex mechanisms are used to govern cellular proliferation. Some molecules, such as growth factors, stimulate cellular growth and division, whereas others, such as the protein products of tumor suppressor genes, are thought to down regulate cell division events. In this section, the cell cycle and the molecular regulation of its events will be described, then the cellular events of mitosis and meiosis will be discussed and, finally, the role of growth factors and protooncogenes will be considered.

Cell Cycle

The Cell Cycle Is Divided into Stages

The cell cycle is defined as the period that extends from the time a cell comes into existence, as a result of cell division, until the instant that the cell divides to give rise to two daughter cells (Fig. 9–1). In rapidly growing cul-

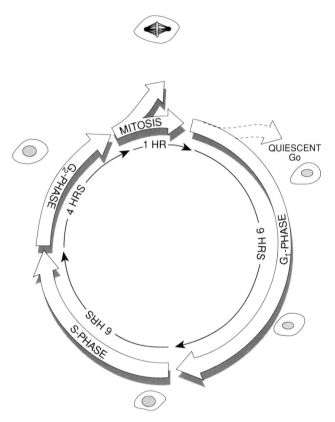

Figure 9–1. The phases of the mammalian cell cycle in culture. Interphase consists of G_1, S, and G_2 phases. The G_1 phase is the most variable stage of the cell cycle. The duration of G_1 can range from nonexistent, as in rapidly dividing embryonic cells, to indefinite, as seen in differentiated cells. Quiescent, nonproliferating cells are said to be in a special G_1 called G_0. During each cell cycle, a cell must decide whether it will differentiate or continue to cycle. The environmental and molecular cues that help a cell make this decision are not understood completely. However, this decision is made very early in G_1. If the growth environment favors proliferation, a cell will continue to cycle. If essential nutrients are lacking or if the culture chamber is overcrowded with cells, the cells will cease to divide. The point during the cell cycle at which a cell decides to continue cycling or to enter the quiescent G_0 state is called the *restriction point*. The S phase includes the entire period of chromosomal replication, and G_2 is a period when the cell prepares for mitosis. The M phase includes the events of spindle formation, chromosomal segregation, and cytokinesis. Two identical daughter cells are formed in each cell cycle. The approximate duration of each phase of the cell cycle for an imaginary mammalian cell line in culture is given in hours.

tures of human cells, the cell cycle generally lasts for approximately 20–24 hours. In the human body, the cell cycle can be relatively short—as with cultured cells—or it can be much longer. For example, liver hepatocytes are thought to divide only once or twice a year. During mitosis, or M phase, cultured cells round up, the chromosomes condense, the mitotic spindle forms, and the chromosomes then separate to opposite spindle poles. The events of nuclear division are sometimes called *karyokinesis*. After the chromosomes segregate to opposite spindle poles, the cell itself is split into two daughter cells by a process called *cytokinesis*. The events of M phase generally occur within 1 to 2 hours.

The remainder of the cell cycle is known as *interphase*. Interphase has been subdivided into the period of DNA duplication (S phase); the interval between the end of M phase and the beginning of S phase, termed G_1 phase (G stands for gap); and the period between the end of S phase and the onset of mitosis, called G_2 phase. The G_1 phase is a period of cellular growth, whereas G_2 is when the cell is preparing for mitosis. Experimentation has demonstrated that the length of time occupied by S, G_2, and M phases is relatively constant. Therefore, variation in the length of the cell cycle generally is confined to the duration of the G_1 phase. In some cells, such as rapidly dividing early embryonic cells, G_1 is virtually nonexistent. In other cells, G_1 may be so long that one has the impression that the cells have completely stopped cycling. Such quiescent cells are said to be in a special G_1 state called G_0.

The Passage Through the Cell Cycle Is Driven by a Small Group of Proteins

Our understanding of the molecular events occurring inside the cytoplasm that result in cells progressing through the cell cycle is advancing rapidly. Experimental studies with yeast mutants and nonmammalian embryos have identified many of the key cell cycle regulatory proteins. Subsequent experimentation has demonstrated that these same proteins also are present and active in mammalian cells. Principal among these important cell cycle regulators are two families of proteins called the *cyclins* and the *cyclin-dependent kinases* (cdks). The cyclins derive their name from the fact that these proteins are cyclically synthesized and degraded each cell cycle, whereas the cdk enzymes are so named because these enzymes are inactive until they bind to a specific member of the cyclin family. The individual cdks that are responsible for driving cells through M phase and S phase are cdk1 and cdk2, respectively. Experimental studies have shown that the enzymatic activity of these two proteins fluctuates during the cell cycle. Although cdk1 and cdk2 are present in constant amounts at all phases of the cell cycle, the enzymatic activity of the two proteins is turned on at key cell cycle transition points. Specifically, cdk1 is activated at the transition from G_2 into M phase, whereas cdk2 is responsible for driving a cell from G_1 into S phase. Current belief is that at these particular points in the cell cycle, the activated kinases phosphorylate the appropriate target proteins resulting in the cell being driven into either M phase or S phase.

The proteins that are largely responsible for activating the cdk enzymes are the cyclins. Different classes of cyclins have been identified, and separate types of cyclins are responsible for turning on the activity of the different cdk enzymes. The mitotic cyclins activate cdk1, allowing cells to be driven into M phase. Mitotic cyclins cannot be detected in cells in early G_1; however, synthesis of the mitotic cyclins begins and continues throughout interphase until a threshold level is reached at the G_2-M transition. Binding of mitotic cyclins to cdk1 induces a series of posttranslational modifications of cdk1 enzyme, inducing cdk1 activation, which results in the onset of mitosis. The actual targets of the activated cdk1 enzyme are not known in entirety, but one can speculate on potential cdk1 substrates when one considers the changes that occur in cells at the G_2-M transition. For example, as cells enter M phase, the nuclear envelope disassembles, the chromosomes condense, the mitotic apparatus forms, and cells round up and lose their substrate attachment. Therefore, logical candidates as substrates for the activated cdk1 enzyme would be nuclear lamins, chromosome proteins, components of the microtubule cytoskeleton, and actin attachment proteins. About midway through mitosis, the mitotic cyclins are proteolytically degraded, cdk1 kinase is turned off,

and the events of early mitosis are reversed by cellular phosphatases, resulting in chromosome decondensation, nuclear envelope reformation, spindle disassembly, and flattening of the progeny cells (Fig. 9–2).

A similar mechanism appears to be at work at the G_1-S boundary, with the key players in the G_1-S transition being cdk2 and a group of G_1 cyclins. The G_1 cyclins are synthesized and then degraded in an oscillatory manner during each cell cycle, with peak levels of the proteins being present late in G_1. Upon binding to cdk2, the enzyme gets activated, and the cell is driven into S phase. Once again, the targets of the activated cdk2 are not known, but logical candidates would be enzymes required for DNA synthesis and the machinery responsible for the production of DNA precursors. Following the completion of DNA synthesis, the G_1 cyclins are degraded, and the cell begins to prepare for M phase.

In addition to the mitotic and S phase-specific cyclins and cdks, recent evidence suggests that other members of the cyclin and cdk families exist in cells. Studies suggest that these additional cell cycle regulatory molecules are involved in the precise advancement of cells through G_1 phase and the integration of extracellular growth regulatory signals with the cell cycle progression machinery. The activity of one of these other cdks, cdk4, will be described in a later section.

In summary, a picture is beginning to emerge concerning mechanisms that result in the progression of cells through the cell cycle. According to current thinking, key events in the cell cycle occur under the direction of enzymes called *cyclin-dependent kinases*. The activity

Figure 9–2. The regulation of cdk1 and cdk2 during the cell cycle. The level of the cdks does not change during the cell cycle, although the activity of the enzymes fluctuates. **Upper panel:** The accumulation of the mitotic cyclins increases until the cyclin levels peak near the G_2–M boundary. The high levels of cyclin activate cdk1, and the enzymatic activity of cdk1 drives the cell through M phase. Late in M phase, cyclin is degraded, and the activity of cdk1 drops to resting levels. **Lower panel:** The regulation of cdk2 by the G_1 cyclins. G_1 cyclin levels increase rapidly until they peak near the G_1–S boundary. High levels of the G_1 cyclins result in activation of cdk2, and the enzyme phosphorylates the appropriate substrates necessary for the induction of DNA synthesis. Again, the G_1 cyclins are degraded, and the kinase activity of cdk2 returns to baseline levels.

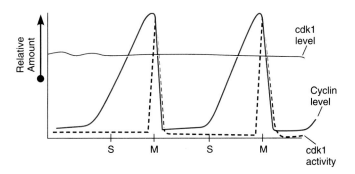

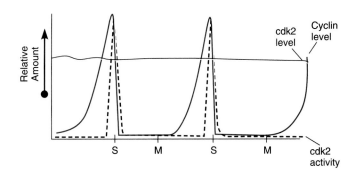

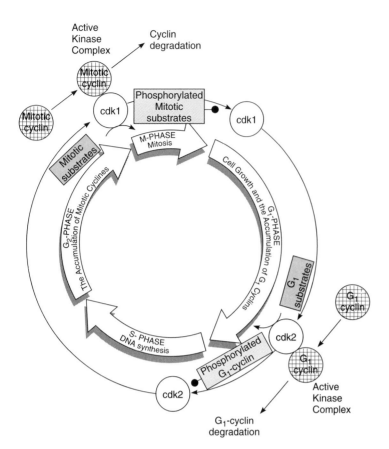

Figure 9–3. A schematic summarizing the biochemical and cellular events of the cell cycle. Early in G_1, cdks are disassociated from cyclins and are inactive. During G_1, cell growth occurs, and the G_1 cyclins accumulate. The level of the G_1 cyclins reaches a threshold, and they bind to and activate the cdk2. The activated enzyme complex phosphorylates appropriate substrates necessary for the induction of DNA synthesis, the G_1 cyclins are then degraded, and cdk2 activity turns off. In G_2, the mitotic cyclins accumulate. The mitotic cyclins bind to cdk1 and activate the enzyme. The activated enzyme complex phosphorylates appropriate substrates, and the cell is driven through M phase. The mitotic cyclins are then degraded, cdk1 kinase is inactivated, and mitosis is completed. Presumably, phosphatases exist that reverse the effects of the cdks following S and M phases.

of the appropriate cdk is controlled by members of a small family of proteins called *cyclins.* The cyclins are synthesized and degraded in an oscillatory manner during the cell cycle, and different cyclins activate different cdks. The molecular changes that occur in a cell which are thought to drive the cell cycle are summarized diagrammatically in Fig. 9–3.

Despite the rapid progress that has been made in our understanding of the cell cycle, several key gaps still exist in our knowledge of cell cycle regulation. For example, key events at the G_1-S and G_2-M boundaries appear to be due to the phosphorylation of target proteins. Many of these potential target proteins have not yet been identified. At a higher level, the involvement of molecules such as growth factors in modulating the behavior of the key cell cycle regulatory proteins is not known in detail. In addition, it is not yet completely clear how the protein products of many protooncogenes might be influencing cell cycle progression. Many of these questions should be resolved over the next several years.

Mitosis Results in the Production of Two Progeny Cells that Are Identical to the Parental Cell

The cell cycle culminates in mitosis, the onset of which is triggered by the activation of cdk1 kinase. This enzyme presumably induces several changes in a cell that result in the cell's progression into mitosis. These changes include, but are not limited to, the condensation of the chromosomes, nuclear envelope breakdown, spindle formation, and alterations in the organization of the actin cytoskeleton.

Mitosis traditionally is divided into five stages: prophase, prometaphase, metaphase, anaphase, and telophase. During *prophase,* the chromatin, which was duplicated during S phase, slowly condenses into chromosomes. The number of chromosomes varies from species to species, but in humans the diploid chromosome number is 46. The mitotic chromosome consists of two chromatids that are connected at a region termed the *centromere.* On the surface of the centromere are two *kinetochores*—one associated with each chromatid. The kinetochore is the region of the chromosome to which the spindle microtubules attach and through which the forces that result in chromosomal movements appear to act (Fig. 9–4). Prophase also is the period when the cytoplasmic microtubules are broken down as the cell prepares for the reorganization of the cellular microtubules into the mitotic apparatus.

Prometaphase begins at the instant the nuclear envelope disassembles. During prometaphase, some of the microtubules of the forming spindle apparatus make contact with and become attached to the kinetochores of the condensed chromosomes. Early in prometaphase, the chromosomes are haphazardly scattered around the mitotic spindle. However, the chromosomes begin to

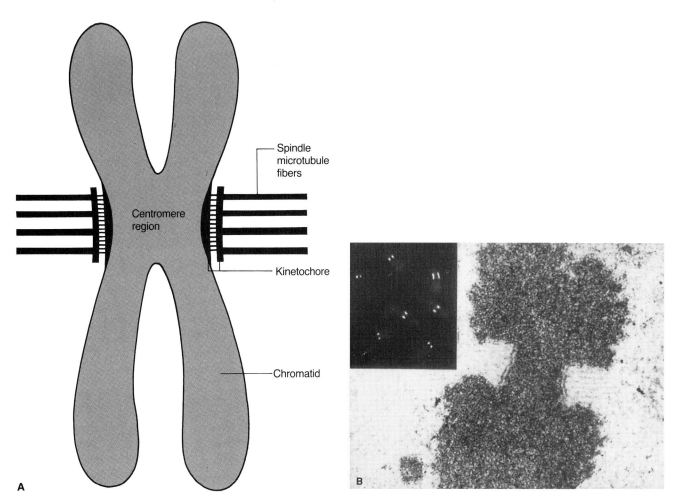

A

B

Figure 9–4. A schematic drawing of a metaphase chromosome. A: Two chromosome arms (chromatids) are attached at the centromere region. Sitting on the surface of the chromosome at the centromere are the kinetochores. The kinetochore is a trilaminar plate that is involved in spindle microtubule binding. The inner plate of the kinetochore rests on the centromeric heterochromatin and is probably involved in attaching the kinetochore to the chromosome. The outer kinetochore plate binds to spindle fibers. The plates of each kinetochore are nucleoprotein aggregates. The two kinetochore plates are separated by a weakly staining region that is probably composed of chromatin that holds the two plates together. B: An electron micrograph of a mitotic chromosome. In this figure, a cultured mammalian cell (*Indian muntjac*) was fixed and processed for electron microscopy. The trilaminar plate kinetochore can be seen at the primary constriction. The inset shows a cultured *I. muntjac* cell stained with an antibody that specifically recognizes kinetochore proteins. (**B**, courtesy of B. R. Brinkley and R. P. Zinkowski.)

undergo a series of movements that result in the alignment of all of the chromosomes at the equator of the spindle. The arrangement of chromosomes at the midzone of the spindle is sometimes termed the *metaphase plate configuration*. Under the microscope, metaphase appears as a period of inactivity or rest as the cell prepares for anaphase.

Anaphase is the period of mitosis when the sister chromatids separate. The onset of anaphase is the time at which the centromere splits in two. At that moment, the chromatids begin their migration, or segregation, toward opposite poles of the spindle. The force-generating mechanism that results in anaphase chromosomal segregation has not been identified, although there are several possible explanations for this motile event. Possibly, one or more of the proteins that compose the kinetochore contain the ATPase activity that allows the movement of chromosomes along the spindle microtubules. Alternatively, force-generating molecules may be associated with the spindle microtubules. Although the details concerning anaphase chromosomal migration are unknown, what is clear is that the microtubules that extend from the spindle pole to the kinetochore must disassemble for anaphase chromosomal movement to occur. If cells are treated with antitumor drugs that inhibit the breakdown of spindle microtubules, anaphase chromosomal segregation is blocked.

The final stage of mitosis is *telophase*. Telophase is identified as the period when the daughter chromatids have completed their segregation to the opposite spindle poles. Other events that occur during telophase are the reformation of a nuclear envelope around each set of chromosomes, the decondensation of chromosomes into chromatin, the dissolution of the mitotic apparatus, and cytokinesis. Cytokinesis is accomplished by a contractile ring of actin microfilaments that encircles the cortex of the dividing cell in an area of the cell surface overlying the area where metaphase plate chromosomes were organized. The end result of mitosis is that two progeny cells are produced that are identical in genetic composition to the original parent cell.

The Mitotic Apparatus and Contractile Ring Are M-Phase-Specific Cytoskeletal Structures

The cellular machine that is responsible for directing the events of chromosomal segregation is the mitotic spindle apparatus. Mitosis occurs with such unerring fidelity because the genes that are contained within the chromosomal arms must be divided evenly. However,

the genes themselves contribute little to the process of mitosis. Instead, the active roles in mitosis are played by the spindle microtubules, the centrosomes, and the kinetochore region of the chromosomes. The chromosomal arms, which contain the genetic material, are thought to be passive participants in the mitotic event.

The centrosome is responsible for nucleating microtubule growth in most human cells. The centrosome is composed of a centriole pair and a surrounding cloud of amorphous substance called the *pericentriolar material* (see Chapter 3). Experimental analysis has demonstrated that the centrosome's microtubule nucleating capacity is contained within the pericentriolar cloud and not the centriole cylinders (Fig. 9–5). As the cell proceeds through interphase, the centrosome is duplicated and, at the onset of mitosis, the daughter centrosomes migrate to opposite sides of the nucleus, where they will serve as the mitotic spindle poles (Fig. 9–6). Along with this migration, the interphase array of cytoplasmic microtubules is disassembled, and the interphase cytoplasmic microtubule complex is then replaced by the mitotic apparatus. The mitotic apparatus is a complex array of microtubules, membranous vesicles, chromosomes, and other cellular proteins (see Fig. 1–3). As in interphase, all of the spindle microtubules appear to originate in the pericentriolar material.

Three classes of microtubules, all of similar biologic composition, can be identified in the mitotic apparatus. The first class, the astral microtubules, are arranged in

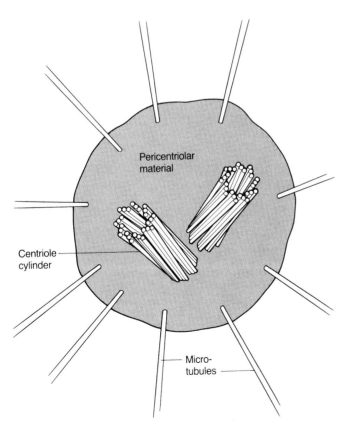

Figure 9–5. A schematic of the centrosome complex. A centrosome is composed of a pair of centriole cylinders arranged at right angles to each other and an amorphous cloud of pericentriolar material. The centrioles are composed of a set of nine short, triplet microtubules. The composition of the pericentriolar material is not known completely, but the pericentriolar cloud serves as the site of microtubule nucleation.

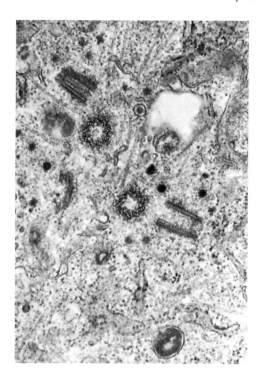

Figure 9–6. The morphology of the centrosome complex. An electron micrograph of a cultured mammalian cell at the G₂–M boundary. The duplicated centrosome can be seen. The centrosome will split, and the two centrosomes will migrate to opposite ends of the cell to serve as the mitotic spindle poles. In each centriole pair of the duplicated centrosome, one of the centrioles is cut in cross section, whereas the other is cut longitudinally. (McGill M, et al. *J Ultrastruct Res* 1976;57:43–53, with permission.)

class of spindle microtubule extends from the centrosome past the metaphase plate region, and these microtubules then overlap with microtubules that extend from the opposite spindle pole. These polar microtubules are thought to be involved in pushing the two spindle poles apart late in anaphase, so that the contractile ring can split the cell in two (Fig. 9–7).

One of the key events in spindle morphogenesis is the attachment of the chromosomes to the spindle microtubules. This occurs only at the kinetochore regions of the mitotic chromosomes (see Fig. 9–4). This has been demonstrated in cells in which the kinetochores have been experimentally detached from the mitotic chromosomes. In these cells, the detached kinetochores are able to attach to the spindle microtubules and to undergo the entire repertoire of mitotic chromosomal movements. The chromosomal arms in these experimentally treated cells do not associate with the spindle and are displaced to the cell periphery. Therefore, the proteins of the kinetochore region are responsible for the attachment of chromosomes to spindle microtubules, as well as for directing most of the chromosomal migrations that occur during prometaphase and anaphase.

The other major cytoskeletal structure that plays an active role in cell division is the contractile ring (Fig. 9–8). The contractile ring is composed of a belt of actin microfilaments, myosin, and other actin cytoskeletal components that are involved in attaching the microfilaments to the plasma membrane. During cytokinesis, the belt of actin microfilaments constricts, which results in the cell being cleaved in half. Molecular biologic studies have demonstrated that cytokinesis is dependent on the activity of myosin, suggesting that the events of cytokinesis may occur by a mechanism that is similar to smooth-muscle contraction. Following mitosis, both the contractile ring and mitotic apparatus are disassembled and replaced by the interphase configurations of microtubules and microfilaments.

a starlike fashion around each spindle pole. These microtubules are thought to be important in orienting the contractile ring. The second type is the kinetochore microtubule that extends from the centrosome to the kinetochore and is important for directing the chromosomal migrations that occur during mitosis. The final

Figure 9–7. A schematic of the organization of the mitotic spindle at metaphase. Three classes of microtubules can be recognized: the astral fibers, the kinetochore microtubules, and the polar microtubules. The polar microtubules extending from one spindle pole overlap with those that are nucleated from the opposite spindle pole. In electron micrographs, the overlapping polar microtubules appear to be linked together. At metaphase, the chromosomes are aligned at the midregion of the spindle.

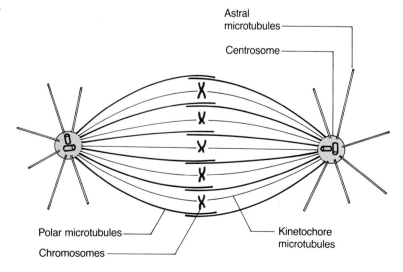

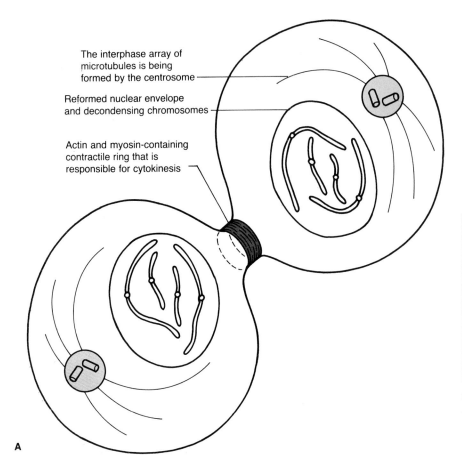

The interphase array of
microtubules is being
formed by the centrosome

Reformed nuclear envelope
and decondensing chromosomes

Actin and myosin-containing
contractile ring that is
responsible for cytokinesis

A

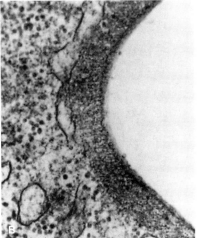

Figure 9–8. The contractile ring is responsible for cleaving a cell in half during
cytokinesis. **A:** The contractile ring is composed of bundles of actin microfilaments
and myosin. **B:** An electron micrograph through a portion of the contractile ring.
The dense band of actin myosin filaments can be seen just inside the plasma membrane. (Beams HW, Kessel RG. *Am Sci* 1976;64:279–290, with permission.)

Meiosis

Meiosis Results in the Production of Haploid Gametes

Unlike mitosis, which results in the production of genetically identical diploid progeny cells, *meiosis* is a specialized form of reductive division that results in the generation of haploid gametes. Moreover, each of the gametes that is produced is genetically distinct from each other and from the original parent cell. This is achieved through a two-step process that has many similarities to mitosis but several very key differences (Fig. 9–9).

Before the onset of the first meiotic division, the parental cell undergoes a round of DNA synthesis, and the cell then proceeds into the first round of meiosis. It is during meiosis I that the key events resulting in genetic variability take place. First, genetic recombination, or crossing over, occurs during the extended prophase I period. Crossing over results in the actual physical exchange of portions of chromosomes between maternally and paternally derived chromosomes of a homologous pair. This exchange results in a mixing of the parental genes that leads to increased genetic combinations. The second event that contributes to genetic diversity is the independent assortment of paternal chromosomes during meiosis I. From independent assortment alone, 2^n different types of gametes could be formed (where n is the haploid number of chromosomes). Because humans have 23 pairs of chromosomes, 2^{23}, or about 8×10^6 genetically unique gametes could be formed from this event. In reality, however, the number of genetically distinct gametes that are produced is much higher, owing to the recombination events that occur during prophase I.

The net result of meiosis I is that two progeny cells are produced that have a diploid amount of DNA. However, unlike mitosis, both of the daughter cells are different genetically from the diploid parent cell. A second round of DNA synthesis does not occur between the first and second meiotic divisions (see Fig. 9–9). Hence, meiosis II resembles mitosis, in that a spindle

Figure 9–9. A schematic summarizing the events of meiosis. During meiosis I, a round of DNA synthesis occurs. In prophase I, the homologous chromosomes pair and genetic recombination (crossing over) occurs. In metaphase I, the homologous chromosome pairs align on the spindle equator such that some of the maternal chromosomes face one pole of the meiotic spindle apparatus, whereas other maternal chromosomes face the other spindle pole (the same is true of the paternal chromosomes). This results in the independent assortment of the maternal and paternal homologues during anaphase I. DNA synthesis does not occur during meiosis II, and the chromatids separate during anaphase II. The net result of one round of meiosis is that four nonidentical haploid cells are produced.

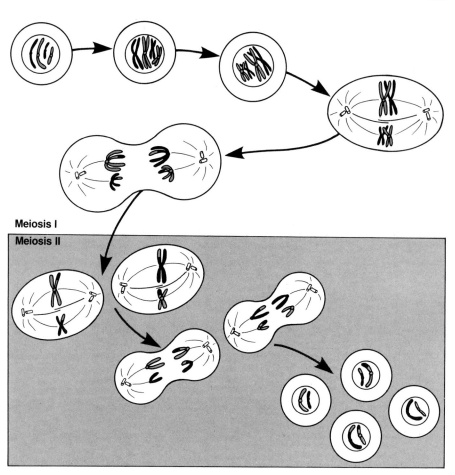

Meiosis I

Meiosis II

apparatus forms, and the chromosomes segregate. However, because no DNA synthesis has occurred, the progeny cells of meiosis II are haploid, rather than diploid. Moreover, because of the specialized events of meiosis I, the cells that are produced following meiosis II are different genetically. On completion of meiosis, the vertebrate egg is fully mature, whereas the sperm must still undergo maturation.

Spermatogenesis Results in the Production of Highly Specialized, Streamlined Cells

The process by which spermatozoa are produced is called *spermatogenesis*. Spermatogenesis occurs in the seminiferous tubules of the testes and continues from the onset of puberty until death. The haploid cells that are produced during meiosis II in the male are known as *spermatids*, and inside the seminiferous tubules the spermatids undergo the morphologic differentiation that produces the highly motile sperm. During this morphogenesis, much of the cytoplasm is shed, the nucleus is condensed, and a large flagellum is assembled. The cells that are produced are highly specialized and streamlined for swimming. From one round of meiosis, four mature spermatozoa are produced.

Egg Maturation Is Tightly Regulated and Occurs in Stages

Oogenesis, or the process of oocyte development, begins *in utero*. Within the embryonic ovary, some of the cells differentiate into primary oocytes, and the meiotic events commence. However, the cells progress only as far as prophase of meiosis I, at which point division arrests until sexual maturity. During this arrested period, the oocyte grows. This growth is due to yolk accumulation and to the production and storage of other proteins that will be needed for early embryonic growth. At sexual maturity, several of the primary oocytes are stimulated every month by hormones to resume progression through the meiotic cycle. The completion of meiosis I is characterized by an asymmetric cell division, such that most of the yolky cytoplasm is retained within the developing oocyte, now called a *secondary oocyte*. The remaining half of the chromosomes are pinched off of the oocyte to form a small residual structure called a *polar body*. The secondary oocyte then continues into meiosis II but is arrested at metaphase.

Experimentation has shown that a substance exists inside of oocytes that is responsible for metaphase II arrest, and this material is referred to as *cytostatic factor* (CSF). The mechanism by which CSF arrests oocyte

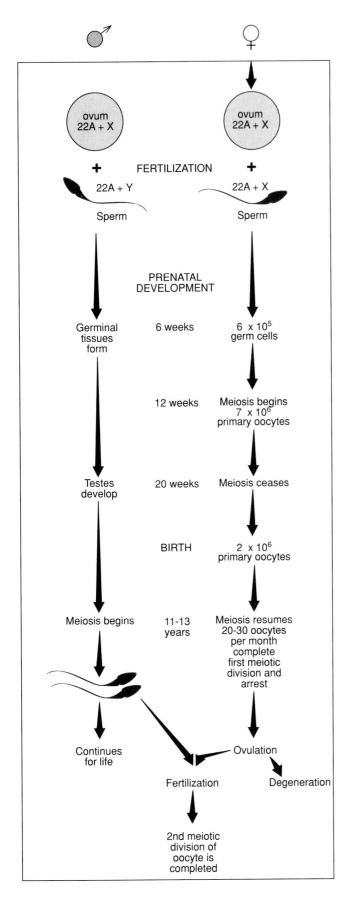

◀ **Figure 9–10.** The timing of meiotic events in humans. (Modified from Widnell, CC, Pfenninger, KH. *Essential Cell Biology.* Baltimore: Williams & Wilkins, 1990.)

development is not clear, but experimental evidence suggests that it involves a signaling pathway that includes the c-mos protooncogene product and a kinase called *MAP kinase.* It is thought that the mos protein, which itself is a protein kinase, is synthesized during oogenesis and activates MAP kinase, either directly of indirectly. The activated MAP kinase, through some unknown mechanism, then arrests the oocyte in metaphase II. At fertilization, the mos protein is degraded, CSF activity is destroyed, and meiosis is completed. A second asymmetric division occurs, generating a mature fertilized egg and second polar body (the polar body that was produced by the first meiotic division may also complete a second meiotic division, producing two polar bodies). The purpose of the asymmetric divisions that occur during oogenesis is to preserve the nutrient-rich cytoplasm that will be necessary to sustain the egg until implantation into the uterus occurs.

Unlike spermatogenesis, in which four functional gametes are produced by each round of meiosis, the meiotic divisions of oogenesis result in the production of only one functional egg and three nonfunctional polar bodies. A time scale for the events of gametogenesis in males and females is provided in Fig. 9–10.

Fertilization Restores the Diploid State

The key event of the process of fertilization is not simply the fusion of the sperm and egg cells, but the mixing of the haploid male and female genomes that restores the diploid condition. Sperm egg fusion occurs in the oviduct, and the zygotic embryo is transported by the cilia of the oviduct cells to the uterus where implantation of the embryo into the uterine wall occurs. During its transport down the oviduct, the metabolism of the developing embryo is driven by the nutrient-rich, yolky cytoplasm. Sometime following implantation, a circulatory connection is made to the maternal blood supply, and the embryo is then nourished by the mother.

As sperm migrate through the female reproductive tract, they undergo a process called *capacitation.* Up until the time of capacitation, sperm are not able to fertilize an egg; following capacitation, mammalian sperm are capable of fusing to an egg. The mechanisms of capacitation are still unknown, but it is thought that secretions in the female genital tract cause an alteration in the sperm plasma membrane that permits sperm egg binding to occur. Sperm egg binding seems to be mediated by the zona pellucida, a glycoprotein-rich extracellular matrix that surrounds the unfertilized oocyte. The zona pellucida in mammalian cells is composed of three proteins, ZP-1, ZP-2, and ZP-3, with ZP-3 serving as the actual sperm receptor. Species-specific molecules on the surface of the sperm bind to a species-specific ZP-3 receptor glycoprotein within the zona pellucida during fertilization. This binding of the sperm to the zona induces the sperm to release digestive enzymes that enable the sperm to bore its way through the zona pellucida (Fig. 9–11). The plasma membranes of the sperm and egg then fuse, and the sperm nucleus enters the egg cytoplasm.

Following sperm–egg fusion, the egg undergoes a series of changes that result in the metabolic activation of the dormant egg cytoplasm. The first identifiable change in the mammalian egg at fertilization is that numerous membranous vesicles that line the cortex of the egg are discharged. These vesicles, the cortical granules, release digestive enzymes that modify the glycoprotein network of the zona pellucida in such a way that additional sperm are no longer able to bind to the egg. In addition, an increase in the intracellular level of Ca^{2+} occurs in the egg cytoplasm. This increase in Ca^{2+} is responsible for the inactivation of cytostatic factor, resulting in the subsequent completion of meiosis. After meiosis is completed, the maternal and paternal genomes are united to form the zygote nucleus, and the fertilized egg then begins a series of very rapid embryonic cell divisions.

Environmental Regulation of the Cell Cycle

Growth Factors Help Regulate Embryogenesis, Development, and Other Processes that Require Cellular Proliferation

The human body is a complex society of billions of cells working together to ensure the survival of the organism as a whole. The cells in the body have the ability to regulate the behavior of one another through complex communication networks involving both cell–cell interactions and the secretion of molecules. Some of these environmental cues are involved in controlling the proliferation of cells.

Cell proliferation is required for several processes, including embryogenesis and development, wound repair, the immune response, and the replacement of cells that are lost through normal wear and tear. Many of these cellular proliferative events occur as responses to any one of a small group of polypeptide growth factors that regulate a given population of cells. Each growth factor has specific target cells and tissues. This specificity is regulated through the presence of growth factor receptors in the plasma membranes of the target cells. After being bound to a growth factor, the activated receptor transduces this signal to the inside of the cell, either directly or through the use of intracellular secondary messengers. Regardless of which mechanism is used, the metabolic activity of the target cells is modified considerably. Following stimulation by a growth factor, a cell can be triggered to exit from the quiescent G_0 phase of the cell cycle and to enter into G_1. The cell then is committed to complete one round of the cell cycle. After completion of M phase, if sufficient growth factor is present in the environment, the progeny cells will continue to cycle. If the stimulatory signal is no longer present, the cells will exit the cell cycle and reenter the quiescent G_0 state. In this manner, the growth and proliferation of cells in a multicellular organism can be regulated by the environment.

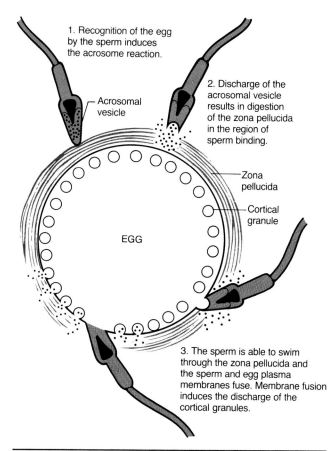

1. Recognition of the egg by the sperm induces the acrosome reaction.

— Acrosomal vesicle

2. Discharge of the acrosomal vesicle results in digestion of the zona pellucida in the region of sperm binding.

— Zona pellucida

— Cortical granule

EGG

3. The sperm is able to swim through the zona pellucida and the sperm and egg plasma membranes fuse. Membrane fusion induces the discharge of the cortical granules.

Figure 9–11. A schematic drawing summarizing the events of sperm egg fusion. Species-specific molecules on the surface of the sperm recognize molecules in the zona pellucida, inducing the acrosome reaction in the sperm. During the acrosome reaction, digestive enzymes are discharged from the acrosomal vesicle that cause a localized digestion of the zona pellucida. This permits the sperm to traverse the zona to the surface of the egg, where membrane fusion occurs between the plasma membranes of the sperm and egg. The fusion of the sperm and egg membranes causes a discharge of the cortical granules and activates the dormant oocyte cytoplasm. The activated oocyte is then able to complete meiosis, and the haploid maternal and paternal nuclei are united to form a zygotic embryo.

Growth factors initially were discovered when it was found that fibroblasts grown in culture would proliferate when grown in a medium that contained serum, but not in a medium that had been supplemented with plasma. The difference between serum and plasma is that serum is prepared by allowing blood to clot, whereas plasma is prepared by centrifuging cells out of blood without allowing clotting to occur. Following wounding, blood platelets are triggered to release the contents of their secretory granules, an event that induces the clotting response. One of the factors in these secretory granules was found to be the component that was responsible for triggering fibroblast growth in culture. This compound, called *platelet-derived growth factor* (PDGF), is a small glycoprotein that binds to PDGF receptors on the surface of connective tissue cells to stimulate wound repair. Similar experimental observations have led to the discovery, purification, and characterization of numerous other growth factors. All known growth factors are small proteins that work either singly or in concert to regulate the proliferation of cells in the human body. Some growth factors, such as interleukin-2 (IL-2), which stimulates T lymphocytes, and nerve growth factor (NGF), which promotes nerve cell growth and survival, have very narrow target cell specificities. Other growth factors, such as PDGF, epidermal growth factor (EGF), and the insulin-like growth factors I and II (IGF-I and II) have broader target cell specificities and can modulate the growth behavior of numerous cell types.

Many growth factors also have a second, very important, physiologic function. Growth factors can also act as chemoattractant agents. For example, the chemoattractant properties of PDGF may be essential for drawing cells to the site of an injury before initiating their proliferation. In addition, the chemoattractant characteristic of growth factors may be very important in guiding cells and cellular extensions, such as nerve growth cones, to the appropriate regions during embry-

onic growth and development. Growth factors, their receptors, and signal transduction events were described in detail in Chapter 8.

Cancerous Cells Have Escaped from Normal Cell Cycle Regulation

For the human body to survive as a healthy entity, the growth and proliferation of cells must be a tightly regulated process. Occasionally, a cell escapes from the normal constraints that are placed on cell growth and begins to divide uncontrollably. Such cells are said to be *transformed*. For years, it was felt that studying transformed cells and identifying the changes in these cells that allow them to escape from normal cell growth regulation could provide important information about the genes and proteins that control the mammalian cell cycle. This experimental strategy came to fruition with the discovery of oncogenes, protooncogenes, and the identification of the proteins encoded by protooncogenes. This has resulted in an understanding of how a mutation can result in the abnormal cellular proliferations that are characteristic of tumorigenic cell growth.

Before discussing how a mutation causes a cell to escape from the constraints that are normally placed on cell growth, we should consider the characteristics of cancer cells. Transformed cells show several distinct phenotypic traits when grown in culture. Unlike normal cells, which cease to grow under culture conditions, transformed cells will continue to proliferate under certain conditions of nutrient or growth factor deprivation. In addition, transformed cells will grow in culture to extremely high densities. Normal cells will continue to divide until they completely cover the substrate on which they are cultured, at which time they will cease to proliferate. This phenomenon is known as *density-dependent inhibition of growth*. Cancer cells, on the other hand, will not cease growing after they have

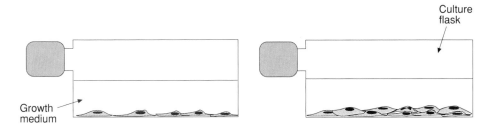

Figure 9–12. Transformed cancerous cells behave differently than do nontransformed cells in culture. Left: A culture flask containing nontransformed mammalian cells. Note that all of the cells are attached to the growth substrate and that the cells ceased proliferating when a confluent monolayer of cells was formed. Right: A culture flask containing transformed cells is shown. Notice that the cells no longer require substrate attachment, and that the cells did not stop proliferating after confluence was achieved.

formed a confluent monolayer on a culture dish. Instead, transformed cells will begin growing on top of one another in layers. This phenomenon highlights a second characteristic of transformed cells. Normal cells require substrate attachment to grow in culture; cancer cells can be grown in suspension, completely unattached to a substrate. Finally, cancers are malignant. That is, cancer cells are invasive—the cells not only proliferate abnormally, but tumor cells also are able to invade other tissues to form secondary tumors, or metastases. These, as well as several other less obvious traits, set cancer cells apart from normal cells and contribute to the development of tumors (Fig. 9–12).

Oncogenes and Protooncogenes

Experimental Assays Have Been Developed for the Identification of Oncogenes

Analysis of the DNA in transformed cells exhibiting aberrant behavior has allowed the identification of the genes that cause the transformed phenotype. For these types of studies, cultured cells are treated with a mutagenic agent such as ultraviolet (UV) light. The cells are then cultured, and the growth characteristics of the cells are monitored (Fig. 9–13). Those cells that show loss of anchorage dependence, loss of density-dependent inhibition of growth, and other traits of cancer cells, are cloned and the DNA is analyzed. For DNA analysis, the DNA is isolated from the mutated cell and digested into very small pieces with restriction enzymes; the cleaved DNA is then used to transform a second normal cell line. Those DNA fragments that are able to transform the recipient cells are then cloned, sequenced, and analyzed.

A second method of identifying cancer-causing genes is to analyze the genomes of tumor viruses. Many types of viruses have been identified that can cause tumorigenic growths in various mammalian species. The most useful viruses for cancer studies are the RNA tumor viruses, or retroviruses. When the genomes of tumor viruses are analyzed, the appropriate cancer-causing gene can be identified. Surprisingly, the cancer-causing genes that are carried by retroviruses and other tumor viruses are not normal parts of the viral genome. Apparently, these viruses have picked up the cancer-causing genes during the normal infection of a healthy cell and modified or mutated the DNA sequences during the incorporation of these transforming genes into the viral genome. To visualize how this could occur, one needs to consider the retroviral infectivity pathway (Fig. 9–14). After infecting a cell, retroviruses convert their RNA genome into a double-stranded DNA molecule by a virally encoded enzyme called *reverse transcriptase*. The viral DNA is incorporated into the host DNA, where it is duplicated as the host cell DNA is doubled. After insertion into the host DNA, a series of genetic rearrangements are thought to occur that result in a portion of a cell regulatory gene being inserted into the viral genome. During these putative genetic rearrangement steps, the normal cellular gene apparently is modified in a fashion that converts it to a gene that results in the loss of growth control. For example, consider an enzyme that is important for cellular proliferation events. Enzymes often have regulatory domains, as well as catalytic domains, so that enzyme activity can be turned on and off at the appropriate time. Suppose that during the genetic rearrangements that were described, the DNA encoding the regulatory domain is damaged in such a way that the cell cannot turn off the enzyme once it is produced. Because the catalytic

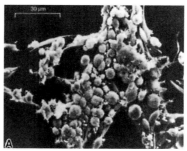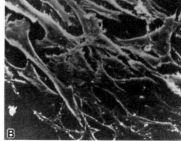

Figure 9–13. The morphology of transformed cells in culture as viewed by SEM. Mammalian cultured cells transformed using a strain of Rous sarcoma virus containing a temperature-sensitive oncogene. **A:** At the permissive temperature, the cells assume a transformed phenotype characterized by a rounded morphologic appearance and loss of substrate attachment. **B:** At the restrictive temperature, the cells return to their nontransformed phenotype, evident by their flattened appearance, which permits strong adherence to their culture dish. (Alberts B, Bray D, et al. *Molecular Biology of the Cell*, 2nd ed. New York: Garland Publishing, 1989, with permission. Photo courtesy of Dr. G. S. Martin.)

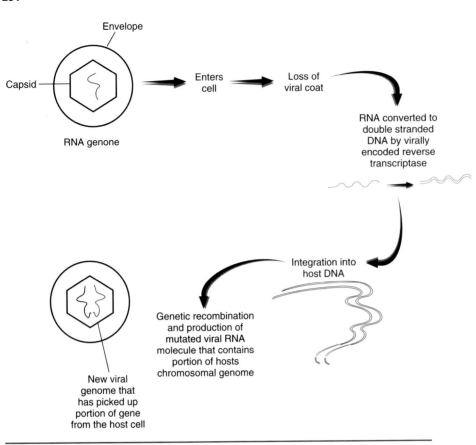

Figure 9–14. A potential mechanism of transformation caused by viral infection. A schematic showing the viral infectivity pathway of a retrovirus that has invaded a mammalian cell. During a subsequent infection, the mutated virus could cause transformation of the infected cell.

domain would remain fully active in this scenario, the modified form of the enzyme would have the ability to stimulate the cell. However, because the regulatory domain would be damaged, the cell would not have the capacity to turn off the excitatory enzyme. As a result, the cell would receive a constant stimulatory signal from the mutated gene product and would proliferate indefinitely. This scheme provides a potential explanation of how both oncogenic viruses and environmental mutagens could convert a normal cell regulatory gene into a cancer-causing gene.

Other Modifications in the Genome Can Result in Cancerous Growth

In addition to direct modification of the coding region of a gene, other alterations in the genome can result in abnormal cellular proliferation. For example, chromosomal translocations can result in the loss of normal growth regulation. The best-studied example of this phenomenon is the so-called Philadelphia chromosome (Ph[1]) that is characteristic of chronic myelogenous leukemia. In this type of cancer, a piece of chromosome 9 has been exchanged with a portion of chromosome

22. It has been shown that a cancer-causing gene called *abl* lies at the breakpoint of chromosome 9. As a result, a truncated form of the *abl* protein product that causes overstimulation of cells is produced.

Another example of how a normal gene can be converted into a cancer-causing gene is by simple overamplification of the gene. By some unknown mechanism, gene duplication events can occur that result in hundreds of copies of a particular gene being produced, and these additional gene copies subsequently will be transcribed and translated. If the gene in question happens to be a gene that is responsible for stimulating cell growth, overproduction of this protein product would lead to overstimulation of cellular activity.

The Protein Products of Cancer-Causing Genes Have Been Identified

The mutated genes that are responsible for the development of a cancer are called *oncogenes*, and the normal cellular counterparts of these genes (i.e., the nonmutated forms) are called *protooncogenes*. Molecular cloning and sequencing of protooncogenes has helped identify and characterize the protein products of the protoonco-

Figure 9–15. A schematic drawing showing the localization of and interplay between many known protooncogenes and protoonco-gene products.

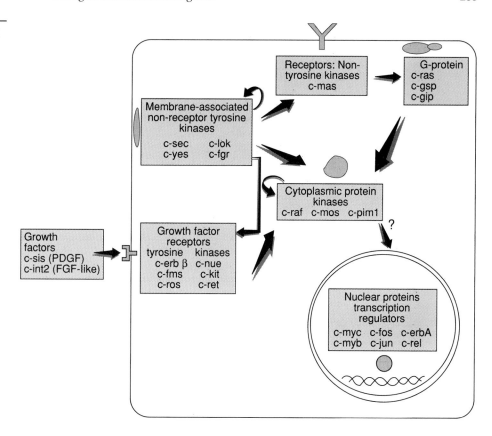

genes. Perhaps not too surprisingly, experimentation has shown that many of the protooncogenes encode proteins that are important for the regulation of cellular growth and proliferation. For example, the protooncogene c-*sis* (the *c* stands for cellular; the oncogene counterpart of c-*sis* is v-*sis*, in which the *v* refers to the viral oncogene) encodes the growth factor PDGF. Cells that contain v-*sis* overproduce a modified PDGF-like molecule that binds to PDGF receptors on the surfaces of cells and results in the constant stimulation of cell proliferation. Likewise, the *ras* protooncogenes encode a class of GTP-binding proteins that are involved in signal transduction events (see Chapter 8). During certain signal transduction events, *ras* proteins bind to GTP and are activated; *ras* proteins are inactivated by cleaving GTP to GDP and, as a result, the stimulatory signal is turned off. Mutated forms of *ras* proteins, encoded by v-*ras*, are able to bind to GTP efficiently but are unable to cleave the molecule enzymatically. Therefore, cells containing this aberrant protein are constantly receiving a stimulatory signal and, thus, proliferate uncontrollably.

Many protooncogenes encode kinases. This seems logical because, as discussed earlier, phosphorylation and dephosphorylation of cellular proteins is a mechanism that is used to control the cell cycle and cellular proliferation. Some protooncogenes, such as c-*mos*, encode common serine-threonine kinases. However, many protooncogenes encode much less common forms of kinases called *tyrosine kinases* (see Chapter 8). Tyro-

sine kinases are so called because these enzymes phosphorylate tyrosine residues on proteins. Many tyrosine kinases are transmembrane cell surface receptor proteins that have their kinase activity on the cytoplasmic domain of the molecules. Normally, these receptors bind to ligand, are activated, then phosphorylate various target substrates. Mutated forms of these receptors maintain kinase activity, even in the absence of bound ligand. An example of this type of oncogene is v-*erb-B*, which encodes a mutated form of the EGF receptor.

Finally, several protooncogenes encode proteins that are involved in regulating transcriptional activity. One of these, c-*jun*, encodes the transcriptional activator AP-1. This intranuclear transcription factor binds to a specific nucleotide sequence found in many promoters and enhancers. The mechanism by which these proteins induce transformation is not clear. Figure 9–15 summarizes the intracellular location and functional activity of many protooncogenes.

Tumor Suppressor Genes Inhibit Cell Proliferation

In addition to protooncogenes, a second type of mutated gene appears to be involved in the onset of certain cancers. These genes, the so-called tumor suppressor genes, or anti-oncogenes, behave in a fashion quite distinct from that of protooncogenes. The protein products of protooncogenes are involved in

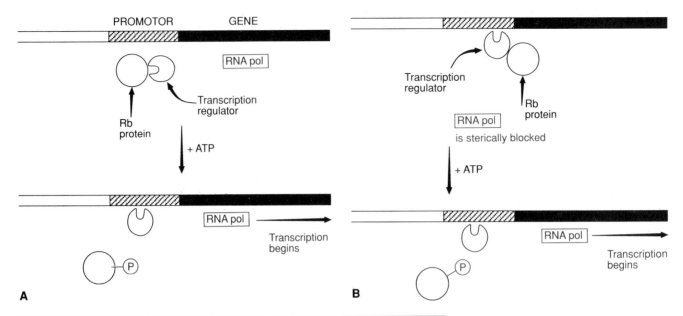

Figure 9–16. The activity of the Rb tumor suppressor protein may be regulated by a cdk-type enzyme. A schematic showing two potential mechanisms of action of the Rb tumor suppressor gene product and the putative regulation of the protein by a cdk family member. **A:** The Rb protein is attached to a gene regulatory protein (a transcription activator), thereby inhibiting the activator from binding to a promotor region and turning on transcription. Phosphorylation of Rb by cdK4 causes the Rb protein to dissociate from the transcription activator, allowing the activator to bind near the promotor region so that transcription of the genes immediately downstream will begin. The gene products presumably would allow the cell to progress deeper into the cell cycle. **B:** The Rb protein is bound to a transcription activator that is sitting near the promotor region. Phosphorylation of Rb by cdK4 causes the Rb protein to dissociate from the complex, which allows transcription to begin. From these schematics, it is clear that the loss of functional Rb protein because of mutation would result in the loss of cell cycle regulation.

growth stimulation; the protein products of tumor suppressor genes, on the other hand, are thought to repress cellular growth and cell division. Therefore, loss of these genes through mutation can lead to cell transformation by removing the restraints that normally regulate cell growth.

The most intensively studied tumor suppressor gene is the gene whose loss results in the onset of the rare eye cancer retinoblastoma, as well as certain breast and lung cancers. It is thought that the retinoblastoma gene (*Rb*) encodes a protein that binds to and regulates proteins that are involved in the transcription of DNA into mRNA. Because it is a growth suppressor, it is thought that the Rb protein binds to transcription factors and inactivates them, thereby inhibiting the mRNA production necessary for the cell to progress through the cell cycle. Loss of the Rb protein through deletion or mutation would result in the constant synthesis of mRNAs required for cell cycle progression. This, in turn, would lead to uncontrolled cell growth. Recent evidence suggests that the activity of the Rb protein may be regulated by one of the cdks called *cdk4*. If this is true, it may pro-

vide an explanation for the growth-suppressive behavior of Rb. In its unphosphorylated state, the Rb protein would bind to transcription regulators and would inhibit transcription of genes necessary for advancement through the cell cycle. Treatment of a cell with a growth factor sets off an intracellular cascade of reactions that results in the activation of cdk4. Following phosphorylation by cdk4, Rb would disassociate from transcription factors and would allow the transcriptional activity necessary for cell cycle progression (Fig. 9–16)

A second important and intensively studied tumor suppressor is a protein called *p53*. The normal function of the p53 protein appears to be as a guardian of the genome. Specifically, the p53 protein seems to have the capacity to sense DNA damage. Cells have developed elaborate mechanisms that allow the repair of mutated DNA. Occasionally, DNA gets damaged, either by an environmental agent or by exposure to UV. If this damaged DNA is not repaired, deleterious mutations can be passed on to the progeny cells during cell division. In theory, these mutations could lead to the development of cancerous cells. However, in normal cells, the p53 protein apparently has the ability to sense DNA damage. When p53 detects mutations, the activated protein arrests the cell until the damage is repaired; p53 achieves this by inducing the synthesis of a class of proteins that can bind to and inactivate cyclin–cdk complexes. This blocks the progression of the damaged cell through the cell cycle and allows the DNA repair machinery to repair the mutated DNA. Once the abnormal DNA is repaired, the cell then resumes progression through the cell cycle. From this scenario, it is relatively easy to see how either a deletion of the p53 gene or a mutation in the p53 gene that results in the production of nonfunctional p53 protein would lead to an increased incidence of cancer. If a cell lacked functional p53 protein, the cell cycle could not be arrested to allow the repair of DNA damage. Deleterious mutations would be passed on to the progeny cells (e.g., mutations in protooncogenes) and, with time, a cell could build up enough mutations to develop into a cancer.

Cancers Arise from Multiple Mutations

The development of a cancer in a human does not appear to be due to a single mutation. Instead, it seems that cancers develop as a result of several independent mutations. Support for this conclusion comes from numerous observations. First, if the development of a tumor would occur following a single mutagenic event, cancer rates would be the same for all age groups. However, the rate of cancer occurrence is much higher in elderly persons, suggesting that transformation events occur as a result of several additive, unrepaired mutational events that allow a cell to escape from normal growth regulatory mechanisms.

This conclusion is supported by work with transgenic mice. If the DNA encoding an oncogene is microinjected into a mouse egg to produce a transgenic mouse, few, if any, tumors develop in the adult animal, even though each and every cell in the adult mouse expresses the oncogene. Likewise, transgenic mice can be produced by injecting two separate oncogenes into a fertilized mouse egg nucleus. In these mice, only a few more tumors develop in the adult animals than if only a single oncogene is present. Again, remember that each cell in the adult mouse would be expressing two mutated genes in this example. These data support the conclusion that the development of a cancer occurs as the result of several mutations in the cell and demonstrate the complexity of cell cycle regulation.

Clinical Case Discussion

We have learned in this chapter that many oncogenes encode mutated forms of tyrosine kinase. An example is the P210 phosphoprotein produced in chronic myelogenous leukemia. Adult chronic myelogenous leukemia

(ACML) predominantly affects persons between the ages of 30 and 60 years. It is a progressive disease that usually begins with a chronic phase, followed by a transitional phase, and terminating in an acute phase called *blastic crises* with conversion to acute myeloblastic or lymphoblastic leukemia. The peak incidence of ACML in the pediatric age group is between 10 and 14 years. Recent studies suggest that the disease results from a postzygotic transformation of the pluripotent stem cell. The onset of symptoms are usually mild or insidious. Common presenting symptoms include malaise and fatigue, weight loss, and bone and joint pain. The most common physical finding is marked splenomegaly (80–95%). Other abnormalities include hepatomegaly, priapism, and papilledema. The hematologic findings at diagnosis: leukocytosis with white blood cell counts above 100,000/mm^3, mild anemia, and thrombocytosis. Circulating blasts are observed in the peripheral blood. The bone marrow is hypercellular, the myeloid:erythroid ratio frequently is in excess of 20:1 (normally 3:1), with normal maturation patterns.

The Ph^1 chromosome is an abnormal chromosome 22 that is present in 85–95% of bone marrow cells in patients with ACML. The Ph^1 chromosome results from a reciprocal translocation between the long arm of chromosome 9 and chromosome 22 t(9:22). The molecular event is the translocation of the *c-abl* protooncogene located on chromosome 9 into a specific breakpoint cluster region on chromosome 22, resulting in the novel chimeric *bcr-abl* gene and producing a tumor-specific 210-kDa phosphoprotein (P210), with augmented tyrosine kinase activity that confers growth factor-independent cell growth. Over 300 variants have been classified. Hydroxyurea is a ribonucleotide reductase inhibitor that blocks DNA synthesis and is effective in ACML. Alternative treatment with Busulfan (Myleran) leads to a median survival of 30–45 months. Intensive combination chemotherapy regimens have been developed, with the objective of eradicating the Ph^1 clone.

Suggested Readings

Cell Cycle

Beach D, Basilico C, Newport J, eds. *Cell Cycle Control in Eukaryotes*. Cold Spring Harbor, NY: Cold Spring Harbor Press, 1988.

Hyams J, Brinkley BR, eds. *Mitosis: Molecules and Mechanisms*. San Diego: Academic Press, 1989.

Murray A, Hunt T. *The Cell Cycle*. Oxford: Oxford University Press, 1993.

Meiosis

Austin C, Short R, eds. *Reproduction in Mammals: I. Germ*

Cells and Fertilization, 2nd ed. Cambridge: Cambridge University Press, 1982.

Sagata N, Watanabe N, Vande Woude G, et al. The c-*mos* protooncogene product is a cytostatic factor responsible for meiotic arrest in vertebrate eggs. *Nature* 1989;342:512.

Wasserman P. Early events in mammalian fertilization. *Annu Rev Cell Biol* 1987;3:109.

Environmental Regulation of the Cell Cycle

Evered D, Nugent J, Whelen J, eds. Growth factors in biology and medicine. *Ciba Foundation Symposium 116*. London: Pitman Publishing, 1985.

Feramisco J, Ozanne B, Stiles C, eds. *Cancer Cells, III. Growth Factors and Transformation*. Cold Spring Harbor, NY: Cold Spring Harbor Press, 1985.

Oncogenes and Protooncogenes

Bishop J. Viral oncogenes. *Cell* 1985;42:23.

Hunter T. The proteins of oncogenes. *Sci Am* 1984;252(2):70.

Weinberg R. Finding the anti-oncogene. *Sci Am* 1988;259(3):44.

Review Questions

1. Which of the following statements concerning meiosis and mitosis is correct?
   a. Meiosis occurs in all cells of the body.
   b. Independent assortment and crossing over occur in both types of division.
   c. Genetically identical progeny are produced from mitosis but genetically different progeny cells are produced during meiosis.
   d. Mitosis produces haploid cells.
   e. None of the above

2. The important cell cycle regulatory molecules that are synthesized and degraded each cell cycle are called:
   a. cdks
   b. Cyclins
   c. Protooncogenes
   d. Growth factors
   e. None of the above

3. The principal environmental cues that regulate cell cycle progression in somatic cells from multicellular organisms are:
   a. Protooncogenes
   b. Hormones
   c. Cytoskeletal proteins
   d. Growth factors
   e. All of the above

4. Which of the following statements is incorrect?
   a. Products of protooncogenes send stimulatory signals to cells.
   b. Cancerous cells usually show several genetic lesions.
   c. All protooncogenes encode kinases.
   d. Tumor suppressor gene products repress cellular proliferation.
   e. None of the above

5. The most common chromosomal translocation in adult chronic myelogenous leukemia (ACML) involves:
   a. t(9:21)
   b. The *bcr* and *abl* protooncogenes
   c. The *c-myc* protooncogene
   d. t(8:14)
   e. c-*jun* and c-*fos* protooncogenes

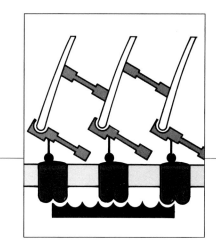

Chapter 10
Cell Motility

Clinical Case

A 3-month-old female, Christine Varsel, presented in the emergency room with a 1-day history of swelling, redness, and pain in the right cheek. Her past medical history included recurrent otitis media and stomatitis. Birth history was normal but she was treated for an umbilical cord infection and had late separation of the cord. Physical examination revealed a small febrile child in mild distress with erythema of her right cheek and eye. The remainder of the examination was normal. Laboratory tests: leukocyte count was 19,500/mm^3 (60% neutrophils, 5% bands, 30% lymphocytes, and 5% monocytes), and hemoglobin and platelets were normal. The child was diagnosed with a facial cellulitis and treated with appropriate antibiotics. She remained febrile after 7 days of treatment, and a repeat complete blood count showed persistent leukocytosis (29,000/mm^3). A hematology consult was obtained. Serum immunoglobulin levels were normal. The nitroblue tetrazolium dye reduction test following phagocytosis was normal. A Rebuck skin window test was recommended, which showed complete absence of neutrophils at the abrasion site. Analysis of leukocyte cell surface markers by flow cytometry using a panel of commercially available monoclonal antibodies was performed, and a complete absence of β$_2$ integrins (CD11/CD18) was demonstrated.

Chemotaxis and Cellular Motility

To appreciate the difficulties in understanding how non-muscle cells move, it is necessary to ask: What are the requisite aspects of cell behavior that contribute to cell movement? If, as an example, we take a cell migrating in a chemotactic gradient, such as might occur when a white blood cell homes in on a bacterium, the following cellular events are assumed to occur. The chemotactic molecule must be received by a cellular receptor, and this information must be transduced across the plasma membrane. The cell then uses the information conveyed by the chemotaxin to become polarized. Implicit in the concept of polarization is that the cell can direct the extension of a leading edge and cause cellular constituents to be moved into appropriate areas of the cell to accommodate forward migration. This process appears to be dependent on the microtubule-organizing center working in conjunction with the Golgi apparatus. Next, the cell must extend a lamellipodium or ruffled edge forward in the general direction of motion. Finally, the cell must establish and break a series of attachments with the extracellular matrix if it is to develop traction against which to propel itself. Thus, for movement to occur, each and every one of these events must be coordinated to effect directed cellular translocation.

Bone marrow-derived cells such as neutrophils must adhere to the vascular endothelium and migrate into the tissues to participate in the inflammatory response.

Neutrophils are the white blood cells responsible for the protection of the body's connective tissues from attack by microbial pathogens. These highly motile cells travel passively in the bloodstream until they exit the circulation and enter the connective tissues. Upon leaving the vasculature, these cells seek out infectious agents and engulf and kill them. The cellular and molecular events that neutrophils carry out to arrive at specific sites of infection comprise the phenomenon of chemotaxis. Chemotaxis relies on another cellular phenomenon, cell motility. Virtually every cellular process described in the preceding nine chapters of this book comes into play as neutrophils extravasate and move up a chemical concentration gradient in carrying out their role as the body's first line of defense against infection.

The key to understanding most of these processes appears to reside in the actin cytoskeleton and the proteins and molecules that regulate its structure. The change of shape of a neutrophil from its hydrodynamically fluid structure to a flow-resisting cell that tumbles within the confines of a capillary depends on changes in the actin cytoskeleton. The adhesion molecules that firmly attach the leukocyte to the vascular wall are not only attached to the actin cytoskeleton but also coupled to various signal transduction systems. Finally, the regulated expression and activation of surface receptors for adhesion, chemotaxis, and phagocytosis are all dependent on links to the actin cytoskeleton of the neutrophil.

Extravasation of White Blood Cells Requires a Change in Cell Shape and Expression of Leukocyte Adhesion Molecules

When neutrophils are in suspension, as occurs during their passage in the circulation, their cytoskeleton tends to be more solated than gelled, resulting in a relaxed shape (Fig. 10–1A). While in this relaxed state, the cells are deformable, and they pass easily through the capillary beds of the various organs. The first event that occurs in the process of margination is a stiffening of the actin membrane skeleton and cytoskeleton (i.e., an increase in the rate of sol-to-gel transformation), which translates into a less deformable cell shape. This change in cellular behavior serves to slow the cell down as it passes through progressively smaller and smaller vascular lumens. The stiffening of the cytoskeleton begins when membrane-

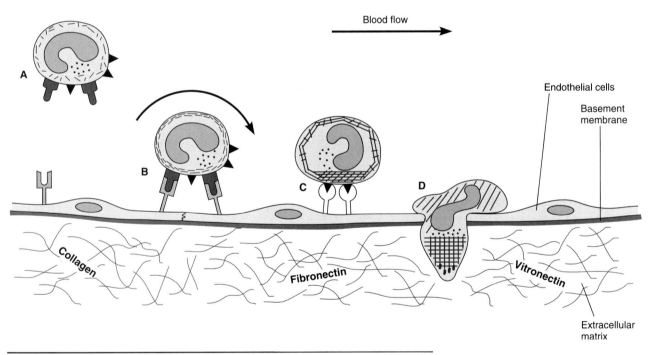

Figure 10–1. Stages of margination and extravasation of a white blood cell from a capillary showing relationships of cytoskeleton and adhesion molecules. **A:** Pliable cell with solated actin cytoskeleton expressing integrins (*wedges*) and glycoproteins (*ovals*) that serve as ligands for selectins on endothelial cells (*bottom of figure*). **B:** In rolling phase, cytoskeleton stiffens as selectins bind the glycoproteins, causing the cell to tumble or roll as it progresses along the endothelial wall. **C:** Integrins engage their counterreceptors on the endothelial cell as the cell establishes a leading edge framework of actin and crawls toward a gap between adjacent endothelial cells. **D:** The cell extends its leading edge through the retracted endothelial cells and diapedeses through the vascular wall into the extracellular matrix.

spanning glycoproteins on the neutrophil surface bind selectin molecules that protrude into the vascular lumen from the surface of the lining endothelial cells. This binding initiates an increase in $[Ca^{2+}]_i$ and/or initiates one or more multiple signaling cascades. The end result of these signals is the generation of an activated cell.

Endothelial cells located in the vicinity of the inflammatory focus express low-affinity adhesion molecules known as *selectins*. The expression of these molecules can be stimulated by either metabolites from pathogenic microbes or extracellular mediators, such as lymphokines and chemokines, produced by cells resident in the connective tissues that happen upon the pathogen. Each selectin has a lectin-like binding domain on its distal end that binds to a specific glycoprotein expressed on the surface of the neutrophil. As neutrophils pass over this bed of selectins, a transient series of loose attachments are made and then broken, causing the neutrophil to "roll" over the surfaces of successive endothelial cells (Fig. 10–1B). A stronger adhesive interaction occurs when integrins such CD11/CD18 appear on the neutrophil surface and bind to specific endothelial cell surface-associated adhesion molecules such as intracellular adhesion molecules (ICAMs) (see below). These adhesive interactions are also considered to be of a low affinity, but integrins are present in sufficient numbers to hold the neutrophil to the endothelial surface. Now the neutrophil can "crawl" over the endothelial surface until it finds a site between adjacent endothelial cells to diapedese into the connective tissues (Fig. 10–1C).

At the same time that neutrophils are adhering to the endothelium, other mediators, such as thrombin and histamine, are also acting on the endothelium, causing cellular retraction to occur, which opens gaps in the vascular wall. The crawling neutrophil will seek out these gaps and diapedese into the connective tissues. Histamine and thrombin exert their effects on the endothelial cell by interacting with a transmembrane glycoprotein called *endoCAM* (designated *CD31*). This specific receptor–ligand interaction stimulates the phosphorylation of serine residues on the cytoplasmic tail of CD31. This phosphorylation ultimately results in a contraction of the actin cytoskeleton causing endothelial cells to loosen their junctional attachments to each other. This contraction not only allows white blood cells easy access to the connective tissue compartment, but it permits antibodies and other blood proteins to leak into the surrounding perivascular connective tissues.

Transmembrane Proteins Mediate Specific Types of Interactions with Vascular Endothelial Cells

The neutrophil surface is covered with many different kinds of integral membrane proteins, and most of these proteins link the inside of the cell to the outside environment. To attach to the vascular wall and leave the circulation, white blood cells use at least two types of adhesion molecules, namely, selectins and integrins (see Chapter 6). The selectins are long peptide chains with terminal lectin binding domains, meaning that their ligands are specific sugar molecules such as a fucose residue. The integrins, of which more than 20 are known, are a more complex adhesion system by far than are the selectins. The β integrins are one-half of a binding pair of molecules expressed on the surface of the white blood cell. The other half of the binding pair to which the neutrophil will attach is expressed on the surface of the endothelial cell. Vascular endothelial cells express a plethora of adhesion molecules, including vascular cell adhesion molecules one and two (VCAM-1 and -2), intracellular adhesion molecules one and two (ICAM-1 and -2), and endoCAM. The integrins on the neutrophil are considered the receptors, and the corresponding ligands on the endothelial cell are the counterreceptors. Different white blood cells express different integrins; so, too, do the endothelial cells. For example, neutrophils express the $β_2$ integrin CD11/CD18 (integrins are heterodimeric members of the immunoglobulin gene super family). CD18 is the common β chain for the $β_2$ integrin family, and CD11 is the α chain). The counterreceptor for CD11/CD18 expressed on the endothelial cell surface is ICAM-1, and the net result of the adhesive interaction of this receptor and counterreceptor pair is the accumulation of large numbers of neutrophils in the connective tissues perfused by this stretch of ICAM-1-expressing cells in the capillary wall. The accumulation of a large focus of neutrophils constitutes the acute inflammatory infiltrate and, along with the extravasated blood proteins, is known as *pus*.

In contrast to neutrophils, eosinophils, another type of white blood cell seen in cases of allergy or asthma, express a β1 integrin known as *VLA-4*. VLA-4 binds preferentially to VCAM-1 on endothelial cells. The release of different inflammatory mediators produced and released by cells residing in the connective tissues that have encountered an inflammatory irritant determines which type of white blood cell would be most appropriate to deal with the particular irritant, and the receptor-counterreceptor specificity ensures that the appropriate cell is preferentially attracted from the circulation.

This system of selective adhesions in response to soluble mediators (i.e., chemokines or cytokines) depends on the presence of even more receptors on the endothelial cell surface. These chemokine receptors are located in the domain of the plasma membrane adjacent to the capillary basal lamina (and, therefore, closest to the source of the chemokine in the extracellular spaces). Occupation of the chemokine receptors leads to upregulation of the expression of genes encoding the ICAM or VCAM molecules; by this mechanism of regulated gene expression, the endothelium can control the qual-

Table 10–1

Proteins Associated with or Localized to Areas of the Focal Adhesion

Cytoskeletal Proteins	Signal Transduction Proteins
Actin	Protein kinase C
Talin	Other kinases
Vinculin	pp60[v-src]
α-Actinin	Calpain II
Tensin	
Zyxin	

From Williams MJ, et al. *Trends Cell Biol* 1994;4:109.

ity and quantity of the ensuing inflammatory infiltrate. Thus, the phenotypic expression of inflammation depends on which chemokine or cytokine receptors are activated, how many endothelial cells respond, and how long the endothelial cells continue to express the adhesion molecules on the luminal domains of their plasma membranes.

The sites where the integrins of the cell actually attach to the extracellular matrix are known as *focal adhesions*. By using fluorescently labeled antibodies, it has been shown that, along with the integrin, many cytoskeletal proteins and/or signal-transducing molecules colocalize to the focal adhesion (see Table 10–1). As the cell moves, the adhesion to the extracellular matrix must be broken, and it has been suggested that the enzyme urokinase might function to catalytically break the interaction between integrin and its receptor. A schematic depiction of the linkage of CD11/CD18, the β_2 integrin, to the extracellular matrix is shown in Figure 10–2.

Receptor Sensitivity During Chemotaxis Must Be Constantly Regulated to Allow the Cell to Adjust to Ever-Changing Extracellular Conditions

Cell adhesion, by its very nature, must be precisely regulated. With too much adhesion (either due to too many integrin molecules or too high an affinity of receptor for ligand), the cell will be stuck firmly in place. Too little adhesion will result in the cell being unable to gain a propulsive advantage over its substrate, and in a sense, it will "spin its wheels." The process whereby the affinity state of the receptor binding site is altered is termed *inside-out signaling*, and it results from conformational changes arising from within the cytoplasmic domain of the integrin molecule. Once the cytoplasmic domain has been altered (e.g. by phosphorylation), the conformational change is transmitted through the cell membrane to the integrin binding site and changes its affinity for its ligand. Phospholipases, intracellular Ca^{2+}, G proteins, and protein kinases have all been shown capable of inducing inside-out signaling in integrins.

The formylated methionine (f-met) receptor is coupled to a G-protein-mediated signal transduction pathway. It is necessary for the neutrophil to adapt its chemotaxic motion to changing conditions—such as stopping when the cell has arrived at the source of the gradient—and initiating other activities—such as phagocytosis—to neutralize the chemotaxic stimulus. To do this, the cell must change the significance that is given to an occupied receptor, be it an adhesion molecule or a chemotaxic one. This is typically done in any of several ways. The cell can decrease the number of receptors on the cell surface, either by not recycling as many as before or by not making any more new receptors (a form of regulated gene expression). The cell can downregulate the receptor, which implies that receptors are actively endocytosed and subjected to autophagia. Another, even more subtle, mechanism to desensitize a G-protein-coupled receptor to a stimulus is to phosphorylate the intracellular domain of the receptor. Phosphorylated receptors show an increased affinity for signal-blocking proteins, which, when bound to the intracellular tail of the receptor, significantly inhibit G-protein-mediated signaling.

It has recently been suggested that the f-met receptor may be both an integral membrane protein and also an actin-binding protein associated with the cytoskeleton. Thus, the receptor has at least two proteins it can be attached to by its intracytoplasmic domains; namely, the guanyl nucleotide binding protein (i.e., the G protein itself) or actin. It further seems that these two proteins actively compete for the privilege of attaching to the formylated tripeptide f-methionine-leucine-phenylalanine (fMLP) receptor. In a fully desensitized neutrophil, the receptor would be anchored to the actin membrane or cytoskeleton and moved laterally through the phospholipid bilayer away from the binding sites on the G proteins. G proteins have been shown to bind to microtubules; thus, they can be moved in a different direction by the action of a microtubule-associated motor protein such as dynein or kinesin. In this way, an isolated receptor located in a plasma membrane domain different than its G-protein partner, that bound its ligand, would be incapable of sending a signal to interior of the cell.

The Initiation of Directed Cellular Migration Requires the Cell to Polarize the Cytoplasmic Constituents and Redistribute Cell Surface Proteins to Different Plasma Membrane Domains

For a cell, like the neutrophil, to sense a chemical gradient and orient itself within the gradient so that it can begin to migrate toward the source, the cell must be able to polarize its insides and to control where on its

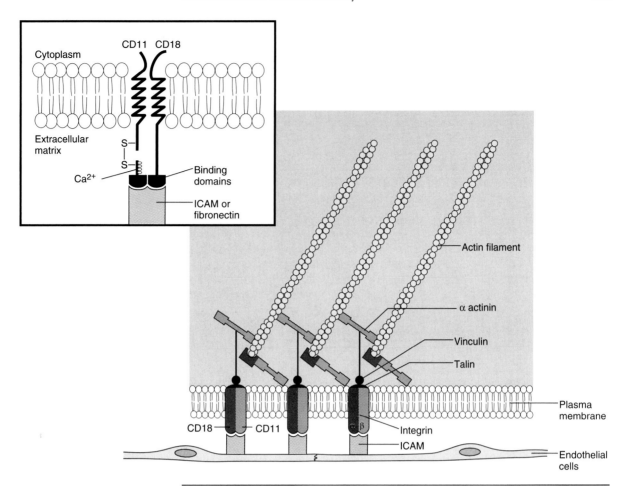

Figure 10–2. Schematic representation showing the β₂ integrin linking the cytoskeleton of a cell to the surface of an endothelial cell. The common β chain is represented as CD18, whereas the α chain, which can be CD11a [lymphocyte-function-associated-antigen (LFA)-1], CD11b (Mac-1), or CD11c (p150,95) is labeled *CD11*. Different white blood cells express in common the β chain, whereas the CD11a, -b, or -c chains are differentially expressed on the leukocytes. The biochemical structure of a β₂ integrin is depicted (*inset*). The α chain contains an intrachain disulfide linkage, a characteristic of the immunoglobulin gene superfamily, and a divalent cation binding domain that typically accepts up to four Ca²⁺ ions, which can contribute to the avidity of the binding site. Both the CD11 chain and the CD18 chain contribute to the formation of binding site for either cellular adhesion molecules or extracellular matrix proteins such as fibronectin.

surface various chemotaxic receptor proteins are located. The redistribution of the cellular internum has been appreciated for some time, whereas the significance of the cell surface domains and the proteins contained in them during the migratory process has only recently been appreciated.

When migrating cells are viewed by phase microscopy, the cell is seen to extend a leading edge, or lamellipodium. Actin filaments are the major cytoskeletal element associated with the leading edge. Most other organelles are excluded from this site and, as the cell locomotes, small particles dropped in the path of the oncoming cell are picked up on the surface membrane and carried backward. This apparent retrograde move-

ment of particles (and, by extension, cell surface receptors) suggests a flow of membrane away from the apex of the leading edge of the cell. Paradoxically, as the cell proceeds to extend a forward protrusion, there is a concomitant rearward migration of surface receptors associated with a simultaneous rearward movement of intracellular actin filaments. At the same time that these rearward movements are taking place, there is a continual polymerization of actin filaments at the cell's leading edge. Collectively, these observations indicate that there is a constant cycling of surface membrane from the front to the rear of the cell; however, it seems more than likely that the movement of surface receptors from front to rear can be accounted for by linkage of the receptor to the actin cytoskeleton, concomitant with a continuous polymerization of new actin filaments at the front and depolymerization somewhere toward the center of the cell.

In addition to the adhesion molecules, migrating leukocytes also express chemotactic receptors, which gather information needed to "aim" or "steer" the cells toward the inflammatory focus. The best characterized chemotaxic receptor on human neutrophils is one that recognizes f-met, which is a characteristic of bacterial proteins. Bacteria formylate the first methionine encoded by the ATG start codon of their genes. Human neutrophils exhibit very strong chemotaxic responses to the fMLP; therefore, the receptor is known as the *f-met* or *fMLP receptor*.

Scientists have devised a clever way to study neutrophils migrating in a gradient of fMLP. A small gelatin plug containing the fMLP is placed in the large tapered end of a glass micropipet that has an orifice of about 5 μm, or the size pore through which neutrophils can crawl. The fMLP then diffuses out the tip of the pipet and establishes a concentration gradient leading back into the pipet. If neutrophils are then placed near the opening of the pipet, they will sense the gradient and migrate into the pipet, as indicated in Figure 10– 3. The major advantage of this system as opposed to migration on glass surfaces is that the cell can interact with the circular wall of the pipet; thus, its entire surface is engaged, as would be expected to occur when the cell is migrating in the extracellular matrix of the body. This setup can be viewed under a microscope and manipulated in a variety of ways to test different parameters of cellular motility. For example, the pipet can be coated with different adhesion molecules or extracellular matrix proteins to study how well neutrophils adhere to them.

Using this pipet system of chemotaxis, neutrophils have been observed moving toward the fMLP at about 0.3 μm/second. When a counterpressure is applied against the migrating cell, a pressure of approximately 17 cm of water is sufficient to completely stop the cell in its tracks. This implies that the adhesive interaction of the neutrophil with the wall of the pipet is at least as strong as the pressure head against which the cell is migrating.

Neutrophils migrating in an fMLP gradient within a glass micropipet exhibit three distinct regions. The region that most closely resembles the leading edge of the cell migrating on a flat surface contains an F-actin gel. This zone is free of other cytoplasmic organelles and granules. The middle zone is short and represents an interface between the first and third regions. It is characterized by granules that exhibit Brownian motion. This rapid motion indicates that the surrounding fluid is of a much lower viscosity than the leading edge zone, suggesting that a wave of actin gel depolymerization is occurring at the boundary of the first and second zones. The third and largest region comprises the posterior half of the cell and contains the nucleus and other organelles. These structures show little or no motion, suggesting that they are being pulled forward by the rear of the cell.

White Blood Cells Can Sense a Chemical Gradient and Migrate Toward the Source of the Chemical in a Process Known as *Chemotactic Migration*

The concept of chemotaxis is predicated on three categorical kinds of motion that can be observed in nonmuscle cells. The first of these motions is designated *random motion* and implies that the cell neither moves toward nor avoids any stimulus in its immediate envi-

Figure 10–3. Diagrammatic representation of the glass pipet model of studying neutrophil migration in response to the chemotaxic molecule fMLP. A gelatin plug containing the fMLP is situated at the beginning of the pipet taper, and, with time, the chemical diffuses out of the narrowed 5-μm opening, thereby establishing a concentration gradient. Neutrophils placed near the tip will migrate toward the gelatin plug, up the gradient of fMLP. Depicted (*cartoon*) is an interaction between an integrin such as CD11/CD18 on the neutrophil surface and its counterreceptor, ICAM-1, which has been applied to the walls of the pipet. Also indicated (*cartoon*) are the zones that include the leading edge [F-actin gel (*1*)], intermediate zone [depolymerizing actin exhibiting Brownian motion (*2*)], and the posterior aspect of the cell containing the nucleus and organelles [*dimension bar (3)*]. **Upper insert:** The hypothesis is that an area of high osmotic pressure forms in response to a Ca^{2+} flux or transient at the leading edge and that this protuberance is stabilized by polymerization of actin into the cytoskeleton, while actin is depolymerizing at the base of the leading edge (which corresponds to *zone 2*, above). **Lower insert:** The transmembrane linkage of the cytoskeleton to the intracellular adhesion molecules (ICAM-1) used to coat the pipet. It should be noted that CD11/CD18 cannot only bind to ICAM-1 but can bind to the extracellular matrix protein fibronectin. (Modified from Skalak et al. *Blood Cell* 1993;19:389.)

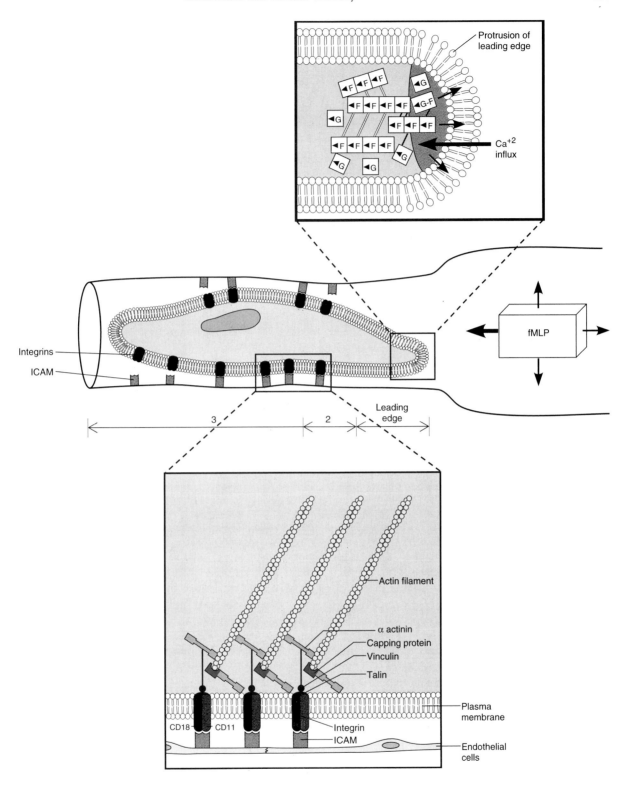

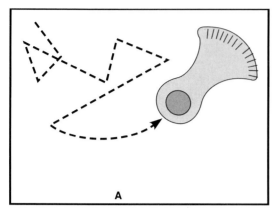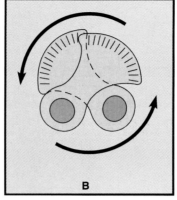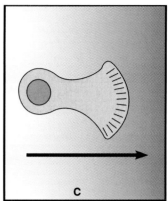

Figure 10–4. Three categories of cellular motion. **A:** Random motion occurs when a motile cell migrates neither toward nor away from anything in its vicinity. **B:** Chemokinesis occurs when a cell is exposed to a chemotactic molecule in a uniform solution, such that receptors anywhere on the cell surface detect equimolar amounts, resulting in the cell moving about on its own axis. **C:** Chemotaxis occurs when a cell has recognized the chemical and actively migrates in the direction of greatest receptor occupancy, which is into the face of the chemical gradient.

ronment and that the long axis of the moving cell is not oriented toward any particular reference point (Fig. 10–4A). The second described motion is called *chemokinesis*. Chemokinesis is defined as an increase in velocity or frequency or magnitude of rotation of the cell on its axis in response to a chemical stimulus in its environment (Fig. 10–4B). *Chemotaxis*, in contrast with chemokinesis, occurs when the cell migrates either to or from the source of the chemical (i.e., purposeful migration) (Fig. 10–4C). Human white blood cells have been described migrating only toward a chemotaxin, never away from it. Bacteria are capable of both types of directed migratory activities. It is likely that, at the molecular level, all these forms of movement are the same. This is because a chemical such as fMLP, that elicits a directed chemotactic migration when presented to the cell in a concentration gradient, generates chemokinetic motion when the cell's receptors are exposed to the chemical in a uniform solution. In addition to fMLP, neutrophils have chemotactic receptors for the C5a split product of the complement cascade, platelet activating factor, LTB$_4$, and various kinds of immunoglobulins, such as aggregated IgG.

The Chemotactic Signal Is Transduced Across the Cell Membrane in the Form of Calcium Ions

How, then, do cells detect the presence of a chemical gradient? Individual neutrophils can detect a concentration gradient from one region of the cell surface to another. To make such a determination suggests that the cell must have numerous receptors present over the entire cell surface. Thus, the cell responds to differences in receptor occupancy and, in the case of the neutrophil, concentration differences of as little as 1% are sufficient

to elicit a chemotactic migration when approximately half of the cell's receptors are occupied. Furthermore, sustained migration necessitates that unbound receptors be made continually available to permit subsequent ligand sensing and receptor binding. Two possible mechanisms have been suggested, and both likely contribute to meeting the cell's need for new receptors at the leading edge of the cell. The cell can synthesize new receptors and, by directed internal movement, insert them into the leading edge. The incorporation of new receptors into the membrane of PMNs is associated with the release of the neutrophil's azurophil granules, which occurs in response to chemotactic stimulation. Alternatively, the cell can recycle receptors that have previously bound ligand and been internalized. Such an event has been demonstrated, using both fluorescently labeled and radioactively labeled chemotactic peptides. In these experiments, the tagged molecules were observed to be internalized into endosomes that then passed through the compartment to uncouple receptor and ligand (CURL). The ligand is released and digested in lysosomal vacuoles, and the receptor molecule is returned to the leading edge of the cell, where it again becomes available to bind ligand.

The binding of chemotactic molecules to the leukocyte's surface not only starts the chemotactic response but initiates several other processes at the same time. Among the processes activated by the transduced signal are the respiratory burst, which is a manifestation of oxidative metabolism; the release of lysosomal enzymes contained within the azurophil granules that are prevalent in granulocytes such as the neutrophil; and the production and release of arachidonic acid metabolites, produced by the cyclooxygenase (prostaglandins) and lipoxygenase pathways (leukotrienes). Given that the cell must have a certain number of receptors occupied by the chemotaxin to begin the process of directed migration, how, then, does the cell get the message into the cytoplasm to initiate the sequence of events necessary to effect this migration? One explanation suggests that, following the binding of the receptor, there is a local increase in Ca^{2+} in the area of greatest receptor occupancy. This divalent cation activates the enzyme phospholipase A_2, the substrate for which is phosphatidylcholine (PC). The lipase converts the PC to arachidonic acid, which is then taken through the lipoxygenase pathway to 5-hydroxyeicosatetraenoic acid (5-HETE) and leukotriene B_4 (LTB_4). Leukotriene B_4 is secreted into the extracellular space, where it acts both as a new chemotaxin, attracting more cells, and as an autocrine signaling factor, further activating the cell that released it when it binds to the LTB_4 receptor on the cell's outer membrane. When this binding event occurs, the amplification of the initial signal is completed because binding the LTB_4 receptor boosts the initial Ca^{2+} flux, which generates even more phospholipase A_2 activity.

The Actin Cytoskeleton Is Involved in Extending and Maintaining the Leading Edge of a Migrating Cell

Based on numerous microscopic observations of migrating cells, the most conspicuous aspect of directed cellular movement is the extension of a pseudopod or lamellipodium, which constitutes the leading edge of the cell. This extension appears to be synonymous with cellular propulsion, although the exact identification of the source of the propulsion is not known with certainty. Regardless of the mechanism for moving the cell forward, the leading edge must protrude beyond its current position relative to the cell's location. It has been suggested that movement occurs because there is an increase in the osmotic pressure in the area where the cytoskeleton approaches the domain of plasma membrane that bounds the current leading edge. The anteriormost aspect of the actin cytoskeleton consists of many short filaments that assemble and disassemble, and an increase in focal osmotic pressure might arise when complexes of G-actin bound to profilin, in response to a transient rise in intracellular free Ca^{2+}, are split, releasing the actin monomers. This focal accumulation of G-actin monomers would attract cellular water through the interstices of the gelled cytoskeleton to the leading edge, and the increasing pressure would push forward a new membrane extension of the leading edge. As the newly liberated G-actin is polymerized (plus end towards the leading edge), the newly extended leading edge would be stabilized.

As stated, there is no consensus as to how moving cells extend their leading edge. Other explanations have also been put forward to explain this aspect of cellular

motility. One alternative to the increasing focal osmotic pressure hypothesis is that the actin cytoskeleton that currently supports the leading edge can contract, thereby exerting a directed squeezing of cellular water forward to "bleb out" a new leading pseudopod. It is equally possible that cellular propulsion can be attributed solely to the force produced against the plasma membrane by newly polymerizing actin. Whereas equally good cases can be made for each of these propulsive mechanisms, it can be clearly seen that the one common feature shared by all three suggested mechanisms is the modification of the actin cytoskeleton. This indicates that the forces that control the various actin capping, bundling, cross-linking, and severing activities of the cell must be exquisitely and precisely coordinated in order for a cell to move forward.

Migrating Cells Are Steered When the Microtubule-Organizing Center Realigns Itself Along an Axis Between the Nucleus and the Leading Edge of the Cell

Observations made on cells that have been inserted into a chemotactic gradient have shown the occurrence of intracellular events that precede the extension of the leading lamellipodium. The most dramatic of these processes—and perhaps the most important for migratory purposes—is the reorientation of the microtubule-organizing center (MTOC) and the Golgi apparatus (GA). It follows that if particles picked up by the leading edge of the cell are observed to flow backward as the cell moves forward, and if receptors are being internalized, there must be a mechanism to replace the membrane at the leading edge. The organelles responsible for directing the insertion of new membrane mass into the migrating cell membrane are the MTOC and GA. This insertion of new membrane was demonstrated in an elegant experiment using vesicular stomatitis virus (VSV)-infected fibroblasts.

Shortly after most cells are infected by viruses, they express on their surfaces viral-encoded proteins that are not present before infection. When fluorescent antibodies against a VSV-encoded protein, known as the *G glycoprotein*, were used to stain migrating cells, the first appearance of the G protein on the cell surface was at the leading edge of the cell. Because the G glycoprotein protein is synthesized by the infected cell, glycosylated in the GA, then transported to the cell surface in large transfer vesicles, it was concluded that new membrane mass was, indeed, inserted into the leading edge of the locomoting cells. In the case of neutrophils, the delivery of the azurophil granules to the cell's leading edge and their subsequent exocytosis of hydrolytic enzymes at the site of inflammation requires the presence of intact microtubules. Movement of the azurophil granules towards the plus end of the microtubule would require kinesin. Additionally, chemotaxic and adhesion molecules are contained on the inner surfaces of these granules, such that, during fusion with the leading edge cell membrane, new receptors are inserted onto the cell surface in the proper orientation. Both azurophil granules and the Golgi vesicles in fibroblasts are steered by microtubules that, in turn, are oriented along the axis of migration by the MTOC. The insertion of new membrane in some cells (e.g., fibroblasts) can be inhibited by treating the cells with colchicine. This agent depolymerizes the microtubules of the cell and results in a cessation of locomotion. Interestingly, this treatment has little or no effect on neutrophils, but agents that depolymerize actin filaments, such as cytochalasin B, can inhibit PMN movement. These disparate observations suggest that different types of cells depend on each of these cellular processes to differing degrees. Additional evidence that microfilament-directed insertion of new membrane mass is necessary for neutrophil migration derives from the observation that monensin, a drug that inhibits membrane insertion and recycling, inhibits chemotactic migration but does not prevent the MTOC and GA from orienting in a chemotactic gradient.

Failure to Properly Express Integrin Gene Products Results in a Clinical Condition Known as *Leukocyte Adhesion Deficiency,* Which Is Characterized by the Inability of White Blood Cells to Migrate from the Vasculature to Sites of Infection

Clearly, the presence of adhesion molecules on the surface of human neutrophils is important in facilitating the passage of these cells from the circulation to the extracellular matrix. People with leukocyte adhesion deficiency (LAD) suffer recurrent bacterial infections because their neutrophils do not express adhesion molecules needed for their recruitment via the chemotaxic pathway to sites of inflammation. Two different forms of LAD have been described. Type I LAD neutrophils are characterized by a congenital absence of the β-chain (CD18) of the integrin molecule. These cells do not bind to cells expressing ICAM-1, and intravital microscopy has suggested that establishment of a firm adherence to endothelial cells and diapedesis does not occur. Chemotactic stimuli that lead to phosphorylation of cytoskeletal proteins in normal neutrophils do not result in phosphorylation of cytoskeletal proteins in type I LAD neutrophils. In contrast, type II LAD neutrophils fail to express the ligands, Lewis X or Siayl 1, for E- or P-selectins on endothelial cells. The primary defect in type II LAD occurs in the rolling phase of intravascular recruitment of neutrophils. People with type II LAD often show developmental abnormalities, including diminished growth and mental retardation. This has led to the suggestion that LAD type II may be a part of a systemic defect in the general metabolism of fucose.

Leukocytes isolated from individuals with LAD can phagocytose and kill bacteria *in vitro* as well as can leukocytes from normal donors. However, when LAD leukocytes are tested in chemotaxis assays in response to such well-characterized chemoattractants as formylated peptides or split products of the complement cascade, they fail to migrate, owing to their inability to adhere to the test substrate, and they do not adhere to cultured endothelial cells expressing the corresponding ligands for the missing adhesion molecules. Additionally, LAD cells do not stain with labeled antibodies specific for integrins. Leukocyte adhesion deficiencies may be ameliorated one day by gene therapy.

Summary

In summary, it can be seen that almost every aspect of cell biology comes into play in the complex act of amoeboid movement. This type of movement is complex and highly integrative, in that it requires coordinated structure and processes in order for directed migration to occur. Transmembrane proteins acting as chemotaxic receptors bind chemotactic molecules. Occupied receptors then transduce a signal across the plasma membrane, most likely in the form of a Ca^{2+} flux through ion channels. Once the Ca^{2+} flux is sufficient to activate the cell, there is a distinct polarization of the cell's internum, which is most strongly reflected in the realignment of the microtubule-organizing center and Golgi apparatus. The MTOC and GA serve to steer new and recycled membrane to the advancing leading edge of the cell. At the same time, a series of membrane-spanning integrins, attached intracellularly to the cytoskeleton, are continuously binding to and releasing from the extracellular environment as the cell advances. Failure to express the molecules that link the cytoskeleton to the outside environment of the cell can have severe clinical consequences, as is seen in the LAD disorders.

Clinical Case Discussion

Leukocyte adhesion deficiency is a rare autosomal recessive disorder, mainly of neutrophils. The neutrophils from such patients are deficient in many adhesive-dependent functions, including spreading on surfaces, aggregation, directed migration, and phagocytosis of opsonized particles. The primary defect in patients with LAD is the lack of production of the β_2 integrin adhesion molecules, which contribute to critically important cell-to-cell and cell-to-extracellular matrix interactions in the immune system. The β_2 integrins are heterodimeric molecules consisting of an α and a β chain found on the surface of leukocytes. These dimeric molecules facilitate adhesion to endothelial cells. The common β subunit CD18 can bind with one of three different α chains designated *CD11a, -b,* or *-c.* The primary defect in patients with LAD is a lack of expression of the gene (on chromosome 21) that encodes the CD18 protein.

Two clinical forms of LAD have been described using flow cytometric analysis. A severe form (LAD type I) results in a complete absence of the CD18 molecule and a deficiency of all three β_2 integrins. In contrast, patients with type II LAD have only a partial absence of the CD18 molecule. The first characteristic sign in the clinical presentation is delayed separation of the umbilical cord and omphalitis (inflammation of the navel). Other features include otitis media, aseptic meningitis, and perianal skin lesions and abscesses. Patients with incomplete CD18 expression usually present in childhood with severe gingivitis and periodontitis. Antibiotics are used during infections, as well as for prophylaxis. Bone marrow transplantation is the only curative therapy. Recent studies using gene transfer techniques have successfully restored CD18 expression to CD18-deficient human bone marrow progenitor cells.

Suggested Readings

Cell Motility

Cramer LP, Mitchison TJ, Theriot JA. Actin-dependent motile forces and cell motility. *Curr Opin Cell Biol* 1994;6:82–86.

Klotz K-N, Jesaitis AJ. Neutrophil chemoattractant receptors and the membrane skeleton. *BioEssays* 1994;16:193–198.

Pavalko FM, Otey CA. Role of adhesion molecule cytoplasmic domains in mediating interactions with the cytoskeleton. *Proc Soc Exp Biol Med* 1994;205:282–293.

Stossel TP. The Machinery of cell crawling. *Sci Am* 1994;271: 54–63.

Review Questions

1. Retraction of adjacent endothelial cells to facilitate diapedesis of neutrophils depends on the interaction of which receptor-ligand pair?
   a. CD11/CD18 with ICAM-1
   b. fMLP and the fMLP receptor
   c. Histamine and endoCAM
   d. VLA-4 and VCAM-1
   e. Siayl X and P-selectin
2. Binding of chemotaxic molecules to their corresponding receptors on neutrophils triggers all of the following **EXCEPT:**
   a. Oxidative metabolism
   b. Intracellular Ca^{2+} transients
   c. Exocytosis of azurophil granules
   d. Mitosis
   e. Cellular polarization
3. Characteristic findings in leukocyte adhesion disorder (LAD) can include all of the following **EXCEPT:**
   a. Abnormal leukocyte chemotaxis
   b. Recurrent bacterial infections
   c. A deficiency of the β_2 integrin molecule CD18
   d. An inability to kill phagocytized bacteria
   e. An inability to metabolize the sugar fucose

Review Questions Answer Key

Chapter 1:
1b, 2a, 3d, 4e, 5c

Chapter 2:
1a, 2d, 3d, 4d, 5e, 6c, 7d, 8b

Chapter 3:
1e, 2c, 3d, 4b, 5b, 6b, 7b, 8a, 9d

Chapter 4:
1c, 2e, 3d, 4a, 5e, 6a, 7e, 8b, 9b, 10b

Chapter 5:
1a, 2c, 3c, 4b, 5b, 6d, 7d

Chapter 6:
1b, 2e, 3b, 4e, 5a, 6e

Chapter 7:
1c, 2e, 3e, 4c, 5a, 6c, 7d

Chapter 8:
1b, 2d, 3c, 4b, 5c, 6e, 7b, 8b

Chapter 9:
1c, 2b, 3d, 4c, 5b

Chapter 10:
1c, 2d, 3d

Index

SELF-STUDY PROGRAM INSTRUCTIONS

The *Medical Cell Biology: Self-Study Program* contains questions and answers keyed to Goodman's *Medical Cell Biology*, 2nd edition. The self-study program enables you to answer questions and receive feedback for each correct answer. All questions in the program are multiple choice.

System Requirements
This electronic self-study program has been designed for use with an IBM or IBM-compatible computer and requires Windows 3.11 or higher, 8 MB RAM, a 3.5-inch DSDD floppy drive, an SVGA graphics card or better, and a mouse. The *Medical Cell Biology: Self-Study Program* application is not designed to work in a network environment; use a local copy of Windows for best results.

Installation Instructions
Insert the floppy into the floppy drive. Go to the file manager and select your floppy drive (usually a:). Double click on the SETUP.EXE file and follow the prompts.

Start-up Instructions
From the program manager, open the program group "Medical Cell Biology" by double clicking on the icon. Now double click on the icon labeled "Medical Cell Biology."

How to Take a Quiz
The *Medical Cell Biology: Self-Study Program* contains eleven quizzes. To choose a quiz simply click on the button to the left of the quiz you want to take.

If you want to set a time limit for the quiz, click on the button labeled "Timed Quiz." The default is 30 minutes; you may enter another time between 1 and 120 minutes.

You are now ready to begin the quiz. But before you start be sure to select either "study mode" or "test mode."

The **study mode** lets you know immediately if your choice is correct or incorrect and provides feedback about the correct answer. If you select an incorrect answer, continue to choose answers until you reach the correct one.

The **test mode** allows you to answer all questions and store your answers in memory. After you select an answer, the quiz will automatically advance to the next remaining question.

Review Your Test Results
Results from a single session will be temporarily stored in memory. Choose "Review Statistics" on the main menu to view the results from this session.

You can also review your quiz results for each question by returning to the quiz in study mode. Each question that you answered will show "Correct" or "Incorrect."

To Exit
Selecting File/Exit will end your session and return you to Windows.